Materials for a Healthy, Ecological and Sustainable Built Environment

Related titles

Cost-Effective Energy Efficient Building Retrofitting: Materials, Technologies, Optimization and Case Studies
(ISBN: 978-0-08-101128-7)

Sustainable Construction Materials: Copper Slag
(ISBN: 978-0-08-100986-4)

Advanced High Strength Natural Fibre Composites in Construction
(ISBN: 978-0-08-100411-1)

Woodhead Publishing Series in Composites Science and Engineering

Materials for a Healthy, Ecological and Sustainable Built Environment

Principles for Evaluation

Emina Kristina Petrović

Brenda Vale

Maibritt Pedersen Zari

ELSEVIER

WP
WOODHEAD PUBLISHING

An imprint of Elsevier • elsevier.com

Woodhead Publishing is an imprint of Elsevier
The Officers' Mess Business Centre, Royston Road, Duxford, CB22 4QH, United Kingdom
50 Hampshire Street, 5th Floor, Cambridge, MA 02139, United States
The Boulevard, Langford Lane, Kidlington, OX5 1GB, United Kingdom

British Library Cataloguing-in-Publication Data
A catalogue record for this book is available from the British Library

Library of Congress Cataloging-in-Publication Data
A catalog record for this book is available from the Library of Congress

ISBN: 978-0-08-100707-5 (print)
ISBN: 978-0-08-100706-8 (online)

Publisher: Matthew Deans
Editorial Project Manager: Tessa De Roo
Production Project Manager: Omer Mukthar
Designer: Ines Maria Cruz

Typeset by MPS Limited, Chennai, India

Contents

List of contributors

Ljiljana Đukanović University of Belgrade, Belgrade, Serbia

Rainer Hirth Hochschule Coburg, Coburg, Germany

Nigel Isaacs Victoria University of Wellington, Wellington, New Zealand

Adele Leah Victoria University of Wellington, Wellington, New Zealand

Guy Marriage Victoria University of Wellington, Wellington, New Zealand

Emina Kristina Petrović Victoria University of Wellington, Wellington, New Zealand

Ana Radivojević University of Belgrade, Belgrade, Serbia

Jaime Rios-Calleja BUAP, Benemérita Universidad Autónoma de Puebla, Puebla, Mexico

Mirjana Roter-Blagojević University of Belgrade, Belgrade, Serbia

John Storey Symbiosis Architects and Consultants, Nelson, New Zealand

Brenda Vale Victoria University of Wellington, Wellington, New Zealand

Robert Vale Victoria University of Wellington, Wellington, New Zealand

Marci Webster-Mannison Melbourne Design Studios, Collingwood, VIC, Australia

Maibritt Pedersen Zari Victoria University of Wellington, Wellington, New Zealand

Preface: A call for ontological consideration of building materials

Consideration of building materials has never been more important than now. This is because materials themselves make an integral contribution to the sustainability of construction and any pro-sustainable improvements should include a careful consideration of these. Furthermore, contemporary architecture and design are experiencing forces which call for a stronger conceptual and theoretical emphasis on materials. This book is focused on proposing principles for evaluation of building materials in relation to their suitability in sustainable construction. Underlying the discussion is a deep concern for the importance of this emphasis on materials.

Current efforts to integrate digital innovation into architecture sharpen the importance of materials as the very aspect of all design which remains grounded in its own physicality and "objectness." Calls are already clear for postdigital considerations (Lund, 2015), more object-oriented approaches (Harman, 2002), and generally there is a greater recognition of the role materials can play in architecture and design (Petrović and Perkins, 2016). After all, the floor supporting our weight is made out of a tangible material and we want that to be so. A virtual floor would not be as practical. The same goes for many other architectural features and furnishings. It is important to recognize that digital manufacturing also uses matter, and that goes for all of the digital processes including 3D printing where although the object seemingly "appears" from a digital idea, it still requires a base material in order for it to be fabricated.

It is this inherently unavoidable necessity of materials in architecture that drives this book. Regardless of how much our command of digital technologies develops in the coming years, the tangible materiality of the created objects, "the thingly character of the thing" (Heidegger, 1971, pp. 19, 32), will continue to be fundamental to all architecture and design. The "objectness" or "thingness" of buildings will remain because it is the physical world in which we live, and the means by which we engage with the world around us.

However, a coherent theory about materials is still largely absent. The works which have considered the ontological and poetic importance of architectural construction, such as those of Kenneth Frampton (1990,1995) and Gevork Hartoonian (1994), have tended to mention materials in passing, acknowledging their role, but not giving them significant attention. Similar are the works which have considered architectural construction from a phenomenological perspective, such as those of Peter Zumthor (2006). These important works have increased emphasis on the importance of architectural construction and detail, often under the banner of tectonics, and they have also explicitly criticized scenographic, representational use of architectural construction (Frampton, 1990, 1995). While this has been important for establishing the need for a greater emphasis on materials, materials themselves are far too often implied rather than actively considered, suggesting that in these works materials maintain a relatively supportive role, even if this is no longer scenographic.

Although it is justifiable to think that materials are not the essence of architectural design, there is room to give them a much more active role. This is especially the case as we move towards considering the totality of the implications buildings pose on the environment. In addition to the sheer physicality of the volume of materials used in construction, the materials play a number of important roles in impacting on the environment. Their harvest or extraction can result in long-lasting ecological damage, they play a significant role in the life-cycle energy and energy efficiency of buildings, they influence indoor air quality, and at the end of their lifespan they often become voluminous waste. In fact, materials impact on the sustainability of the whole building more directly and more intensely than many other aspects of architectural construction and design, and represent one of the central issues for efforts to improve the sustainability of building construction. This puts pressure on architects and related professionals to search for better ways of managing and mitigating the impact their work poses on ecosystems and climate. This is not a question of choice. Our professions, as it happens, and whether wanting to or not, is forced to lead the way in pro-sustainable change, and materials play one of the central roles in this process.

It is from this position that an ontological importance of materials for sustainable construction presents itself. Taking ontology as a branch of philosophy "dealing with the nature of being" (Oxford Dictionaries, 2016), it is precisely the totality of the "being" of the materials which is relevant for sustainable construction. It is impossible to attribute instrumental and supportive, and often consequently relatively invisible roles to the material, if they are making one of the most significant contributions to adverse environmental impacts. Rather, if we consider materials in the totality of their impacts, they start to reveal themselves in terms of their sustainable purity or multilayered corruption.

After all, what is the design relevance of a visually beautiful material, if it causes devastation when harvested or extracted, has to be transported long distances, and is toxic while in use and when disposed of? The "beauty" of such a material quickly perishes. But this perishing only takes place upon the more detailed consideration of the totality of impact the material carries within itself, as its ontological totality. Once the material is selected as suitable in the totality of its impacts, it becomes possible and desirable to celebrate it visually and through design, making apparent the ontological beauty that it presents within itself and lends to the project as a whole.

Unfortunately, the works which have considered building construction in relation to ontology, such as those of Frampton and Hartoonian, have already been criticized for their emphasis on the haptic and visual aspects of materials, without sufficient acknowledgment of materials science and engineering (Fernandez, 2006, pp. 10–11). Sustainability was not part of such considerations. Thus, it is essential to call for a new, deeper and more contemporary definition of an ontological approach to building materials, and one which includes the widest possible definition of sustainability.

This book can be seen as one call for a more complete ontological consideration of building materials, including reflection upon a number of criteria discussed in the first three parts of the book. Finally, from this perspective, this book proposes that in order for architecture and the built environment to be more sustainable, materials should be given a more central and active position in contemporary building.

Emina Kristina Petrović

References

Frampton, K., 1990. Rappel a l'ordre: the case for tectonic. Arch. Design. 60 (3/4), 19–25.

Frampton, K., 1995. Studies in Tectonic Culture: The Poetics of Construction in Nineteenth and Twentieth Century Architecture. MIT Press, Cambridge, MA.

Fernandez, J., 2006. Material architecture: emergent materials for innovative buildings and ecological construction. Architectural Press, Boston.

Harman, G., 2002. Tool-Being: Heidegger and the Metaphysics of Objects. Open Court Publishing, Chicago.

Hartoonian, G., 1994. Ontology of Construction: On Nihilism of Technology in Theories of Modern Architecture. Cambridge University Press, Cambridge.

Heidegger, M., 1971. The origin of the work of art. In: Heidegger, M. (Ed.), Poetry, Language, Thought. Harper and Row Publishers, New York, pp. 15–87. First published in German in 1950.

Lund, C., 2015. If the future is software-defined, what to do with our hands?: post-digital commons and the unintended from a design perspective. In: Kulle, D., Lund, C., Schmidt, O., Ziegenhagen, D. (eds.). Post-digital Culture. Available from: http://post-digital-culture.org/clund/ (accessed September 2016.)

Oxford Dictionaries, 2016. Oxford University Press. Available from: www.oxforddictionaries.com (accessed September 2016.)

Petrović, E., Perkins, N., 2016. Materials in furniture design: towards a new conceptual framework. Inter. J. Interior Arch. Spatial Design, ii J. 4, 56–63.

Zumthor, P., 2006. Thinking Architecture. Birkhauser, Basel. First published in 1998.

Acknowledgments

The authors wish to acknowledge the significant contribution the following people have made to this book. Without them it would not have been possible:

Ljiljana Đukanović, University of Belgrade, Serbia
Rainer Hirth, Hochschule Coburg, Germany
Nigel Isaacs, Victoria University of Wellington, New Zealand
Adele Leah, Victoria University of Wellington, New Zealand
Guy Marriage, Victoria University of Wellington, New Zealand
Ana Radivojević, University of Belgrade, Serbia
Jaime Rios-Calleja, BUAP, Benemérita Universidad Autónoma de Puebla, Mexico
Mirjana Roter-Blagojević, University of Belgrade, Serbia
John Storey, Principal Symbiosis Architects and Consultants, New Zealand
Robert Vale, Victoria University of Wellington, New Zealand
Marci Webster-Mannison, Melbourne Design Studios, Australia

Introduction

Maibritt Pedersen Zari, Brenda Vale and Emina Kristina Petrović
Victoria University of Wellington, Wellington, New Zealand

The designing and making of built environments requires professionals to have considerable knowledge of building materials. Traditionally, this has included cost and how materials perform over time, including ease of maintenance. What has changed is that this knowledge now needs to expand to include consideration of the impacts materials have on ecosystems, the global climate, the health of people, as well as ethical considerations, such as whether exploitation occurred in producing the material. Broadly, all these issues are part of what "sustainability" might mean in relation to the design of the built environment. The change is driven by the growing awareness of building professionals, clients, and the general public that these issues are important. Signals, such as the 2015 Paris Agreement, suggest that change in the way buildings and their materials are procured and produced is imminent.

Materials are the physical substance of buildings and built infrastructure. The fundamental nature of architecture and design is the materiality of the created objects. As a result, buildings form the tangible world inhabited by humans, and their materials give the means by which people experience physical space (Petrović and Perkins, 2016). However, three key reasons have now emerged for rethinking the relationship between materials and architecture.

The first is the need to respond to climate change (Walther et al., 2005) and with it the need to change buildings and the materials they are made of (Altomonte, 2008). This implies the need to withstand more extreme weather events and also minimize the use of fossil fuels through energy efficiency measures (Steemers, 2003, Camilleri et al., 2001). Apart from the energy that goes into making and using buildings (see Chapter 3), the sourcing, processing, and transporting of materials also has implications for climate change (see Chapter 2). Mitigating and adapting to climate change requires profound changes in how individual buildings and whole urban areas are constructed and used (Hunt and Watkiss, 2011, Pedersen Zari, 2010, Hamin and Gurran, 2009). This means that change must also occur in what the built environment is made of—its materials.

The second reason for building professionals to increase their ability to select appropriate and sustainable building materials is current global urban growth. Building materials account for around half of all materials extracted from all biological ecosystems for human activities (Murray and Cotgrave, 2007; Roodman and Lenssen, 1995), while at least 40% of what society deems to be "waste" comes from construction and demolition activities (UNEP, 2007). This puts pressure on built environment professions to search for better ways of managing the impact their work poses on ecosystems, global climate, and human health.

The third major reason for change in how building materials are selected and used is the ever-increasing number of new materials and building technologies available. It has recently been estimated that more new materials have been developed in the last 20 years than in the rest of history combined (Schörpfer, 2011; Brownell, 2006). What is often not clear in advertisements for these are their origins, how they are made, and the consequences of their use, including their health impacts. This last is of great importance because the indoor environment is largely a product of its materials. Further, people now spend up to 80%–93% of their time indoors (Delgado-Saborit et al., 2011; Liu and Little 2012; Guieysse et al., 2008; Sexton and Dyer, 1996), again highlighting the responsibility that built environment professionals have for ensuring occupant health through the materials they specify.

This book is a response to these issues and the changes that must occur in building materials. It is focused on proposing principles for their evaluation for their suitability in sustainable construction.

The aim of this book is to examine the issues involved in choosing materials from an ecosystem services perspective, from the viewpoint of the design stage including lifecycle assessment, and the impact of materials on the health of building users. The core proposition of the book is that none of these aspects should be considered in isolation, but rather that a full evaluation of the appropriateness of a material for sustainable construction requires engagement with all concurrently. The book is divided into four parts. Parts I, II, and III propose principles for the evaluation of the appropriateness of various building and furnishing materials for sustainable construction. Part IV consists of a wide range of specific case studies illustrating these principles and providing specific examples.

Part I, made up of two chapters, examines materials as a part of ecosystems and provides an understanding of how typical building materials impact on, or could contribute to ecosystem services. Chapter 1, introduces the concept of ecosystem services in relation to the built environment and discusses the importance of understanding that materials were parts of whole living ecosystems, and remain part of ecosystems through their use and eventual end of life. Being aware of this helps people to realize that environmental impacts are more than calculating how much energy or water is used to process or create a material, but should also include thinking about, and if possible calculating, how using a material over its whole lifecycle impacts on what ecosystems do. These functions of ecosystems, which provide conditions suitable for life on the planet, are termed ecosystem services and are identified in the first chapter. In order to apply this knowledge of ecology to built environmental design, a list of key places for change is presented, along with explanations of which ecosystem services should be focused on in this context. The chapter concludes that the potential for profound positive change in terms of the ecological impact of built environments may be apparent, if an understanding of ecosystem services is integrated into built environment design and materials selection.

Chapter 2, expands on the method of ecosystem services analysis providing an example of how the concept of ecosystem services can be practically applied to built environment design and materials selection. A series of quick reference tables are provided that examine a selection of materials that are grown, extracted, or that are made, considering their impact on or contribution to ecosystem services. The main concluding idea of the chapter is that if built environments are to change over time so that they potentially have positive rather than damaging impacts on ecosystems, there is a need to move to a kind of building design (including materials selection procedures) that recognizes built environments as interdependent systems. This highlights a central theme of this book: a material is not "sustainable" in its own right, but can only be considered to be sustainable (or not) in a specific design, climate, ecological and cultural context over its whole lifecycle.

Part II, again in two chapters, deals with the implications of materials choice at the design stage. Chapter 3, is a discussion of materials used in buildings, grouped as materials which grow, those which are extracted and processed, and those which are manufactured. Building timber is thus discussed twice, first as a grown material and then as a manufactured material by looking at engineered wood products in their many forms. Chapter 3, broadly discusses when materials changed from being used in a form much closer to how they appear in the natural world to their present-day manifestations. The three categories of materials discussed also reflect the increasing use of energy, normally from fossil fuels, in processing materials. Whereas human labor was used for felling and shaping timber in the past, and it would be left stacked to dry out over time, now machinery is used and energy expended to kiln dry timber to shorten the process from harvesting to use. In a world that believes "time is money" such changes are inevitable. In contrast, in a world where making sustainable buildings is the goal and where the one thing in plentiful supply is human labor, different priorities could prevail.

Materials are considered separately in Chapter 3, which is problematic because it is the combination of materials that makes a building. To deal with the issue of which materials might best make a sustainable building, a subject covered from the viewpoint of ecosystem services in Part I, Chapter 4, deals with energy lifecycle assessment, recognizing that materials have other environmental impacts, such as water use, in their manufacture. The advantage of focusing on lifecycle energy assessment is that it includes the impact of the building in use, and necessitates consideration of the life of materials and their maintenance and refurbishment. The lifecycle assessment process is then explained, and the impact of the different phases in the life of a building (construction, use and maintenance, demolition and disposal) are discussed.

Because Part II is written for designers, a scenario is created to set out the energy lifecycle implications of different approaches to making a hypothetical building more sustainable. These include selecting materials which grow, using recycled materials and building smaller rather than larger spaces, in order to compare the relative impact of these approaches. Based on this lifecycle exercise some clear directions for making sustainable buildings emerge. Part II thus ends by setting out principles for guiding the choice of materials to lessen environmental impact.

Part III proposes principles for evaluating the impact of materials on the physical health of building users. It develops an argument based upon examining the problem of the significant increase in the total production and use of synthetic chemicals since World War II (Saunders, 2002), and how these may impact on human health. Chemicals used in building and furnishing materials are part of a larger problem, but there is currently insufficient information for human health impact assessments for an astounding 95% of chemicals used in construction products (Pacheco-Torgal, 2012). While research leading to regulative changes would help to solve this problem, there are concerns over the relatively slow development of new knowledge, and subsequent slow regulative changes. Part III problematizes the issues of slow development in these areas, proposing to consider current developments in both knowledge and regulations as fluid, and provide strategies to navigate this flow of information. Historical and current regulations and knowledge developments are used to identify key patterns in the recognition of health issues, the way such information is communicated, how progress is made towards changes in regulation and implementation of those changes, and the key obstructions to such processes. These are discussed as parts of a complex but nevertheless clear system. The aim is to empower the reader to anticipate and identify a range of likely future regulatory changes, and to be able to act on those even before legally binding regulative changes appear.

Chapter 5, introduces the problem of recognizing health impacts, and sets the tone of this part. Chapter 6, uses the examples of lead and asbestos as the most recognized and regulated against materials in the construction industry, and evaluates how successful regulation was in their elimination, and which steps were needed in that process. It highlights the patterns of very slow recognition and action towards their reduction and elimination. This is followed by Chapter 7, which discusses formaldehyde and phthalate plasticizers in PVC and vinyl, as chemicals currently in the process of being regulated against. The core question of the chapter is whether the current procedures of elimination can be seen as better and faster than the slow process of elimination of lead and asbestos. Chapter 8, discusses a wider range of new chemicals in order to demonstrate typical actions taken to evaluate risk to human health at the point of introduction of these chemicals into prolific use. Part III concludes with Chapter 9, which provides quick and indicative reviews on where many chemicals and materials sit in terms of a progression of recognition of their potential harmfulness. This highlights the importance of including systematic examination of the impacts of building materials on human health as a fundamental requirement for the development of a more sustainable built environment.

Part IV provides a mixture of historical and contemporary case studies. All nine case studies examine the same core question: are there practical examples which respond well to or illustrate the assessment criteria and issues raised in the first three parts of the book? The historical case studies illustrate the richness of solutions that existed before the rise of contemporary construction and to a large extent these are consistent with a sustainable approach to creating built environments. Chapter 10, illustrates traditional building materials and techniques in vernacular Serbian housing, as an example which is similar not only to many parts of Europe but also to vernacular architecture in many other parts of the world. Due to necessity and the limitations of preindustrial production, materials tended to be local, and their repeated use led to a thorough traditional understanding of their physical properties. Chapter 11, expands upon this general discussion, by focusing more specifically on traditional palm thatching in the west coast and highlands of Mexico. A detailed evaluation of the application of palm thatching techniques to a range of small contemporary buildings is given and an embodied energy and ecological footprint assessment of their environmental impact undertaken. The results show that using palm thatch has a low impact where the ways of building follow the traditional vernacular methods, but that impact is much higher when the same materials are used together with modern materials. The conclusion is that it is not the material which makes a sustainable building but how the material is used. In Chapter 12, the difficulty of calculating embodied energy values is discussed due to the complexity involved in knowing where materials come from. Using the New Zealand building industry and six specific materials as examples, it shows that the international trade in both building materials and the raw materials that go into them makes knowing what exactly has gone into a specific building very difficult.

Next follows a series of case studies of contemporary buildings which reflect some, if not all, of the principles and issues discussed in Parts I–III. Chapter 13, presents the experience of designing the Thurgoona Campus of Charles Sturt University in Australia, where rammed earth was used as the main wall material. The numerous challenges of such construction and how these were overcome are discussed. A series of buildings were created that, although first viewed with suspicion by the university, are now an essential part of a demonstrably sustainable campus. Chapter 14, discusses the Hockerton Housing Project in the United Kingdom. It shows how a material that is not thought to be sustainable—concrete—if integrated into the design of a zero fossil fuel and autonomous series of dwellings can lead to a low overall lifecycle environmental impact. In Chapter 15, a detailed description is provided of efforts to recycle or reuse as many materials and components as possible while remodeling an existing house in New Zealand. Issues arising from builders and contractors being uncommitted to the process, despite having committed clients, are discussed.

The final three case studies that make up Part IV concern the impact of building labeling schemes as a way of fostering sustainable architectural practices. In Chapter 16, the Meridian Building is examined, which was the first certified Green Star building in New Zealand. The process and difficulties of making choices to achieve such a building are discussed in this context. In Chapter 17, existing green building labeling systems in Germany are considered alongside issues that arise when designing for people with allergies. Two examples are used to showcase differences in possible emphasis, and challenges with achieving more comprehensive sustainability. Part IV concludes with Chapter 18, and a discussion of the Bullitt

Center in the United States, which is a certified "Living Building." This case study describes the challenges with specifying building materials which could be seen as being truly sustainable, meeting the stringent "Living Building Challenge" criteria.

Finally, underlying all discussions in this book is a deep understanding that materials offer opportunities for a deeper engagement with all construction-related activities. Materials can help facilitate a more ontological engagement with the everyday, which by necessity has to be sustainable.

References

Altomonte, S., 2008. Climate change and architecture: mitgation and adaptation strategies for a sustainable development. J. Sustain. Dev. 1 (1), 97–112.

Brownell, B. (Ed.), 2006. Transmaterial: A Catalog of Materials That Redefine Our Physical Environment. Princeton Architectural Press, New York, NY.

Camilleri, M., Jaques, R., Isaacs, N., 2001. Impacts of climate change on building performance in New Zealand. Build. Res. Inf. 29 (6), 440–450.

Delgado-Saborit, J.M., et al., 2011. Relationship of personal exposure to volatile organic compounds to home, work and fixed site outdoor concentrations. Sci. Total Environ. 409, 478–488.

Guieysse, B., et al., 2008. Biological treatment of indoor air for VOC removal: potential and challenges. Biotechnol. Adv. 26 (5), 398–410.

Hamin, E.M., Gurran, N., 2009. Urban form and climate change: balancing adaptation and mitigation in the U.S. and Australia. Habitat Inter. 33 (3), 238–245.

Hunt, A., Watkiss, P., 2011. Climate change impacts and adaptation in cities: a review of the literature. Climatic Change. 104 (1), 13–49.

Liu, Z., Little, J.C., 2012. Materials responsible for formaldehyde and volatile organic compound (VOC) emissions. In: Pacheco-Torgal, F., Jalali, S., Fucic, A. (Eds.), Toxicity of Building Materials. Woodhead Publishing, Cambridge, pp. 76–121.

Murray, P., Cotgrave, A.J., 2007. Sustainability literacy: the future paradigm for construction education? Struct. Survey. 25 (1), 7–23.

Pacheco-Torgal, F., 2012. Introduction: types of potential toxic building materials. In: Pacheco-Torgal, F., Jalali, S., Fucic, A. (Eds.), Toxicity of Building Materials. Woodhead Publishing, Cambridge, xv-xvii.

Pedersen Zari, M., 2010. Biomimetic design for climate change adaptation and mitigation. Arch. Sci. Rev. (ASR). 53 (2), 172–183.

Petrović, E., Perkins, N., 2016. Materials in furniture design: towards a new conceptual framework. Inter. J. Interior Arch. Spatial Design, ii J. 4, 56–63.

Roodman, D.M., Lenssen N., 1995. A Building Revolution: How Ecology and Health Concerns Are Transforming Construction, Washington, Worldwatch Institute.

Saunders, T., 2002. The Boiled Frog Syndrome: Your Health and the Built Environment. Wiley, Chichester.

Schörpfer, T., 2011. The alternative approach: observation, speculation, experimentation. In: Schörpfer, Thomas (Ed.), Material Design: Informing Architecture by Materiality. Birkhäuser GmbH, Basel.

Sexton, K., Dyer, R.S., 1996. Effects of indoor air quality on human health: setting strategic research directions and priorities. In: Gammage, Richard B., Berven, Barry A. (Eds.), Indoor Air and Human Health. CRC Press, Boca Raton, Florida.

Steemers, K., 2003. Towards a research agenda for adapting to climate change. Build. Res. Inf. 31 (3/4), 291–301.

UNEP, 2007. Buildings and Climate Change: Status, Challenges and Opportunities. United Nations Environment Program, Paris.

Walther, G.-R., Hughs, L., Vitousek, P., Stenseth, N.C., 2005. Consensus on climate change. Trends Ecol. Evol. 20 (12), 648–649.

Part I

Selecting Building Materials for Reduced Impacts on Ecosystem Services: Ecosystem Services Analysis

Utilizing relationships between ecosystem services, built environments, and building materials

1

Maibritt Pedersen Zari
Victoria University of Wellington, Wellington, New Zealand

1.1 Introduction: reducing the environmental impact of built environments

It is well known that the built environment has a large negative effect on ecosystem services (Rees, 1999). More than a third of all of the materials that are harvested, dug up, mined, and processed on the planet end up in built environments, while concurrently approximately a third of all of the waste that is buried, burnt, and dumped comes from construction and demolition activities (UNEP, 2007). In light of this, it is essential to understand the impact of this activity on the ecosystems we inhabit and rely upon for survival, not just in terms of energy use or pollutants (including the emission of carbon), but also in relation to complex and interconnected networks of ecosystem services (Zhang et al., 2010).

One way to reduce or to reverse the negative environmental impact of the built environment may be to create or redesign such environments so that they provide, integrate with, or support ecosystem services, and therefore reduce pressure on ecosystems. This is important as urban environments continue to grow and as the climate continues to change (Eigenbrod et al., 2011, McKinney, 2002). Ecosystem services analysis (ESA), developed by Pedersen Zari (2012a,b, 2014a,b, 2015a,b) is a means by which the concept of ecosystem services is specifically applied to built environment contexts. Part I of this book (this chapter and Chapter 2) expands these concepts to include the selection of building materials. Employing an understanding of ecosystem services in materials selection is a way to consider the contributions ecosystems make to providing materials, as well as the impacts that extracting, harvesting, processing, and using materials can have on ecosystems. By calculating the impact various materials have on ecosystem services at different spatial and temporal scales, a complementary set of decision-making criteria can be devised in addition to life cycle assessment (Part II) and health impact ones (Part III). Perhaps by analyzing the impact of certain building materials in relation to how ecosystems function, steps can be taken towards the creation of a built environment where

Materials for a Healthy, Ecological and Sustainable Built Environment: Principles for Evaluation.
DOI: http://dx.doi.org/10.1016/B978-0-08-100707-5.00001-0

positive integration with, and restoration of, local ecosystems can be realized (Olgyay and Herdt, 2004, Pedersen Zari, 2012a).

1.2 Ecosystem services: definitions and boundaries

Ecosystem services are the benefits that humans derive, either directly or indirectly, from the functions of ecosystems (Costanza et al., 1997). They are fundamental to basic human survival and wellbeing. Human use of ecosystem services is expanding due to human population increases as well as rises in per capita rates of consumption including building materials. In their seminal paper, Costanza et al. (1997) estimated that although humans would not be able to replace the ecosystem services they utilize, if these services had to be paid for in monetary terms, the cost would have been almost twice the entire global gross national product at the time of the research. Many of these services simply cannot be replaced with current human technology (Norberg, 1999), though approximately 60%, including 70% of all regulating and supporting ecosystem services, have now been degraded by human activities (Millennium Ecosystem Assessment, 2005b). The global condition of most ecosystem services except for the provisioning of food and raw materials has declined significantly over the past 60 years (Carpenter et al., 2009). In fact, ecosystems have changed more in the last 60 years than in any other period of human history (Millennium Ecosystem Assessment, 2005b). Several studies have also proven that in most cases, the costs of ecosystem conservation are far outweighed by the net benefits generated by healthy intact ecosystems (Turner and Daily, 2008).

Ecosystem services have been described as "the aspects of ecosystems consumed and utilised to yield human well-being" (Turner and Daily, 2008). These are aside from the intrinsic value some people believe ecosystems have, independent of the ability of humans to exploit them. Farber et al., (2002) state: "as humans are only one of many species in an ecosystem, the values they place on ecosystem functions, structures and processes may differ significantly from the values of those ecosystem characteristics to species or the maintenance (health) of the ecosystem itself." Proponents of deep ecology also believe that ecosystems have value aside from their functional or economic value to humans (Drengson and Inoue, 1995).

Up to 80% of the economic values of ecosystem services are currently not captured in traditional market-based mechanisms (de Groot et al., 2010). The ecosystem services concept is the subject of intensive international research and policy development initiatives in an attempt to define values with the intention that this may lead to preservation or restoration of ecosystems (Naidoo et al., 2008). Understanding ecosystem services aids in the setting of policy and development of tools to determine how much polluters should pay or how much protectors and regenerators of ecosystem services should be compensated (Carpenter et al., 2009, TEEB Foundations, 2010, TEEB in Business, 2011, TEEB in Local Policy, 2011, TEEB, 2011b). Examples of the many schemes incorporating an ecosystem services approach that have emerged since 2000 include: direct compensation paid to landowners for ecosystem services; conservation banking; tradable habitat rights; debt-for-nature swaps; insurance schemes; and tax relief programs. Some proponents of such schemes point out that it is unlikely that economic incentives alone will change human behavior so that people value and adopt environmental stewardship practices (Turner and Daily, 2008).

A focus on ecosystem services has been widely adopted internationally among ecology and policy professionals (Carpenter et al., 2009) and increasingly governments (Executive Office of the President of the United States, 2015), and even large financial institutions (Henry, 2016). This has occurred particularly since the concept was examined and formalized by the United Nations' Millennium Ecosystem Assessment of Ecosystems and Human Wellbeing (Alcamo et al., 2003, Millennium Ecosystem Assessment, 2005a,b, de Groot et al., 2010). Applying the concept of ecosystem services to architectural and urban design, landscape design, and particularly to building materials selection is still being developed however, and faces challenges (de Groot et al., 2010). When it comes to specifying materials for use in built environments the implications for ecosystem services are rarely taken into account (Zhang et al., 2010).

Although ecosystem services are inherently interconnected and interdependent, several ecologists define and list individual ecosystem services (Daily et al., 2000, de Groot et al., 2002, Millennium Ecosystem Assessment, 2005a, Cavanagh and Clemons, 2006, Costanza et al., 1997). The services that humans receive from ecosystems can be divided into: provisioning services such as food and medicines; regulation services such as pollination and climate regulation; supporting services such as soil formation and fixation of solar energy; and cultural services such as artistic inspiration and recreation (see Table 1.1). Daily (1997) provides scientific details of each function and the Economics of Ecosystems and Biodiversity studies (TEEB Foundations, 2010, TEEB, 2011a,b) provide further insights, case examples, and specific details of each ecosystem service.

Certain ecosystem services were only recognized when ecological knowledge increased, therefore it is likely that the list of known ecosystem services will evolve in the future. Lists of ecosystem services could develop further because they are determined not just by the functions of ecosystems but also by human ingenuity in deriving benefit from them (Heal et al., 2005).

Disaggregating the functioning of ecosystems into discrete "services" is philosophically as well as practically difficult, and a concentration on parts of ecosystems, rather than ecosystems as whole entities, could undermine the holism that is central to sustainable or regenerative design. Therefore, ESA should be used in concert with other aspects of design and materials selection that focus on the development of relationships between humans, the built environment, and wider ecosystems (Hoxie et al., 2012, Mang and Reed, 2012, du Plessis, 2012, Cole et al., 2012, Cole, 2012).

1.2.1 Provisioning services

Provisioning services are those which provide some kind of tangible resource to humanity. They provide for the nutritional, medicinal, shelter, and evolutionary (through genetic material) needs of people to survive at the most basic level, and for the great variety of technologies, products, chemicals, and energy needs or wants of humans also. This great range of goods and services extracted from ecosystems is largely based on the diverse amounts of biomass found on the planet,

Table 1.1 Ecosystem services

1. Provisioning services	2. Regulating services (*human time scale*)	3. Supporting services (*long time scale*)	4. Cultural services
1.1 Food • Human (land/fresh water/marine) • Forage	**2.1 Pollination and seed dispersal**	**3.1 Soil** • Formation • Retention • Renewal of fertility • Quality control	**4.1 Artistic inspiration**
1.2 Biochemicals • Medicines • Other	**2.2 Biological control** • Pest regulation • Invasive species resistance • Disease regulation	**3.2 Fixation of solar energy** • Primary production/plant growth (above ground, below ground, marine, fresh water)	**4.2 Education and knowledge**
1.3 Raw materials • Timber • Fiber • Stone • Minerals/ores	**2.3 Climate regulation** • GHG regulation • UV protection • Moderation of temperature	**3.3 Nutrient cycling** • Regulation of biogeochemical cycles • Retention of nutrients	**4.3 Aesthetic value**
1.4 Fuel/energy • Biomass • Solar • Hydro • Other	**2.4 Prevention of disturbance and moderation of extremes** • Wind/wave force modification • Mitigation of flood/drought • Erosion control	**3.4 Habitat provision** • Refugium function[a] • Nursery function[b]	**4.4 Cultural diversity and history**
1.5 Fresh water • Consumption • Irrigation • Industrial processes	**2.5 Decomposition** • Waste removal	**3.5 Species maintenance** • Biodiversity • Natural selection • Self-organization	**4.5 Recreation and tourism**
1.6 Ornamental resources	**2.6 Purification** • Water/air/soil		**4.6 Spiritual inspiration**
1.7 Genetic information			**4.7 Creation of a sense of place** **4.8 Relaxation and psychological wellbeing**

[a]Defined as suitable living space for wild plants and animals (de Groot et al., 2002).
[b]Defined as suitable reproduction habitat (de Groot et al., 2002).

formed through photosynthesis (de Groot et al., 2002). These ecosystems "goods" are a familiar and integral part of traditional forms of human economic systems and tend to be easier to comprehend and measure.

1.2.2 Regulating services

Regulating services relate to the cycling of nutrients at a local and global scale, and the filtering and transformation of pollutants and wastes. These services ensure the regulation and maintenance of the essential functions of ecosystems that support health and growth, and involve interactions between biotic (living) and abiotic (nonliving) elements of ecosystems. Many of these services are less direct in nature than provisioning services in that they do not lead to tangible products for human consumption, and consequently are more difficult to assign value to (TEEB Foundations, 2010), or perhaps to understand. However, these services are fundamental to the survival of all species (including humans). Because of their interlinked nature small changes in one can affect others in a myriad of ways. The multiple effects of climate change are an example of this phenomenon.

1.2.3 Supporting services

Supporting services are ones which enable the on-going structure of ecosystems to endure, as well as the living creatures, plants, and elements within them. Such supporting services ensure the continuation of provisioning and regulation services (Daily and Matson, 2008). Supporting services tend to be less understood or valued by humans than provisioning services or even regulating services, but remain crucially important to the on-going health of ecosystems, and therefore to humanity.

1.2.4 Cultural services

Many aspects of ecosystems are consumed not just for physical survival or to create functional objects but also for psychological or cultural benefits. An example is the creation of gardens for pleasure and relaxation. Such designed landscapes rely on the on-going survival of ecosystems to provide and maintain populations of plants, animals, and insects that are then arranged by humans for their own benefit or enjoyment.

Although social and ecological systems are intimately linked (Berkebile and Hoxie, 2012, Folke, 2006, Young et al., 2006), and proponents of the biophilia hypothesis argue that ecosystem health both affects and is affected by human psychological health (Kellert et al., 2008, Pedersen Zari, 2009), for the purposes of this research cultural services are thought to have less direct or obvious links to improved ecosystem health than the other categories, and therefore the actual provision of regulating, supporting, and provisioning ecosystem services. Also, many of these cultural services are not fulfilled by ecosystems exclusively. Artistic inspiration may equally be found in a city of character as in a forest, for example. For these reasons the category of cultural services and the provisioning service of ornamental resources are not considered further in this context.

1.3 Relationships between ecosystem services

Showing where ecosystem services are potentially linked and how they interact over time may give designers and other building professionals insights into how to select building materials to avoid or minimize negative impacts, or to support multiple ecosystem services that positively reinforce each other. For further examination and diagrams of the relationships between ecosystem services, see Pedersen Zari (2012b). Simplistically, all ecosystem services are related to each other in some way. It is apparent that provisioning services are dependent on both regulating and supporting services, but supporting or regulating services tend not to be dependent on provisioning services. This is because the very idea of "provisioning" involves some kind of human intervention or disturbance (even temporarily) of an ecosystem, such as the harvest of timber or the hunting of animals, for example. De Groot et al. (2010) state:

> *With increasing intensity of use, the production of goods and their benefits can only increase by adding human inputs such as fertilizer, water, pest control and labour. To maximize yield, usually of only one service, ecosystems are reduced to a substrate for production of biomass or other single use purposes. The most extreme types of land use are built up areas and areas covered by concrete or asphalt, where production of natural ecosystem goods approaches zero.*

This demonstrates that it is of great importance that any ecosystem service-based design methodology or evaluation technique does not ignore regulating or supporting services, particularly over long time periods, despite these being more difficult to measure in some cases (Zhang et al., 2010). The provision of food, e.g., is the provisioning service that appears to have the most dependent links to other ecosystem services (Pedersen Zari, 2012b). If people want to be able to continue to grow and harvest food in its many forms, it is vital that the ecosystem services of pollination and seed dispersal, climate regulation, decomposition, soil health, and nutrient cycling, along with most of the others remain functional. Because the ecosystem service of provision of food is fundamental to the continuation of the human species, this highlights how crucial it is to consider individual ecosystem services in the context of their relationships to other ecosystem services, which function at various temporal and spatial scales (de Groot et al., 2010), rather than just in isolation. Ecosystem processes that underpin ecosystem services range from global scales, such as climate regulation, to localized scales, such as nutrient cycling, and can be significant at more than one scale at the same time (de Groot et al., 2010). This is explained further in Table 1.2.

In the case where a particular building or design project has a focus on one ecosystem service only (such as a "zero-energy" home or a "water-positive landscape"), it is still of importance that other ecosystem services are considered, so that when improving or conserving one, another is not degraded (Daily and Matson, 2008). For example, increasing the carbon sequestration potential (climate regulation ecosystem service) of an area can impact negatively on biodiversity (species maintenance ecosystem service) if both services are not considered in tandem. This could happen, e.g., if trees are planted in an attempt to sequester carbon without an understanding

Table 1.2 Ecosystem services in a built environment context

Ecosystem services in a built environment context		Applicability to the built environment	Ecological significance[a]	Negative impact of built environment[a]
Supporting services	1. Habitat provision (*including: genetic information; biological control; fixation of solar energy; and species maintenance*)	Medium	High at biome and ecosystem scales	High at biome and ecosystem scales
	2. Nutrient cycling (*including: decomposition, soil building; and raw materials*)	Medium	High at the biome scale	High at biome and global scales
Regulating services	3. Purification of air, water, and soil	High	High at biome and ecosystem scales	High at biome and ecosystem scales
	4. Climate regulation	High	High at both global and biome scales	High at a global scale
Provisioning services	5. Provision of fuel/ energy for human consumption	High	Medium at a global scale	High at a global scale
	6. Provision of fresh water	High	High at the biome scale	High at the biome scale
	7. Provision of food (*including: provision of biochemicals*)	Medium	Medium at ecosystem scale	High at a global scale

[a]Global scale ≥ 1,000,000 km^2, Biome scale = 10,000–1,000,000 km^2, Ecosystem scale = 1–10,000 km^2 (based on definitions given by de Groot et al., 2010).

of the tree species necessary in that area to provide habitat for other locally occurring organisms (Gitay et al., 2002). This is relevant when selecting timber for construction. It could be argued that using *Pinus radiata*, the most common exotic plantation species used for timber production in New Zealand is a good choice because planting these plantations in place of pastoral land sequesters more carbon in the trees themselves and the soil beneath them for as long as the plantation stands. However, these plantation forests contain only approximately one-tenth of the biodiversity found in indigenous New Zealand forests (Brockerhoff et al., 2003). Thus the ecosystem service of habitat provision is negatively affected by selecting *P. radiata* timber. In *P. radiata* forests, the provision of water service (the flow of streams in and near plantations that are between 3 and 40 years old) can decrease significantly, and nutrient cycling can also be altered negatively (Mead, 2013). Erosion control, a regulation ecosystem service, can be significantly increased with the planting of pine plantations if planted on unstable pastoral grass slopes however (see Fig. 1.1). This changes obviously during harvest times (Mead, 2013).

The example of issues with *P. radiata* timber given above shows that there often is not a simple single solution to selecting materials when concerned about the health of ecosystem services. Often some trade-offs may need to be considered. Understanding the various issues and being aware of them can help to make well-informed decisions however. While there is existing research that can contribute to making these important decisions regarding materials selection, much of it requires patience to access, sift through, and then apply to the world of building and urban design. An initial attempt to bring some of this research together for some of the most common building materials is made in Section 2.5. The tables in that section also identify where some of the major gaps in research are, and indicate that more research needs to be done in some areas to make the idea of ecosystem services applicable to building materials selection. This is true particularly if the methodology is to be developed enough so that people can begin to measure and quantify the impacts of various materials on ecosystem services in order to make more accurate comparisons between various materials.

Figure 1.1 Pine plantation in Marlborough, New Zealand. By P. Capper, New Zealand—Logging tracks, Marlborough, New Zealand, 11 March 2006, CC BY 2.0, https://commons.wikimedia.org/w/index.php?curid = 3251478.

1.4 Defining ecosystem services for a built environment context: key places for change

Information about the negative environmental impact of the built environment is often relative to other human endeavors rather than ecosystems themselves. For example, the United Nations Environment Program states that 40% of all global energy and material resources are used to build and operate buildings (UNEP, 2007). Such a figure is useful in setting an agenda for future research and for establishing the urgency of the need to change urban environments and their use, but has no relationship to how much energy is available, what level of use would be sustainable, or what the environmental impact of this use is. A typical goal that fits into this way of thinking related to water consumption for a building might be "to reduce water use by 10%." This is based upon human-defined goals related to economic, political or convenience factors. It does not give information relative to an example of a successful and sustainable system, nor does it relate to what could be physically possible at a given site. A common reaction to such information is to reduce, remove, or stop certain behaviors or ways of constructing the built environment. ESA aims to enable built environments to move beyond that and towards creating health and wellbeing rather than simply reducing damage. A goal, again related to water consumption but based on ESA, might be to "tailor water use within a given site to its annual rainfall budget." This second kind of target is based upon the physical possibilities the site affords, can be clearly measured, and enables a development to be understood in the wider context of its ecosystem. Examples of this kind of design include the Lloyd Crossing Project proposed for Portland, Oregon (Portland Development Commission, 2004), and the Autonomous House built in the British Midlands in the 1990s (Vale and Vale, 2000). Pedersen Zari (2012a) expands on the concept of ESA in the creation of more sustainable or regenerative urban environments. ESA can be extended to carefully considering the suitability of building materials relative to their impact or relationship with ecosystem services at various life cycle stages. This is the focus of the remainder of this chapter.

Ecological economists and accountants have sought to place value on ecosystem services in different ways as the concept has been developed to fit into policy frameworks (Turner and Daily, 2008, TEEB, 2011b). This work is still in progress and in many cases the information needed to complete the task does not yet exist (ten Brink et al., 2011). In the past two decades many studies have attempted, with increasing success, to place economic value on ecosystem services (ten Brink et al., 2011, Daily et al., 2000) and The Economics of Ecosystems and Biodiversity (TEEB) group have collected more than 1100 different economic values for 10 different biomes and 22 different ecosystem services (TEEB Foundations, 2010).

"Value" is ultimately still a human subjective measure and attempts to determine value weightings for ecosystem services often attract debate (Rosemond and Anderson, 2003, Carpenter et al., 2009). While all aspects of ecosystem functioning are important to the system as a whole, and it is not perhaps even possible to assign value to discrete parts of an ecosystem, it is important to try to identify which of the 26 ecosystem services initially identified (Table 1.1) are the most suitable for inclusion in an urban built environment context. This is because trying to use all 26 ecosystem services is likely to be too cumbersome and complicated in a design or built environment evaluation context, or when specifying materials.

Seven ecosystem services were identified as most suitable for building materials selection and regenerative design methodologies in a built environment context, and are listed in Table 1.2. The rest of Section 1.4 explains how each ecosystem service was ranked. Section 1.5 provides brief descriptions of each ecosystem service.

The seven ecosystem services most suitable for use in a built environment context were deduced by using three ranking criteria:

1. The ability of an ecosystem service to be physically mimicked by or integrated with the built environment.
2. The impact of an ecosystem service on the maintenance of overall ecosystem health.
3. The relative negative impact that the urban environment has on the ecosystem service in question and the scale this relates to in terms of a local, regional, or global context.

Although the second and third criteria are more important in terms of the regeneration of ecosystem health, the physical constraints of the built environment are an important limiting factor in the context of built environment design and materials selection.

1.4.1 Ranking criterion one: integrating ecosystem services into the built environment

Some ecosystem services, e.g., pollination and regulation of species diversity, cannot easily be integrated with or mimicked in a built environment context. Others, such as the provision of energy or fuel are more conducive to integration (Bolund and Hunhammar, 1999). Contribution to or positive integration with some of these ecosystem services could become goals for how materials are selected or could determine overall performance targets for new developments. Following a methodology similar to that described by Shelton et al. (2001), three levels (low, medium, and high) were used to rank the ease of addressing each ecosystem service by the built environment. Criteria used to determine an ecosystem service's applicability to the built environment were: the ease of integrating an ecosystem service into built structure, infrastructure, and its surrounding context; and a consideration of existing technologies or design methods that already support or mimic the service. This allowed services to be prioritized and enabled an initial refinement of the list of ecosystem services suitable for developing ecosystem services-oriented design in the built environment.

Where an ecosystem service is thought to have low suitability for being considered in the context of the built environment, it has been combined with another service as a secondary consideration rather than eliminated altogether. For example, the provision of biochemicals (pharmaceuticals or medicines) is not easily met by the built environment, although a small contribution to this service could be made in a similar way to the provision of food service. The latter could entail deliberately growing edible plants in an urban context. Contributing to the service of provision of biochemicals would mean ensuring some of these plants were chosen for their desirable biochemical properties. Bundling the services together encourages designers initially to consider how to provide food in an urban context but to also be aware of the secondary consideration of providing medicines, using the same methods. Several architectural projects have been designed to incorporate medicinal plants in the landscaping, such as the 1990 Birley Health Centre in Sheffield and the Carolyn Stolman Healing Garden at the Avon Foundation Breast Center in San Francisco, designed by Topher Delaney to incorporate plants traditionally used to treat cancer (Therapeutic Landscapes Network, 2012). Therapeutic Landscapes Network (2012) gives further case studies and listings of landscapes that incorporate medicinal plants and other aspects of therapeutic landscapes. Many of the ecosystem services that cannot be physically provided by an urban environment, such as pollination, provision of genetic material, and species maintenance, can be provided if considered as part of the effort to increase provision of habitat in an urban setting.

The following four ecosystem services are considered to have high applicability to the built environment: provision of fuel/energy, provision of fresh water, purification, and climate regulation (in terms of carbon storage or sequestration). This is because some buildings are already able to perform these functions. For example, generating energy within an urban environment or on buildings has already been achieved in built form with established technologies such as photovoltaic panels. Case study evaluations of urban and building-integrated energy generation are provided by: Bahaj et al. (2007), Droege (2008), Omer et al. (2003), Storey et al. (2002), and Zahedi (2006). It should be noted that several building-integrated energy generation systems are not necessarily energy-effective when analyzed from a lifecycle perspective (Mithraratne, 2009, Omer et al., 2003). This again illustrates the importance of thorough, site- and design-specific environmental performance analysis of any building technology or material before it is integrated into building projects. In a similar way to generating energy in urban settings, the provision of fresh water, purification of air and water, and climate regulation (in terms of carbon sequestration) in urban environments and buildings, has been the subject of extensive research and testing. References for techniques to enable effective urban rainwater harvesting, water conservation, stormwater pollution mitigation, and stream restoration, along with case studies are given in Pedersen Zari (2012a, Table 9, Section 4.5.2.4). The same source also shows similar urban-focused research and case studies for the ecosystem services of purification and climate regulation (in terms of carbon sequestration).

Three ecosystem services may have medium applicability to a built environment context: nutrient cycling, provision of habitat, and provision of food. These may be more difficult to integrate into urban settings than the four services described earlier, but are likely to remain highly relevant when considering the selection of building materials. In the case of nutrient cycling and the provision of habitat, while the built environment can contribute to both, the regeneration of ecosystems that are not part of a built environment may be a more effective long-term means of restoring these services. For example, if the built environment can contribute to the provision of habitat this may be of benefit in terms of maintaining or restoring local biodiversity health. Many species of both plants and animals need areas of undisturbed and nonfragmented habitat (nonurban or human-dominated areas) in order to thrive however. Some plants, and particularly some animal species, are not compatible with humans in close quarters, in cities (such as large carnivores or herd animals), or with human pets, particularly cats, dogs, and rats (Wolch et al., 1995). This means efforts to increase the provision of habitat, and thus maintain or increase biodiversity health, will primarily need to focus on nonurban or periurban areas (Clarkson et al., 2007). Despite this, several researchers argue for increased habitat provision within urban environments and describe numerous benefits (Clarkson et al., 2007, Duncan et al., 2011, Given and Meurk, 2000, Goddard et al., 2010, Loram et al., 2007, Lundholm, 2006, Mathieu et al., 2007). These same sources are useful starting points for further research into how professionals of the built environment can effectively and successfully increase habitat in urban areas, suggesting that efforts to increase habitat provision in an urban setting can and should be made. In terms of nutrient cycling, much of this ecosystem service is due to interactions between organisms that enable decomposition, those that use certain nutrients to grow, and abiotic elements of soil, water, and atmosphere. Because the built environment is largely made up of impervious surfaces (Herald, 2003), and because of the way landfills are constructed and managed, altering the capacity for normal decomposition of organic products (Komilis et al., 1999), this natural kind of nutrient cycling is made more difficult.

The built environment holds great stores of materials in its very fabric along with vast amounts of embodied energy and carbon (Alcorn, 2003, 2006). Because of the quantity of materials used by the built environment and because of the considerable waste associated with construction and demolition activities (UNEP, 2007), the built environment might contribute to the ecosystem service of nutrient cycling by more carefully considered use and reuse of building materials and the provision of nutrient cycling of organic materials (through facilitating composting, for example). The reuse or transformation of "waste" as building materials, or the design of buildings to enable them to be reused or recycled at the end of their life is not a new idea, but could be much more carefully integrated in most urban contexts. Employing such strategies may take pressure off ecosystems to provide more raw materials and may also provide new sources of materials (though see discussion of potential issues with this in Part II (Section 3.3.4)). With few exceptions, buildings are generally demolished rather than deconstructed so that their materials can be reused (Storey et al., 2003, Kibert et al., 2002, p. 24). This is discussed in Chapter 15, in more detail. Several authors discuss ideas for increasing deconstruction and reuse of building materials and provide case studies (Guy and Shell, 2002, Kibert et al., 2000, Krook et al., 2012, McDonough and Braungart, 2002, pp. 92–117). Oliver (1997) provides vernacular examples of building materials reuse or recycling and design for deconstruction. Brown and Jameton (2000) and Pankhurst et al. (2011) give examples of urban composting. In a built environment context, the ecosystem service of decomposition is most closely linked with how waste is created and treated. Although the built environment itself is not capable of decomposing matter, the way it is built and organized, and the way materials are used within it and treated at the end of the life of a building, can support or hinder this service as part of a wider consideration of nutrient cycling.

The provision of food is considered to be of medium applicability to the urban environment. Although it is vital that urban communities begin to grow some of their own food (Pfeiffer, 2003, Koc et al., 1999, Deelstra and Giardet, 2000), devising urban agricultural systems that can feed large proportions of a city's residents may be difficult (Pedersen Zari, 2012a, Section 5.3.7). Food provision might be developed in concert with periurban areas and agriculturally productive hinterlands. Despite this, and like the earlier provision of habitat example, there is no lack of research advocating the need and benefits of urban and even building-integrated agriculture or food growing, with case study examples of how some cities (particularly in non-Western countries) already achieve high rates of urban food production (Koc et al., 1999, Deelstra and Giardet, 2000, Brown and Jameton, 2000). This obviously has implications for how materials might be specified.

Although the provision of habitat, nutrient cycling, and the provision of food are possible in urban settings, in order for these three services to become embedded into cities, particularly in industrialized nations, large changes in behavior and ways of living in urban environments need to take place. This may make the realization of such ideas more difficult. For this reason, these services are listed as being of medium applicability to a current built environment context.

1.4.2 Ranking criterion two: a hierarchy of ecosystem services relative to overall impact

An attempt is made here to examine the importance of each ecosystem service in terms of its physical, biological, and chemical contribution to maintaining the functional health of global ecological systems (after the definition of ecological importance found in Costanza and Folke, 1997). Any discussion of importance, significance, or value of ecosystem services is relative to the specific goals and perspective of whoever is making the definition. Value in terms of ecological importance is defined as "intrinsic" because it does not rely on human preferences, while value related to economic or social factors is defined as "instrumental" because it is inherently anthropocentric in nature (Farber et al., 2002).

Several authors discuss the need for ecological importance factors rather than just economic ones to be taken into account when valuing ecosystem services (Heal et al., 2005, ten Brink et al., 2011). Despite this, a thorough ranking of ecosystem services based on ecological importance has not been conducted to the author's knowledge. Such a ranking is problematic for a number of reasons. It is difficult because of gaps in knowledge about ecosystems, and also because of the nonlinear complexity found in ecosystems and in current ecosystem modeling (Peterson, 2002, pp. 128, 137). This, combined with the fact that ecosystems provide both direct and indirect, as well as tangible and more difficult to quantify benefits, over various temporal and spatial scales, makes it hard to predict ecosystem dynamics and assign value to them (Chee, 2004). Furthermore, although different categories of services can be listed, the interlinked and interdependent nature of ecosystem services often makes it practically difficult to assign value to them as discreet, independent services (Patterson, 2002, Chee, 2004).

Although there are difficulties in providing rankings of services related to ecological value, Costanza and Folke (1997) examine ways scientists can and do assign value independent of human preferences, such as "survival value" and "energy theory of value" based on thermodynamic principles (see also Patterson, 2002). Chee (2004) also discusses an ecosystem service's contribution to ecosystem "resilience" as a possible measure of value independent of human concerns. Patterson (2002) shows ecosystem services provided by the atmosphere, followed by the ocean, are larger in magnitude than those derived from terrestrial ecosystems.

Another way to gauge importance is to examine how certain ecosystem services are related to others. Pedersen Zari (2012a, Figure 63) illustrates that provisioning services are entirely dependent upon regulation and supporting services. The regulation services with the most connections are climate regulation and biological control, while the supporting service with the most connections is provision of habitat. This indicates that materials selection with the increased health of these services in mind may have the most secondary benefits in terms of supporting other ecosystem services. For example, a development that successfully regenerates habitat will have positive effects on species maintenance (biodiversity), nutrient cycling, and pollination services if designed carefully. Regulation services directly support or enable the greatest number of provisioning services. This demonstrates that regenerating or maintaining regulation services should be a priority even if provisioning services provide the most tangible, measurable, and economically profitable results to humans. Without regulation services all of the provisioning services (with the exception of ornamental resources) are negatively affected or disappear completely.

Shelton et al. (2001) found that all ecosystem services were ranked as having "high" ecological importance because decline in one would impact on future production capabilities in another. The research presented here follows a similar pattern. The exception is found in some of the provisioning services. These services, while of obvious importance to human wellbeing, tend to be related to human preferences and do not necessarily have high importance in terms of maintaining ecological health. This is why the provisioning service of fuel/energy is listed as being of "medium" ecological significance in Table 1.2. The provisioning service of fresh water is of great importance to humans but also to all flora and fauna, so has a high rating.

1.4.3 Ranking criterion three: the built environment's impact on ecosystem services

To determine which ecosystem services might be most appropriate to focus on in the context of the urban built environment and what it should be made of, the environmental impact that the built environment has on each ecosystem service was also considered. The underlying rationale was that if the built environment has a large negative effect on a particular ecosystem service, it would benefit that particular service if the built environment could mitigate or reverse this by trying to avoid damage to, integrating with, or working to restore the service. This provided a further and final set of criteria for ranking the ecosystem services in Table 1.2. The basis of the ranking of the impact of the built environment on particular ecosystem services was drawn from a number of sources (such as Doughty and Hammond, 2004, McDonald and Patterson, 2004, Graham, 2003, Newman, 2006, Wilby and Perry, 2006, Kibert et al., 2002).

The seven services (or bundles of services) presented in Table 1.2 suggest that in a similar way to the functioning of an ecosystem, an urban environment (and individual buildings as part of it) could be designed with a deliberate focus on generating a system that: provides habitat for species suitable for coinhabitation with humans in the urban built environment; contributes to soil formation and fertility through careful cycling of biodegradable wastes and recycling of nonbiodegradable wastes; purifies air, water, and soil; contributes to regulating climate through mitigating GHG emissions and the heat island effect, and by sequestering carbon; produces renewable energy; collects and distributes fresh water; and produces human food. In order to achieve this, a focus on designing systems must be employed. The careful selection of materials to create these systems and spaces is also paramount so that ecosystem services can potentially be supported and so that the built environment can perhaps become a positive integral part of ecosystem services. This must occur while the built environment is functioning, but also so that damage or destruction to the same ecosystem services at harvesting/extraction and processing stages of a material's life cycle does not occur.

The ranking results in Table 1.2 suggest that an initial focus on the three ecosystem services of regulation of climate, purification, and provision of fresh water might be easiest to focus on when selecting materials and will have a high number of additional beneficial outcomes for ecosystems and the species within them. The remaining four sets of services can be considered as having an equal ranking. This may change depending on local site-specific climatic and ecological conditions.

In the absence of rankings of ecosystem services provided by ecologists, the methods used to create the list in Table 1.2 have produced a more manageable set of ecosystem services for a design and materials selection context, while still capturing a wide range of ecological considerations. It should be remembered that this is not a ranking of the ecological importance of the ecosystem services alone, but considers if each ecosystem service might be suitable for consideration in a built environment and building materials context. Such a list should be revisited as knowledge gaps are filled in the field of ecology and the understanding of ecosystem services.

1.5 Descriptions of ecosystem services most applicable to a built environment context

A brief description of each ecosystem service most applicable to a built environment context follows, using the numbering system in Table 1.2. When descriptions of each ecosystem service are available, built environment professionals may be better able to utilize the concept of providing ecosystem services through the development or regeneration of the urban environment at early design stages and may also be able to apply the concept to materials selection. A full explanation of the current state of each ecosystem service is beyond the scope of this chapter, however The Economics of Ecosystems and Biodiversity studies (TEEB, 2011a,b, TEEB Foundations, 2010) provide further insights, case examples, and specific details of each ecosystem service. The Millennium Ecosystem Assessment publications provide detailed information about the current state of each ecosystem service (Millennium Ecosystem Assessment, 2005a,b).

1.5.1 Habitat provision

Habitat provision allows for the shelter and protection of organisms and often provides access to shelter and nutritional needs. Habitat provision also has a nursery function in terms of the protection of young organisms. Habitat provision services are relevant to both permanent as well as transient populations of organisms (de Groot et al., 2002). The provision of habitat is of extreme importance to maintaining biodiversity (Krauss et al., 2010).

1.5.2 Nutrient cycling

Nutrient cycling underpins all other ecosystem services because it provides balanced supplies of elements that are necessary for life (Millennium Ecosystem Assessment, 2005b, p. 333). Nutrient cycling services include the regulation of biogeochemical cycles (such as those of oxygen, water, potassium, nitrogen, phosphorus, sulfur, and carbon), as well as the retention of nutrients within a system. Provision for storage, internal cycling, processing, and acquisition of nutrients are parts of this service (de Groot et al., 2002).

1.5.3 Purification

Ecosystems allow for the purification or detoxification of air, water, and soil through assimilation and transformation of toxic compounds (through the bodies of plants or animals), decomposition, and filtering through abiotic elements (filtering of water through soil or sand, for example). Purification also allows the recovery of mobile nutrients and removal or breakdown of excess or toxic nutrients and compounds (Costanza et al., 1997). While the purification ecosystem service may act to reduce concentrations of substances that are harmful to humans, the capacity of ecosystems to do this is finite, and has been exceeded in many locations (Hinga et al., 2005).

1.5.4 Climate regulation

Climate regulation is the ecosystem service that regulates processes related to atmospheric chemical composition, the greenhouse effect, the ozone layer, precipitation, air quality, and moderation of temperature and weather patterns (including cloud formation), at both global and local scales (Costanza et al., 1997). In terms of GHG regulation at the global scale, this may include the ability of ecosystems both to emit and absorb chemicals (Forster et al., 2007). At a local scale vegetation lowers temperatures in urban environments and removes pollutants from the air (TEEB, 2011b). Climate change is one of the most important issues humans must consider (IPCC, 2007, pp. 8–22). This service is of extreme importance in terms of human health and economic prosperity (Alcamo et al., 2003), but this does not eclipse its significance in terms of its impact on biodiversity and the continuation of ecosystems in forms similar to what currently exists (Gitay et al., 2002).

1.5.5 Provision of fuel/energy source

Fuels and energy derived from biomass, biofuels, plant oils, and wind, hydro, and solar energy are an essential ecosystem service. As human technology changes, different kinds of energy sources are needed and can be accessed, expanding the scope and scale of this ecosystem service. Emerging technologies in the development of ocean-powered energy generation (including wave, tidal, and ocean thermal energy conversion) (Benelghali et al., 2007, Pelc and Fujita, 2002), and artificial photosynthesis (Moore et al., 2004, Ron, 2006) are examples.

It is currently difficult for renewable energy sources to compete with cheap and established fossil fuel energy sources (Pelc and Fujita, 2002) due to the on-going subsidization of conventional energy, which encourages greater use of fossil fuels and consequent emissions of GHGs and further climate change (Millennium Ecosystem Assessment, 2005c). Such subsidies were estimated to be $US250–300 billion in the mid-1990s (Millennium Ecosystem Assessment, 2005c).

Most research teams, including the Millennium Ecosystem Assessment, that provide lists of ecosystem services do not include nonrenewable energy sources (fossil fuels) as part of provisioning ecosystem services because they are not dependent on the living components of current ecosystems. Hodas (2007) suggests that the ecosystem service of collecting and storing solar energy as fossil fuels over long time periods should be acknowledged and factored into legal and economic frameworks of valuation, because fossil fuels are nonrenewable and currently are not "managed sustainably." He argues: "It is necessary to value the ecosystem services that manufactured fossil fuels, and to find a legal mechanism to internalize that value into the marketplace, either as a cost on the fossil fuel resource or a subsidy on renewable energy alternatives that seek to collect solar energy and convert it into a usable form." "Geosystem services" which are a consequence of "geodiversity" are terms that have been used since 1993 to describe the abiotic (nonliving) components and resulting services of ecosystems (Gray, 2011). Gray (2011) considers the provision of fossil fuels to be a part of the provision of geosystem services and discusses the interdependent nature of ecosystem and geosystem services. Some researchers include the provision of fuel with the provision of raw materials (see, e.g., TEEB, 2011b, Costanza et al., 1997).

1.5.6 Provision of fresh water

Ecosystem services related to water include the regulation of hydrological flows as well as the storage, purification, and retention of water (de Groot et al., 2002). Water is consumed for human needs by both humans and their livestock to ensure survival and health. It is also consumed in large amounts for crop irrigation (Hoff et al., 2010) and to a lesser extent by industrial processes, such as milling (Oki and Kanae, 2006).

Water can be considered as both a provisioning and a regulating service (Costanza et al., 1997). In terms of regulation, local ecosystems have a large influence on retaining water, managing the volume and timing of eventual run off, recharging of aquifers, control of flooding, and quality of drinking water (Winz, 2005). All of these regulation aspects also impact directly on the provisioning service of water.

1.5.7 Provision of food

The provision of food includes sources from terrestrial, marine, and aquatic ecosystems for humans directly, and also as forage or feed for domesticated livestock. It includes hunting, gathering, subsistence farming, and larger-scale agricultural, horticultural, and commercial fishing activities (Costanza et al., 1997). While the provision of food has increased greatly since the late 1960s due to the "green revolution" (Evenson and Gollin, 2003), this has come at the expense of a decrease in regulation and supporting services (Mooney et al., 2009, Carpenter et al., 2009). Food production rose by 45% between 1992 and 2009. This exceeded the increase in population for the same period which was only 26% (UNEP, 2011).

Although natural ecosystems provide the conditions for agricultural food growing by providing the ecosystem services of pollination, biological pest control, maintenance of soil structure and fertility, nutrient cycling, and hydrological services (Power, 2010), the current form of global food production based on transported products of large-scale agricultural businesses, rather than the more traditional small-scale local farms that existed before the "green revolution," is entirely dependent on the use of fossil fuels. Over time this diminishes the ability of land to produce food without them (Pfeiffer, 2006), while impacting negatively on other ecosystem services (such as habitat provision, water provision and regulation, and soil fertility) and often on biodiversity (UNEP, 2011). For plant-based foods, fossil fuels are used to plant seeds (through the use of fuel-consuming machinery), promote growth (through the use of fertilizer), maintain crops (through the use of pesticides and herbicides), harvest them (using fuel-consuming machinery), process them (again by use of energy-intensive machinery), and finally to transport foods to market (using fossil fuel-consuming vehicles).

The calculated energy content of food based on the fossil fuels used to produce it, exceeds by more than seven times on average the calorific energy content of the food (Eshel and Martin, 2006). Food production accounted for at least 10.5% of total global fossil fuel use in the mid-2000s and 17% in the United States in 2002.

This proportion of energy use is growing and is projected to peak in 2020 (Eshel and Martin, 2006). This means food production and the dietary choices of individuals contribute significantly to climate change. Without the input of fossil fuels the current global model of food production would cease to be viable (Pfeiffer, 2006). Discussion about specific dietary choices in relation to an individual's ecological footprint and contribution to climate change can be found in Vale and Vale (2009, pp. 40, 46), and Eshel and Martin (2006). Essentially, plant-based diets rather than a diet based on meat have a significantly lower overall ecological and climatic impact.

1.6 Conclusion: ecosystem services and the built environment. Moving towards a more positive relationship

The concept of ecosystem services is increasingly being applied to many fields of human endeavor and the potential for profound change in how built environments and the materials within them are designed, valued, built, and used is apparent. This is not just in terms of reducing environmental impacts but also in bringing ecological knowledge into built environment design to the point that the evolving creation of built environments can actually contribute positively to ecosystems (Pedersen Zari, 2012b). As Olgyay and Herdt (2004) state:

> *As we increase the positive impacts of our buildings beyond their negative impacts, we will have a net positive change on our ecosystems structure. This is a profound change in thinking... where we are continually investing in and profiting from our environmental stewardship. The implication of this information is that as the value of our ecosystem services become socially recognised, it will be well within our technical means to design buildings to create an ecological profit.*

Several attempts to integrate ecosystem services into built environment design including materials selection have been made, but currently these are not fully developed or widely used. Several challenges remain before this can happen including: reaching consensus about ecosystem services definitions, boundaries, metrics, and hierarchies; understanding the mechanisms of how ecosystems provide services; defining indicators and benchmarks for measuring the capacity of ecosystems to provide services or recover from damage over time; investigating the use of aggregate metrics and relative importance of different aspects of services when considering substitutions; integrating spatial and temporal issues of both ecosystem services and built environment design; exploring how to spatialize, map, model, and visualize ecosystem services in relation to urban environments; integrating social and economic concerns into ecosystem services methods; better understanding the effect of scale and location on ecosystem services in relation to the materials market; and defining and communicating values of ecosystem services (Koellner and

Geyer, 2013, de Groot et al., 2010). In spite of these challenges and gaps in knowledge, the research described in this chapter provides a basis for further exploration, experimentation, and application to real-world situations. Chapter 2, gives more detail of practical applications of the ESA concept to built environment design and materials selection.

References

Alcamo, J., Ash, N., Butler, C., Baird Callicott, J., Capistrano, D., Carpenter, S., et al., 2003. Ecosystems and Human Wellbeing. Millennium Ecosystem Assessment. Island Press, Washington, DC.

Alcorn, A., 2003. Embodied Energy and CO2 Coefficients for NZ Building Materials. Centre for Building Performance Research, Victoria University of Wellington, Wellington.

Alcorn, A., 2006. Energy and Carbon Coefficients of Building and Related Materials. Centre for Building Performance Research, Victoria University, Wellington.

Bahaj, A.S., Myers, L., James, P.A.B., 2007. Urban energy generation: influence of micro-wind turbine output on electricity consumption in buildings. Energy Buildings. 39 (2), 154–165.

Benelghali, S., Benbouzid, M., Charpentier, J.F., 2007. Marine tidal current electric power generation technology: state of the art and current status. In: IEEE IEMDC'07. Antalya, Turkey, pp. 1407–1412.

Berkebile, B., Hoxie, C., 2012. Implementing regenerative design through community dialogue. Building Res. Inf. 40 (1), 81–94.

Bolund, P., Hunhammar, S., 1999. Ecosystem services in urban areas. Ecol. Econ. 29 (2), 293.

Brockerhoff, E.G., Ecroyd, C.E., Leckie, A.C., Kimberley, M.O., 2003. Diversity and succession of adventive and indigenous vascular understorey plants in *Pinus radiata* plantation forests in New Zealand. Forest Ecol. Manag. 185 (3), 307–326.

Brown, K., Jameton, A., 2000. Public health implications of urban agriculture. J. Public Health Policy. 21 (1), 20–39.

Carpenter, S.R., Mooney, H.A., Agard, J., Capistrano, D., DeFries, R.S., DÃaz, S., et al., 2009. Science for managing ecosystem services: Beyond the Millennium Ecosystem Assessment. Proc. Natl. Acad. Sci. 106 (5), 1305–1312.

Cavanagh, J.-A.E., Clemons, J., 2006. Do urban forests enhance air quality? Aust. J. Environ. Manage. 13 (2), 597–604.

Chee, Y.E., 2004. An ecological perspective on the valuation of economic services. Biol. Conserv. 120, 549–565.

Clarkson, B., Wehi, P., Brabyn, L., 2007. Bringing Back Nature into Cities: Urban Land Environments, Indigenous Cover and Urban Restoration, Report No. 52. Centre for Biodiversity and Ecology Research, University of Waikato, Hamilton.

Cole, R.J., 2012. Regenerative design and development: current theory and practice. Building Res. Inf. 40 (1), 1–6.

Cole, R.J., Busby, P., Guenther, R., Briney, L., Blaviesciunaite, A., Alencar, T., 2012. A regenerative design framework: setting new aspirations and initiating new discussions. Building Res. Inf. 40 (1), 95–111.

Costanza, R., Folke, C., 1997. Valuing ecosystem services with efficiency, fairness, and sustainability as goals. In: Daily, G. (Ed.), Nature Services. Island Press, Washington, DC.
Costanza, R., d'Arge, R., de Groot, R., Farber, S., Grasso, M., Hannon, B., et al., 1997. The value of the world's ecosystem services and natural capital. Nature. 387 (6630), 253–260.
Daily, G.C., 1997. Nature's Services Societal Dependence on Natural Ecosystems. Island Press, Washington DC.
Daily, G.C., Matson, P.A., 2008. Ecosystem services: from theory to implementation. Proc. Natl. Acad. Sci. 105 (28), 9455–9456.
Daily, G.C., Soderqvist, T., Aniyar, S., Arrow, K., Dasgupta, P., Ehrlich, P.R., et al., 2000. The value of nature and the nature of value. Science. 289, 395–396.
de Groot, R., Wilson, M.A., Boumans, R.M.J., 2002. A typology for the classification, description and valuation of ecosystem function, goods and services. Ecol. Econ. 41, 393–408.
de Groot, R.S., Alkemade, R., Braat, R., Hein, L., Willemen, L., 2010. Challenges in integrating the concept of ecosystem services and values in landscape planning, management and decision making. Ecol. Compl. 7 (3), 260–272.
Deelstra, T., Giardet, H., 2000. Urban agriculture and sustainable cities. In: Bakker, N., Dubelling, M., Gundel, S., Sabel-Koschella, V., Zeeuw, A. (Eds.), Growing Cities: Urban Agriculture on the Policy Agenda. Food and Agriculture Development Centre (ZEL), Feldafing, Germany, pp. 43–65.
Doughty, M., Hammond, G., 2004. Sustainability and the built environment at and beyond the city scale. Building Environ. 39, 1223–1233.
Drengson, A., Inoue, Y. (Eds.), 1995. The Deep Ecology Movement: An Introductory Anthology. North Atlantic Publishers, Berkeley, CA.
Droege, P., 2008. Urban Energy Transition. From Fossil Fuels to Renewable Power. Elsevier, Oxford.
du Plessis, C., 2012. Towards a regenerative paradigm for the built environment. Building Res. Inf. 40 (1), 7–22.
Duncan, R.P., Clemants, S.E., Corlett, R.T., Hahs, A.K., McCarthy, M.A., McDonnell, M.J., et al., 2011. Plant traits and extinction in urban areas: a meta analysis of 11 cities. Glob. Ecol. Biogeogr. 20 (4), 509–519.
Eigenbrod, F., Bell, V.A., Davies, H.N., Heinemeyer, A., Armsworth, P.R., Gaston, K.J., 2011. The impact of projected increases in urbanization on ecosystem services. Proc. R. Soc. B Biol. Sci. 278 (1722), 3201–3208.
Eshel, G., Martin, P.A., 2006. Diet, energy, and global warming. Earth Interact. 10 (9), 1–17.
Evenson, R.E., Gollin, D., 2003. Assessing the impact of the green revolution, 1960 to 2000. Science. 300 (5620), 758–762.
Executive Office of the President of the United States, States, E.O.o.t.P.o.t.U, 2015. Memorandum for Executive Departments and Agencies (M-16-01). US Government, Washington, DC.
Farber, S., Costanza, R., Wilson, M.A., 2002. Economic and ecological concepts for valuing ecosystem services. Ecol. Econ. 41 (3), 375–392.
Folke, C., 2006. Resilience: The emergence of a perspective for social-ecological systems analyses. Glob. Environ. Chang. 16 (3), 253–267.
Forster, P., Ramaswamy, V., Artaxo, P., Berntsen, T., Betts, R., Fahey, D.W., et al., 2007. Changes in atmospheric constituents and in radiative forcing. In: Solomon, S., Qin, D., Manning, M., Chen, Z., Marquis, M., Averyt, K.B., Tignor, M., Miller, H.L. (Eds.),

Climate Change 2007: The Physical Science Basis. Contribution of Working Group I to the Fourth Assessment Report of the Intergovernmental Panel on Climate Change. Cambridge University Press, Cambridge, United Kingdom and New York, NY, USA.

Gitay, H., Suárez, A., Watson, R., Jon Dokken, D., 2002. Climate Change and Biodiversity. IPCC. Available at: http://www.coastman.net.co/publicaciones/cc/(0065).pdf.

Given, D., Meurk, C., 2000. Biodiversity of the urban environment: The importance of indigenous species and the role urban environments can play in their preservation. In: Urban Biodiversity and Ecology as a Basis for Holistic Planning and Design Workshop. Lincoln University, Wickliffe Press, Christchurch, New Zealand, 28–29 October. pp. 22–33.

Goddard, M.A., Dougill, A.J., Benton, T.G., 2010. Scaling up from gardens: biodiversity conservation in urban environments. Trends Ecol. Evol. 25 (2), 90–98.

Graham, P., 2003. Building Ecology - First Principles for a Sustainable Built Environment. Blackwell Publishing, Oxford.

Gray, M., 2011. Other nature: geodiversity and geosystem services. Environ. Conserv. 38 (3), 271–274.

Guy, B., Shell, S., 2002. Design for deconstruction and materials reuse. In: Chini, A., Schultmann, F. (Eds.), Design for Deconstruction and Materials Reuse. Report No. 272. CIB, Rotterdam.

Heal, G.M., Barbier, E.B., Boyle, K.J., Covich, A.P., Gloss, S.P., Hershner, C.H., et al., 2005. Valuing Ecosystem Services: Toward Better Environmental Decision-Making. The National Academies Press, Washington, DC.

Henry, K., 2016. Dr Ken Henry AC Speech - Advancing Australia's Natural Capital. 27 May. In: Address to The Fiona Wain Oration, Sydney.

Herald, J., 2003. Pointing the Way to Reducing Stormwater Impacts on Urban Streams. In: Third South Pacific Conference on Stormwater and Aquatic Resource Protection, Auckland, New Zealand.

Hinga, K., Batchelor, A., Ahmed, M., Osibanjo, O., Lewis, N., Pilson, M., 2005. Waste processing and detoxification. In: Hassan, R., Scholes, R., Ash, N. (Eds.), Millennium Ecosystem Assessment: Current State and Trends. Island Press, Washington, DC.

Hodas, D., 2007. Ecosystem subsidies of fossil fuels. J. Land Use Environ. Law. 22 (2), 599–622.

Hoff, H., Falkenmark, M., Gerten, D., Gordon, L., Karlberg, L., Rockstrom, J., 2010. Greening the global water system. J. Hydrol. 384 (3–4), 177–186.

Hoxie, C., Berkebile, R., Todd, J.A., 2012. Stimulating regenerative development through community dialogue. Build. Res. Inf. 40 (1), 65–80.

IPCC, 2007. Climate Change 2007: Impacts, Adaptation and Vulnerability. Contribution of Working Group II to the Fourth Assessment Report of the IPCC. Cambridge University Press, Cambridge.

Kellert, S.R., Heerwagen, J.H., Mador, M.L., 2008. Biophilic Design. John Wiley & Sons, New Jersey.

Kibert, C.J., Sendzimir, J., Guy, B., 2000. Construction ecology and metabolism: natural system analogues for a sustainable built environment. Constr. Manage. Econ. 18 (8), 903–916.

Kibert, C.J., Sendzimir, J., Guy, G.B., 2002. Construction Ecology. Spon Press, New York.

Koc, M., MacRae, R., Mougeot, L., Welsh, J., 1999. For Hunger Proof Cities. Sustainable Urban Food Systems. International Development Research Centre, Ottawa.

Koellner, T., Geyer, R., 2013. Global land use impact assessment on biodiversity and ecosystem servic. Int. J. Life Cycle Ass. 18 (6), 1185–1187.

Komilis, D.P., Ham, R.K., Stegmann, R., 1999. The effect of landfill design and operation practices on waste degradation behavior: a review. Waste Manage. Res. 17 (1), 20–26.

Krauss, J., Bommarco, R., Guardiola, M., Heikkinen, R.K., Helm, A., Kuussaari, M., et al., 2010. Habitat fragmentation causes immediate and time-delayed biodiversity loss at different trophic levels. Ecol. Lett. 13 (5), 597–605.

Krook, J., Svensson, N., Eklund, M., 2012. Landfill mining: a critical review of two decades of research. Waste Manage. 32 (3), 513–520.

Loram, A., Tratalos, J., Warren, P., Gaston, K., 2007. Urban domestic gardens (X): the extent & structure of the resource in five major cities. Landsc. Ecol. 22 (4), 601–615.

Lundholm, J., 2006. Green roofs and facades: a habitat template approach. Urban Habitats. 4 (1), 87–101.

Mang, P., Reed, B., 2012. Designing from place: a regenerative framework and methodology. Build. Res. Inf. 40 (1), 23–38.

Mathieu, R., Freeman, C., Aryal, J., 2007. Mapping private gardens in urban areas using object-oriented techniques and very high-resolution satellite imagery. Landsc. Urban Plan. 81 (3), 179–192.

McDonald, G.W., Patterson, M.G., 2004. Ecological Footprints and interdependencies of New Zealand regions. Ecol. Econ. 50 (1-2), 49–67.

McDonough, W., Braungart, M., 2002. Cradle to Cradle - Remaking the Way We Make Things. North Point Press, New York.

McKinney, M.L., 2002. Urbanization, biodiversity, and conservation. Bioscience. 52 (10), 883–890.

Mead, D.J., 2013. Sustainable Management of Pinus Radiata Plantations. FAO Forestry Paper No. 170. FAO, Rome.

Millennium Ecosystem Assessment, 2005a. Ecosystems and Human Well-Being: Biodiversity Synthesis. World Resources Institute., Washington, DC.

Millennium Ecosystem Assessment, 2005b. Ecosystems and Human Well-being: Current State and Trends, 5 vol. Island Press, Washington, DC.

Millennium Ecosystem Assessment, 2005c. Ecosystems and Human Well-Being: Synthesis. Island Press, Washington, DC.

Mithraratne, N., 2009. Roof-top wind turbines for microgeneration in urban houses in New Zealand. Energy Build. 41 (10), 1013–1018.

Mooney, H., Larigauderie, A., Cesario, M., Elmquist, T., Hoegh-Guldberg, O., Lavorel, S., et al., 2009. Biodiversity, climate change, and ecosystem services. Curr. Opin. Environ. Sust. 1 (1), 46–54.

Moore, T., Moore, A., Gust, D., Hambourger, M., Brune, A., 2004. Artificial photosynthesis and hydrogen production: Strategies for sustainable energy production. In: 13th International Congress on Photosynthesis, Montreal, Canada, Aug 31st -Sept 3rd.

Naidoo, R., Balmford, A., Costanza, R., Fisher, B., Green, R.E., Lehner, B., et al., 2008. Global mapping of ecosystem services and conservation priorities. Proc. Natl. Acad. Sci. 105 (28), 9495–9500.

Newman, P., 2006. The environmental impact of cities. Environ. Urban. 18 (2), 275–295.

Norberg, J., 1999. Linking Nature's services to ecosystems: some general ecological concepts. Ecol. Econ. 29 (2), 183–202.

Oki, T., Kanae, S., 2006. Global hydrological cycles and world water resources. Science. 313 (579), 1068–1072.

Olgyay, V., Herdt, J., 2004. The application of ecosystems services criteria for green building assessment. Solar Energy. 77 (4), 389–398.

Oliver, P., 1997. Encyclopedia of the Vernacular Architecture of the World. Cambridge University Press, New York.

Omer, S.A., Wilson, R., Riffat, S.B., 2003. Monitoring results of two examples of building integrated PV (BIPV) systems in the UK. Renew. Energy. 28 (9), 1387–1399.

Pankhurst, L., Akeel, U., Hewson, C., Maduka, I., Pham, P., Saragossi, J., et al., 2011. Understanding and mitigating the challenge of bioaerosol emissions from urban community composting. Atmos. Environ. 45 (1), 85–93.

Patterson, M.G., 2002. Ecological production based pricing of biosphere processes. Ecol. Econ. 41 (3), 457–478.

Pedersen Zari, M., 2009. An architectural love of the living: Bio-inspired design in the pursuit of ecological regeneration and psychological wellbeing. In: Brebbia, C.A. (Ed.), Sustainable Development and Planning IV. Wessex Institute of Technology, Southampton, pp. 293–302.

Pedersen Zari, M., 2012a. Ecosystem services analysis for the design of regenerative built environments. Build. Res. Inf. 40 (1), 54–64.

Pedersen Zari, M., 2012b. Ecosystem Services Analysis for the Design of Regenerative Urban Built Environments. PhD, Victoria University of Wellington, Wellington, New Zealand [Online] Available at: http://researcharchive.vuw.ac.nz/handle/10063/2491.

Pedersen Zari, M., 2014a. Ecosystem processes for biomimetic architectural and urban design. Archit. Sci. Rev. 58 (2), 106–119.

Pedersen Zari, M., 2014b. Ecosystem services analysis in response to biodiversity loss caused by the built environment. S.A.P.I.EN.S. 7 (1), 1–21.

Pedersen Zari, M., 2015a. Ecosystem services analysis: mimicking ecosystem services for regenerative urban design. Inter. J. Sust. Built Environ. 4 (1), 145–157.

Pedersen Zari, M., 2015b. Mimicking ecosystems for bio-inspired intelligent urban built environments. Intell. Build. Inter. 8 (2), 57–77.

Pelc, R., Fujita, R., 2002. Renewable energy from the Ocean. Marine Policy. 26 (6), 471–479.

Peterson, G., 2002. Using ecological dynamics to move toward an adaptive architecture. In: Kibert, C.J., Sendzimir, J., Guy, G.B. (Eds.), Construction Ecology. Spon Press, London.

Pfeiffer, D.A., 2003. Eating Fossil Fuels. From The Wilderness Publications.

Portland Development Commission, 2004. Lloyd Crossing: Sustainable Urban Design Plan and Catalyst Project.

Pfeiffer D.A., Eating Fossil Fuels, 2003, From The Wilderness Publications.

Power, A.G., 2010. Ecosystem services and agriculture: tradeoffs and synergies. Philos. Trans. R. Soc. B. 365 (1554), 2959–2971.

Rees, W., 1999. The built environment and the ecosphere: a global perspective. Build. Res. Inf. 27 (4/5), 206–220.

Ron, J.P., 2006. An integrated artificial photosynthesis model. In: Anthony, F., Collings, P.C.C. (Eds.), Artificial Photosynthesis. Wiley-VCH Verlag, Weinheim, pp. 13–34.

Rosemond, A.D., Anderson, C.B., 2003. Engineering role models: do non-human species have the answers?. Ecol. Eng. 20, 379–387.

Shelton, D., Cork, S., Binning, C., Parry, R., Hairsine, P., Vertessy, R., et al., 2001. Application of an ecosystem services inventory approach to the Goulburn Broken Catchment. Third Australian Stream Management Conference, Brisbane, August 27-29. Cooperative Research Centre for Catchment Hydrology, pp. 157–162.

Storey, J.B., Donn, M., Osterhaus, W., Jackson, Q., 2002. Urban buildings as energy generators. In: Sustainable Building 2002 Conference. Oslo, Norway.

Storey, J.B., Gjerde, M., Charleson, A., Pedersen, M., 2003. The state of deconstruction in New Zealand, Report to Task Group 39 of The International Council for Research and Innovation in Building and Construction (CIB). In: Meeting of the International Research Group CIB TG39. Gainesville, Florida, April.

TEEB, 2011a. TEEB Manual for Cities: Ecosystem Services in Urban Management. TEEB.

TEEB, 2011b. The economics of ecosystems and biodiversity in national and international policy making (P. ten Brink, Ed.). Earthscan, London.

TEEB Foundations, 2010. The economics of ecosystems and biodiversity: ecological and economic foundations (P. Kumar, Ed.). Earthscan, London.

TEEB in Business, 2011. The economics of ecosystems and biodiversity in business and enterprise (J. Bishop, Ed.). Earthscan, London.

TEEB in Local Policy, 2011. The economics of ecosystems and biodiversity in local and regional policy and management (H. Wittmer, H. Gundimeda, Eds.). Earthscan, London.

ten Brink, P., Berghöfer, A., Neuville, A., Schröter-Schlaack, C., Vakrou, A., White, S., et al., 2011. Chapter 10: Responding to the value of nature. In: ten Brink, P. (Ed.), The Economics of Ecosystems and Biodiversity for National and International Policy Makers. Earthscan, London.

Therapeutic Landscapes Network, 2012. Medicinal Plants. Available at: http://www.healinglandscapes.org/resources/plants/ (Accessed: August 8 2016).

Turner, R., Daily, G., 2008. The ecosystem services framework and natural capital conservation. Environ. Resour. Econ. 39 (1), 25–35.

UNEP, 2007. Buildings and Climate Change: Status, Challenges and Opportunities. United Nations Environment Program, Paris.

UNEP, 2011. Keeping Track of Our Changing Environment: From Rio to Rio + 20 (1992–2012). Nairobi, Division of Early Warning and Assessment (DEWA), United Nations Environment Programme (UNEP).

Vale, B., Vale, R., 2000. The New Autonomous House: Design and Planning for Sustainability. Thames and Hudson, New York.

Vale, R., Vale, B., 2009. Time to Eat the Dog? The Real Guide to Sustainable Living. Thames and Hudson, London.

Wilby, R., Perry, G., 2006. Climate change, biodiversity and the urban environment: a critical review based on London, UK. Prog. Phys. Geogr. 30 (1), 73–98.

Winz, I., 2005. Assessing sustainable urban development using system dynamics: the case of New Zealand's urban water systems. In: Ecological Economics in Action, Massey University, Palmerston North, 11–13 December. Institute for the Study of Coherence and Emergence.

Wolch, J., West, K., Gaines, T., 1995. Transspecies urban theory. Environ. Planning D: Soc. Space. 13, 735–760.

Young, O.R., Berkhout, F., Gallopin, G.C., Janssen, M.A., Ostrom, E., van der Leeuw, S., 2006. The globalization of socio-ecological systems: an agenda for scientific research. Glob. Environ. Chang. 16 (3), 304–316.

Zahedi, A., 2006. Solar photovoltaic (PV) energy; latest developments in the building integrated and hybrid PV systems. Renew. Energy. 31 (5), 711–718.

Zhang, Y., Singh, S., Bakshi, B.R., 2010. Accounting for ecosystem services in life cycle assessment, Part I: a critical review. Environ. Sci. Technol. 44 (7), 2232–2242.

Ecosystem services analysis: incorporating an understanding of ecosystem services into built environment design and materials selection

2

Maibritt Pedersen Zari
Victoria University of Wellington, Wellington, New Zealand

2.1 Introduction: a wider perspective on sustainability and the built environment

Ensuring that the built environment becomes more sustainable or perhaps even regenerative, must include careful consideration of which materials to build with. There is no such thing as a "sustainable material" without understanding the context of where and how it is used. It could be useful therefore to incorporate knowledge of ecosystem services into designing, planning, and decision making, which is by necessity site-specific and grounded in the physical ecological reality of the planet. Although this book is about building materials specification, in the case of using ecosystem services as a basis for improving the sustainability of the built environment, it is important to first understand how this concept relates to the wider task of designing buildings and cities before defining in detail how the same ideas can relate to specifying individual materials. A building material can only be considered sustainable (or not) when understood in the context of the design, climate, and culture it is used within over time.

It is complex and in some cases currently impossible to quantify the impacts on each different ecosystem service that each kind of building material has, and to then compare or rank which materials are better from an ecosystem service impact perspective. This is due to some information from the fields of ecology or biology not being available, the vast differences in metrics used to examine different ecosystem services, the complexities of a globalized trade in building materials products where manufacturers and retailers often do not know the source or composition of their own products (see Chapter 12), and the inherent complexities in ecosystem services and the relationships between them. Despite this, summaries of some of the major impacts, both positive and negative on ecosystem services that some common building materials have are provided in Section 2.5. These impacts can then be part of considerations when selecting building materials.

Materials for a Healthy, Ecological and Sustainable Built Environment: Principles for Evaluation.
DOI: http://dx.doi.org/10.1016/B978-0-08-100707-5.00002-2

2.2 Ecosystem services analysis and whole building or urban design

When applying the concept of ecosystem services analysis to whole building or urban design, the first step is to determine if there is an adequately healthy existing ecosystem in the locality that can be studied. If not, basing design targets on an ecosystem that existed prior to development on the site could become the focus of study. Measurable rates of ecosystem service provision that exist (or existed) on a site can then be determined. For example, specific figures such as annual rainfall and water retention in a particular place relate to the ecosystem service of provision of fresh water, and can be calculated with accuracy. Although as discussed, there are knowledge gaps in the field of ecology related to measuring ecosystem services (ten Brink et al., 2011), each ecosystem service has aspects that can be measured and are useful in setting initial design targets for regeneration. These targets can then be used to determine the optimal environmental performance of the built environment that is (or will be) on the same site as the ecosystem studied. For instance, the level of habitat provision in a new (or retrofitted) development should ideally be equal to the level of habitat provision in the original ecosystem that existed on the same site. In examining climate regulation, one aspect that could form a design goal would be to determine how much carbon was contained in and sequestered by the original ecosystem. This would set the optimal level as a goal for the new development. Although it may be difficult to accurately determine certain rates or figures related to ecosystem services, an approximate figure is still useful in determining site-specific ecological regeneration goals for built environments. This process is summarized in Fig. 2.1.

An ecosystem services analysis methodology could be employed by teams of designers and ecologists as well as urban planners, policy makers, and ecological economists at local and regional levels. An incremental process that focuses on improving existing ecosystem services in a specific location to an optimal or preexisting level could be a tangible starting point in the process of regenerating ecosystem services in the urban built environment. The next stage would be to initiate measures to reintroduce ecosystem services that may be absent in urban areas due to past degradation and removal of ecosystems, or because of conventional ways of constructing urban environments. This suggests that a regenerative built environment, if it can exist at all, will need to evolve over time rather than be expected to be fully functional after the initial realization of a design.

When applying ecosystem services analysis just to materials selection the process is slightly different. Despite this, determining overall goals for the performance of the building, landscape, or urban system that the materials will make up, is a necessary first step to determining which materials are most appropriate in that specific climatic and cultural context. This is why it is never possible to say with absolute certainty if a material is sustainable or not, because it always depends on the context a material is used in.

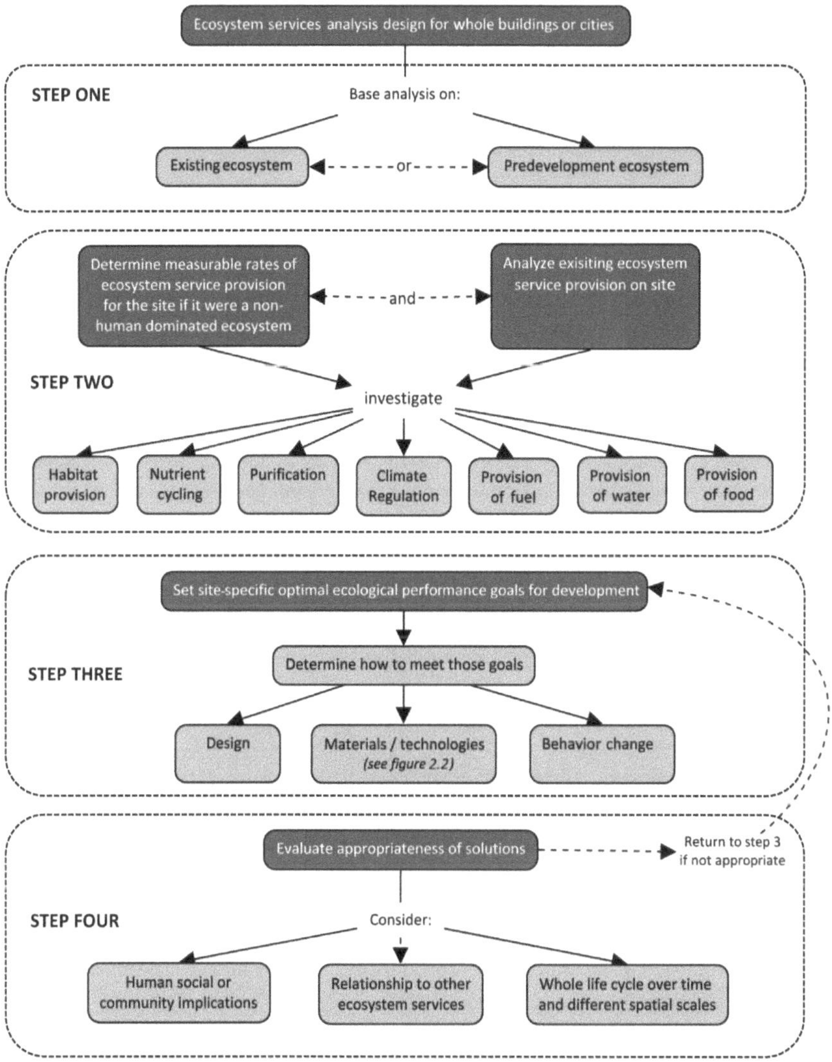

Figure 2.1 Stages of ecosystem services analysis for built environment design.

2.3 Ecosystem services analysis and materials selection

The built environment and the materials that it is made of vary greatly according to different climatic, ecological, economic, political, and cultural contexts. Systemic approaches that are appropriate to each place will therefore also vary. Although it has been established that focusing on climate regulation, purification, and water provision services are perhaps the most important services to consider initially (see Table 1.2), using ecosystem services in a materials selection context requires design teams to consider which ecosystem services are most important to consider for a particular site before selection begins, both in terms of where the materials come from and where they will be used. Discussions with ecologists who have knowledge of local ecosystems may further define the hierarchy of importance of the ecosystem services for a specific site and identify a set of appropriate ecological considerations. This suggests that wider disciplinary inputs into the process would be necessary than those normally found in traditional design team configurations. There is a need to determine to what extent a rural hinterland must be considered in tandem with an urban counterpart if regenerative goals are to be achieved across multiple ecosystem services. This implies a need to understand ecosystem services at a larger scale (city, region, or ecosystem boundary) when devising goals and targets for specific individual buildings or small developments. Careful thought in each case needs to be put into whether it is more appropriate to use human-defined urban boundaries, such as city limits or suburb boundaries, or those related to ecosystems themselves, such as habitat type demarcations, or water catchment zones, when employing ecosystem services analysis.

Although quantifiable ecosystem service metrics for each material may not be possible, except when calculated for specific designs in specific locations, formation of criteria or a methodology to increase the possibility that materials, technologies, and systems in a building development are appropriate in terms of overall environmental impact on multiple ecosystem services may be possible. This is true even though ecological "trade-offs" (as well as social and economic ones) will exist for most materials. For example, increasing the provision of metals through mining (increasing the ecosystem service of provision of raw materials) due to their ability to be recycled many times (engaging more effectively with the ecosystem service of nutrient cycling) could be deemed inappropriate, given the negative impacts increased mining could have on the ecosystem services of habitat provision and climate regulation. As discussed in Section 1.3, all ecosystem services are intimately linked. An important part of an ecosystem services methodology therefore, is to ensure that strategies to address one ecosystem service do not impinge negatively on the health of others at both the level of whole building design and materials selection. An analysis of associated impacts or benefits of design decisions on ecosystem services is necessary.

A proposed process of using ecosystem services analysis for materials selection is summarized in Fig. 2.2. When using ecosystem services analysis for materials selection the first step is to understand and if possible calculate the inputs from ecosystem services into a specific material, and the ecosystem service impacts caused by the harvesting, processing, and use of that material in the built environment.

This is likely to require research to ascertain how each material being considered is made and where it comes from. This research can initially be at a general level, e.g., understanding how all steel is made, but will also need to include research into specific types or brands of materials. This is because the way that each brand of material is harvested or extracted and processed is likely to vary considerably depending on where it is from and how it has been made. This means that each

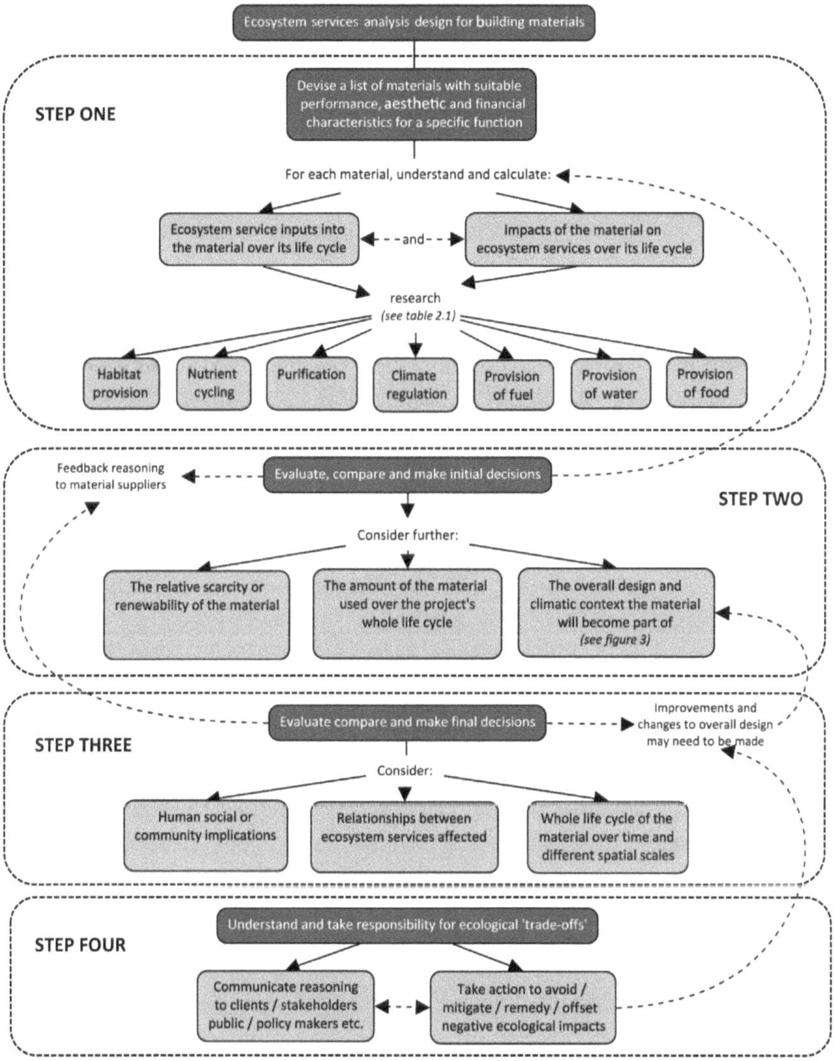

Figure 2.2 Stages of ecosystem services analysis for building materials selection.

brand may have very different impacts on ecosystem services even if the end products essentially are the same in terms of performance. Unfortunately, at present much of this information is often not readily available from materials suppliers, so extra time is required for materials specifiers to find these data. Although this may initially be somewhat frustrating it should be remembered that a huge proportion of materials that are harvested, extracted, or created end up being used in the built environment. Materials specifiers therefore have a great deal of power to change common practices and whole markets if they collectively begin to ask questions about where materials are form and how they are made and then purchase according to high ecosystem service impact standards. A crucial step in affecting change is to communicate why certain decisions have been made back to manufacturers and suppliers of materials. A selection of questions that may be useful to consider at this stage of the process of working with ecosystem services in materials specification is given in Table 2.1.

The second step is to consider the ecosystem services impact of each material in the context of how the material will be used in a whole building design, and how much of it will be used over the life time of the building. For example, using a material that has high ecosystem service impacts at harvest, but that contributes to producing ecosystem services while in use could be considered appropriate when whole lifecycle impacts are analyzed. The silica in sand, e.g., is a primary constituent, often two-thirds of the raw materials used, in the production of silicate glass (the most commonly used kind of building glass) (Bansal and Doremus, 1986). Sand is sourced by dredging seabeds or rivers, using sand dunes or beach sands, or by open pit mining depending on the type and origin of the glass. In 2015, 7% of sand mined in the United States, which is the world's largest supplier of silica sand, was used for glass-making purposes (U.S. Geological Survey, 2016, p. 144).

Sand mining often can have significant negative impacts on local ecosystems (Crawford, 2011, Erftmeijer and Robin Lewis, 2006) (see Fig. 2.3). Glass is a vital part of passive solar design however, and its appropriate use in combination with high-mass materials and insulation can contribute to buildings being highly energy-efficient, as well as to energy being generated in certain design situations (Bainbridge and Haggard, 2011). Carbon dioxide (CO_2) emissions, which are a major contributor to climate change, are reduced by using certain kinds of glass (e.g., low-emissivity glass in double or triple glazing units) even when taking into account higher CO_2 emissions during manufacture compared to production of standard glass for single-pane windows (Carrillo Usbeck et al., 2011, TNO Built Environment and Geosciences, 2008). One study estimated that nearly 100 million tonnes of CO_2 emissions could be reduced in Europe annually if low-emissivity insulated double-glazed window units were used in all European buildings (TNO Built Environment and Geosciences, 2008), compared to annual CO_2 emissions from the architectural glass manufacturing industries in Europe of up to 5 million tonnes (GEPVP, 2005). Glass, and the sand used to make it, can therefore contribute positively to the ecosystem services of energy provision, and climate regulation while in use in buildings. Without a given specific design context, and without knowing the exact composition of glass and the glass-making method in question, it is difficult to quantify this further. Several resources exist to assist in quantification of the lifecycle of glass products and can aid in choosing which kind of glass and which kind of glass production methods to choose. See, e.g., Grbeš (2015), TNO Built Environment and Geosciences (2008), and Carrillo Usbeck et al. (2011).

Specifying a material with high ecosystem service impacts at harvest or extraction may be deemed necessary because it has specific performance characteristics that cannot easily be substituted with another material. Copper wiring, e.g., which is the material of choice for electrical wiring in buildings, is made by mining copper ore and then refining it through a series of processes as described in detail by

Table 2.1 Ecosystem services and their relationships to building materials selection

Ecosystem services in a built environment context		Questions to consider when selecting building materials
Supporting services	1. Habitat provision	• How does the harvesting and processing of the material impact on habitat for nonhumans in the short and long terms? • What are both local and distant habitat impacts of using the material? • Can the material be appropriately integrated into built environment systems to provide habitat for flora or fauna?
	2. Nutrient cycling	• Can the material be biodegraded or recycled in closed loops? • What is the impact of using the material on soil formation and the renewal of fertility over time? • How does the method or length of time between harvests/extraction affect nutrient cycling for the ecosystem the material is from? • How can the material be integrated into built environments to ensure future reuse, recycling, or biodegradation of it and/or other materials?
Regulating services	3. Purification	• How does harvesting/extraction and processing of the material affect water, air, and soil quality and purification over time? • Are there different sources or processes for creating the same material that have different purification outcomes for ecosystems? • How is the site of harvest/extraction remediated or cared for? Does it recover over time? • How can the material be integrated into built environments to facilitate purification of air, water, and soil?
	4. Climate regulation	• How does the material contribute to or disrupt regulating climate over time? Consider: sequestering of carbon; GHG emissions during all stages of lifecycle; if the material provides protection from decreased ozone during use; and how the material lessens or contributes to the heat island effect

(*Continued*)

Table 2.1 (Continued)

Ecosystem services in a built environment context		Questions to consider when selecting building materials
		• How can the material be integrated into buildings to reduce GHG emissions? • Can the material be integrated into buildings to store or sequester carbon over time?
Provisioning services	5. Provision of fuel/ energy for human consumption	• What is the embodied energy of the material over its lifecycle? • Does creation of the material affect the ability of the ecosystem it is from to provide energy for other uses? • How can the material be integrated into buildings to encourage effective energy use? • Can the material be integrated into buildings to facilitate energy production?
	6. Provision of fresh water	• How does the extraction/harvesting, etc., of the material affect local or wider-scale water storage and movement patterns over time? • How does extraction/harvesting of the material affect local or wider-scale fresh water quality? • What is the embodied water of the material over its lifecycle? • How can the material be integrated into a building to encourage the conservation or filtering and purifying of water?
	7. Provision of food	• How does the extraction/harvesting, etc., of the material affect local or wider-scale food production over time? • Does the harvesting, extraction, or making of the material compete with food provision in terms of land use or energy use? • How can the material be integrated into a building or landscape to encourage the production of food?

the US Congress Office of Technology Assessment (1988). Some copper is made into new components from recycled sources (termed secondary copper). The proportion of this varies from country to country but in the United States and Europe the figure may be approximately 40% (Joseph, 1999, p. 22). Approximately half of all copper mined becomes copper wire and approximately 15% of copper used in the

Figure 2.3 Sand mine on river edge, USA. By E. Possardt, U.S. Fish and Wildlife Service. *Source*: Sand mine on river edge, CC0. https://commons.wikimedia.org/w/index.php?curid=28642462.

United States becomes building wire (Joseph, 1999, pp. 21, 37). The impacts of copper mining depend largely on the country of origin. For example, strict environmental laws governing air, water, and land pollution are in place in many developed nations while they are not in many developing countries. It should be noted that presently at least two-thirds of copper comes from mines in developing countries (U.S. Geological Survey, 2016, p. 55). In 2015 the top six copper-producing countries in the world were: Chile (5,700,000 tonnes or 30% of global total); China (1,750,000 tonnes or 9.4% of global total); Peru (1,600,000 tonnes or 8.6% of global total); United States (1,250,000 tonnes or 6.7% of global total); Congo (990,000 tonnes or 5.3% of global total); and Australia (960,000 tonnes or 5.1% of global total) (U.S. Geological Survey, 2016, p. 55). Copper mining often negatively affects the ecosystem services of purification (in terms of air, ground and surface water, and soil quality), climate regulation (due to the release of greenhouse gases (GHGs) and acid rain causing substances such as sulfur dioxide), provision of fresh water (related to the impact on ground water quantities and quality, as well as pollution of surface water due to leachates from tailings), and habitat provision (particularly in the case of open-pit mines which disrupt the landscape, and remove vegetation and topsoil. Aquatic life is also negatively affected by heavy metal pollution from mine tailings, spoils and leachates or by acid mine drainage (Greenspec, 2016; Northey et al., 2013; US Congress Office of Technology Assessment, 1988) (see Fig. 2.4).

Figure 2.4 Open pit copper mine, Pima County, Arizona, USA, 2015. The mine is 4 km long by 2.4 km wide and is 366 m deep. Benches are 12.2 m high. By Docentjoyce, USA. *Source*: Open pit copper mine, Pima County, Arizona, USA, 2015, CC BY 2.0. https://commons.wikimedia.org/w/index.php?curid=3251478.

It is of great importance that the relative scarcity of a material is also taken into account when specifying materials (Klinglmair et al., 2014; Zhang et al., 2010b). It is well known that various constituents of some building materials are becoming more scarce (Heinberg, 2013). The implications for ecosystems (including all life within them) of these materials being mostly used up by humans for their own purposes and being made unavailable in ecosystems is difficult to quantify or perhaps even imagine. The implications of resource depletion for human economies and human wellbeing are more easy to anticipate and obviously are very negative. Biotic resources and their renewal rates have not received much attention within lifecycle assessment (LCA) research (Klinglmair et al., 2014). There is currently no clear consensus of how to incorporate the issue of resource depletion (both biotic and abiotic) into LCA methods, though Zhang et al. (2010a) present an approach to take this into account. Again using the example of copper, Sverdrup et al. (2014) estimate that the peak of copper production will occur between 2031 and 2042 but that the supply of copper to society will likely continue until approximately 2400. Harmsen et al. (2013) showed that copper scarcity in the next three decades will likely result in deteriorating ore quality. This will increase the amount of material that needs to be mined in order to produce the same amount of copper, and will lead to a higher gross energy requirement for copper production in the future. This will affect standard LCA calculations. This same study also estimated that the gross energy requirement of copper in a 2050 100% renewable energy system will probably be 2–7 times larger than it was in 2013 (Harmsen et al., 2013).

The third step when using ecosystem services analysis to select materials is to evaluate the best choices and to again consider if the material in question is the best fit for the project in terms of its overall ecological performance goals. Considering the relationships between ecosystem service impacts when the material is made, and ecosystem impacts when the material is being used in a building, landscape, or urban setting is an important part of this. This step may mean alterations to the overall design of a building are necessary.

The final step of the process of using ecosystem services analysis for materials selection is to be aware of, and if possible to take some responsibility for, ecological "trade offs." If it is deemed that a particular material must be selected, understanding that using it has negative effects on the ecosystem it has been extracted from, such as habitat provision in the examples of glass and copper as discussed above, could help design teams realize that their own development may need to provide extra habitat to alleviate this damage, or perhaps that donations to ecosystem restoration in the country of origin of the material could be appropriate and built into building costs. Ethically, economically, and ecologically it is preferable to avoid damage in the first place to ecosystems and therefore to the services they provide to humans, rather than to try to remediate damage after it has been done however (Turner and Daily, 2008).

While it may prove to be too complicated to accurately quantify the tradeoffs between ecosystem services that are affected by materials when they are made compared to when they are in use (for reasons described earlier), understanding that there are in fact tradeoffs can enable design teams to decide how they will address these. In an ideal world people would not specify materials that have significant negative impacts on ecosystem services, however if it has been decided that a material that does have negative impacts on ecosystem services when it is harvested, extracted, processed, etc., has to be used, certain actions can be taken to potentially begin to address this. A short selection of strategies follows:

1. Ensure that the building development creates or regenerates the ecosystem service negatively affected within the development itself. For example, if fresh water quality is negatively affected by specifying a certain material, special attention could be given to designing systems within the development that contribute to fresh water quality.
2. Ensure that the developers contribute to the restoration of the ecosystem service negatively affected by paying for this work to be done in that particular place. For example, contributing funds to conservation or restoration groups in the location that has been damaged by the harvest/extraction/processing of the material.
3. Ensure that the developers contribute to the restoration of the ecosystem service negatively affected by paying for this work to be done in a different location. For example, if a small amount of material is used that negatively affects distant marine ecosystems, the developers could sponsor or organize a clean-up of a local beach, or something else appropriate.
4. Pay for "offsets." This is most relevant in terms of carbon or other GHG emissions, but may also be appropriate in the case of loss of habitat, or loss of provision of food services. For example, if the habitat loss caused by the materials chosen is calculated for a particular project, the same amount of land (either in pristine condition or in need of restoration) could be purchased and set aside as a habitat reserve. There are several ethical considerations that should be taken into account when purchasing offsets, particularly in terms of

carbon. The effectiveness of offsetting carbon is also controversial for several reasons. Discussing effective and appropriate options with experts is recommended.

Employing some of these ideas may affect the overall design of a building project. With each of these strategies care needs to be taken to be as accurate as possible with calculations and to effectively communicate the intentions of the developers or design team to both materials suppliers, and clients, stakeholders, and the public to avoid tokenistic responses or accusations of "green-washing" (the act of describing a behavior, group, concept, or product as being more environmentally sustainable than it actually is to enhance its public reputation, and/or to increase financial profits). It should be remembered that no action is likely to completely undo or make up for damage done to ecosystem services so these strategies are intermediate steps to be taken while the harvesting, extraction, and processing methods of materials change so that negative ecosystem service impacts are avoided in the first place.

2.4 Benefits and difficulties of applying the ecosystem services concept to built environment design and materials selection

Aside from general potential ecological benefits of more sustainable or regenerative built environments, there are significant social and economic benefits as described by Pedersen Zari and Jenkin (2009) such as: more resilient communities as climate changes; more equitable communities; potential new revenue streams from buildings; and increased financial value of buildings. Elaboration upon these will not be repeated here, but there are several additional advantages when adding ecosystem services analysis to a materials selection process that will be discussed.

Using an understanding of ecosystem services when selecting building materials is conducive to a sustainable or regenerative approach, but enables a more specific targeted and potentially measurable design response than simply aiming to "regenerate ecosystems" or be "more sustainable." Understanding ecosystem services fits well into the philosophy of encouraging people to move beyond the typical "reduce, recycle, efficiency" sustainability paradigm (Reed, 2007, McDonough and Braungart, 2002), and to grasp the potential for increased ecological and human health when adopting a regenerative approach (Olgyay and Herdt, 2004). For example, if materials were selected that are known to aid in the conservation or creation of ecosystem services when grown or extracted in certain ways, rather than using materials where the opposite is true, the building industry can make use of its huge potential to change materials markets positively to aid in ecological restoration. This ability to change markets, and therefore affect real change, comes about from the vast amount of materials humans source from the earth that become building materials as discussed earlier.

The Economics of Ecosystems and Biodiversity group (TEEB, 2011) provides clear reasons why an understanding of ecosystem services should be integrated into urban-level decision making. This should include how designers select materials. They point out that when the benefits derived from local ecosystems are understood or become apparent these are more valued and perhaps therefore preserved. For example, understanding that purer water is a result of nearby forests in a particular city could mean it is easier to convince people of the need to conserve the forest for that purpose rather than to use it just for timber. This has the potential to contribute to prioritizing or preventing certain urban development projects in particular areas, and therefore to long-term effective spatial planning right down to the detail of materials selection (providing the overall economic values of society change to allow this). In addition, by considering impacts on ecosystem services at a materials selection level, the impacts of decision making can be understood across various spatial boundaries, timescales, and multiple interconnected environmental issues, and can therefore be communicated to clients, city managers, and residents. This means more accurate planning and budgeting and in some instances could lead in time to a reduction in a city's overall ecological footprint (refer to TEEB (2011) for case studies). Other benefits of incorporating an understanding of ecosystem services into architectural and urban design identified by TEEB (2011) include increased human health, and increased biodiversity in urban areas.

Working with the concept of ecosystem services requires a multidisciplinary and multiscale approach and requires complex modeling (Zhang et al., 2010a). This is a challenge when trying to use an ecosystem services approach to design and materials selection, made even more difficult by the fact that ecologists still do not have absolutely certain or comprehensive knowledge about all ecosystem services and are still devising typologies and metrics associated with them (Zhang et al., 2010b). There is also a need to develop internationally agreed definitions of ecosystem services typologies, values, and metrics, along with methodologies and reporting mechanisms, and to understand how these relate to social ecologies and economic systems at various spatial and temporal scales. This is vital in order to facilitate learning and sharing of the concepts, transferability of results, and to enable comparisons to be made between different decisions that have been made (de Groot et al., 2010).

2.5 Potential impacts on ecosystem services of common building materials

As de Groot et al. (2010) point out:

> *The quantitative relationship between biodiversity, ecosystem components and processes and services is still poorly understood. The specific nature of interdependencies between the structure and diversity of biotic communities and the functioning of ecosystems remains one of the most important unresolved questions in ecology... Criteria and indicators are needed to comprehensively describe the interaction between the ecological processes and components of an ecosystem and their services.*

A thorough and detailed discussion and list of quantifiable metrics or ratings of how ecosystems and then ecosystem services are affected by the extraction or harvesting of all common materials used in the built environment is therefore beyond the scope of this chapter. However, a list of potential indicators and metrics for various ecosystem services is provided by de Groot et al. (2010), part of which forms the basis for Table 2.2. Additional metrics for some land-based ecosystem services may be provided by LCA research detailed in Koellner and Geyer (2013) and Zhang et al. (2010a).

Quick reference tables are provided in this section along with references if further research is required for selecting common building materials. These tables are not designed to be exhaustive and do not investigate every material that is used in the built environment. Instead a selection of some of the most commonly used materials are examined. The intention of the quick reference tables is to serve as an example of how researchers or design teams can approach investigating the impacts of specific materials from an ecosystem services perspective, and to enable materials specifiers to be aware of some of the major issues associated with each material in terms of ecosystem services. This may enable them to formulate specific questions to ask of manufacturers. Other environmental impacts and human health implications of some of these materials are discussed in Part II (Chapter 3 and Chapter 4) and Part III (Chapter 5 to Chapter 9).

The quick reference tables are divided into three categories:

1. Materials that are grown;
2. Materials that are extracted or mined;
3. Materials that are made or processed.

A series of questions to work through for each type of material when considering them for selection is given after each set of tables to summarize the major issues to be aware of.

2.5.1 Materials that are grown

Materials that are grown include timber, straw, bamboo, and other plant-based building materials. Less traditional grown building materials also include some algae products, fungi, animal products, and even materials made through the process of biomineralization. Materials that are grown are often renewable but the health of the soil or growing medium needs to be taken into account over multiple cycles of harvest. The addition of chemicals such as fertilizers, pesticides, and herbicides, which can negatively affect multiple ecosystem services, is also a common concern associated with grown materials. Three quick reference tables are provided in this section: straw, timber, and bamboo.

Table 2.2 Indicators and potential metrics for ecosystem services

Ecosystem services in a built environment context		Ecological process and/or component providing the service (or influencing its availability)	State indicator (how much of the service is present)	Performance indicator (how much can be used/provided in a sustainable way)
Supporting services	1. Habitat provision	Importance of ecosystems to provide breeding, feeding, or resting habitat for transient species. Maintenance of a given ecological balance and evolutionary processes	Number of transient species and individuals. Natural biodiversity (esp. endemic species). Habitat integrity (min. critical size)	Dependence of other ecosystems (or "economies") on habitat. "Ecological value" (i.e., the difference between actual and potential biodiversity value)
	2. Nutrient cycling	Role of biota and abiotic processes in transformation, removal, or breakdown of organic matter, xenic nutrients, and compounds. Movement of minerals and nutrients through an ecosystem	Denitrification (kg N/ha/year). Immobilization in plants and soil. Rates of nutrient movement	Maximum amount of chemicals or nutrients that can be recycled or immobilized on a sustainable basis
Regulating services	3. Purification	Capacity of ecosystems to extract, filter, and/or transform aerosols, chemicals, and other toxins from the atmosphere, soil, or water	Immobilization in plants and soil. Leaf area index, NO_x fixation, etc.	Amount of aerosols or chemicals "extracted" and the effect on air, water, or soil quality
	4. Climate regulation	Influence of ecosystems on local and global climate through land-cover and biologically mediated processes	GHG balance (esp. carbon sequestration). Land cover characteristics, etc.	Quantity of GHG, etc., fixed and/or emitted and their effect on climate parameters
Provisioning services	5. Provision of fuel/energy for human consumption	Presence of species or abiotic components with potential use as fuel/energy	Total biomass (kg/ha). Total potential energy (kJ)	Net productivity (kg/ha/year). Energy potential (kJ/ha/year)
	6. Provision of fresh water	Presence of fresh water reservoirs. Role of forests in water infiltration and gradual release of water	Total amount of water (m^3/ha). Water retention capacity in soils, etc., or on the surface	Max. sustainable water extraction (m^3/ha/year). Quantity of water retention and influence of hydrological regime (e.g., irrigation)
	7. Provision of food	Presence of edible plants and animals	Total or average stock in kg/ha	Net productivity (in kcal/ha/year or another unit)

Partially adapted from de Groot et al. (2010).

Many different species of trees are used as timber in buildings so ecological impacts will vary depending on the type of logging, the size of the operation, its location, and whether the forest was old growth or plantation grown.

The tables examining straw, timber, and bamboo (Tables 2.3–2.5) indicate that when investigating grown materials for use in buildings, it is important to know not just where materials come from but also how they have been grown and in what specific context in order to consider impacts on ecosystem services. This will vary between brands or types of similar materials. The most important issues to be aware of with grown materials include impacts on soil health and nutrient cycling, interaction with the carbon cycle (because most grown materials store carbon), and implications for biodiversity of plantation growing. In general, old growth materials should be avoided.

To summarize, the following aspects should be investigated:

1. Where has the material been grown? Are there environmental laws in that place that indicate good ecological practice?
2. How exactly has the material been grown? Which chemicals have been used to promote growth or reduce "pests" and what are the impacts of those? Are certain practices better from an ecosystem impact point of view?
3. How has the material been processed or treated? Which chemicals have been used and what are the impacts of those on ecosystems over time and at end of life of the material? What are the in-use implications on indoor air quality?
4. How has the growth and harvest of the material contributed to or reduced GHG emissions?

   How has the growth and harvest of the material affected local biodiversity in the short and long terms? Does it increase or diminish habitat? How has this been addressed by the company? Are certain practices less damaging?
5. How does the growth and harvest and use of the material interact with fresh water?
6. How has the growth and harvest of the material affected local air, water, and soil health? What are the in-use and end-of-life implications?
7. How does the growth or harvest of the material interact with nutrient cycles? Are there issues with end-of-life disposal or reuse of the material that need to be considered?
8. Does the growth and harvest of the material compete with food growing or energy production potentials?
9. Are there groups of people (particularly indigenous) that oppose the harvest of the material? What are the main issues? Are there associated ecological or cultural issues that need to be considered?

2.5.2 Materials that are extracted or mined

Materials that are extracted or mined are those which most typically are dug out of the earth as raw materials which are then minimally processed into various products. Most extracted materials are not renewable but many are abundant, such as earth and sand. The relative scarcity of the material needs to be taken into account.

Table 2.3 Potential impacts on ecosystem services of common building materials that are grown: Straw

	Straw
Climate regulation *High priority*	*At harvest*: There may be a negative effect on carbon stores in soil when straw is removed for use as a building material. No-till strategies may mitigate this in part (Blanco-Canqui, 2012). Using straw as a building material rather than burning it is preferential in terms of climate regulation. Burning increases CO, O_3, and particulates (P_{10}) + SO_2 (Li et al., 2008) *In use*: Straw stores carbon during its life as a building material. It does this more effectively than biochar (Alcorn, 2003; Mattila et al., 2012)
Purification *High priority*	*At harvest*: It is better to use straw as a building material than to burn it in terms of air quality (purification) (Andreae, 1991) *In use*: The use of straw (and several other natural materials) can have a positive effect on indoor air quality (humidity and air quality) depending on how it is used in a design (Korjenic et al., 2011; Ashour et al., 2011)
Provision of water *High priority*	*At harvest*: Leaving the straw in place on fields rather than removing it for building materials (or burning) reduces evapotranspiration and the need for irrigation. This can increase yield significantly in subsequent crops (Huang et al., 2005)
Habitat Provision *Medium priority*	*At harvest*: Leaving straw on a field versus taking it away for use as a building material can alter the balance between predators and "pests" in terms of habitat provision. This is site- and climate-specific (Landis et al., 2000)
Nutrient Cycling *Medium priority*	*At harvest*: There is potential removal/alteration of soil nutrients, particularly nitrogen and carbon when removing straw from fields for use as a building material. Various farming techniques impact on figures over time. Generally, research indicates leaving straw in place is better for soil health and nutrient cycling (Malhi et al., 2006). Long-term productivity of medium to heavy textured soils may not be affected providing that <40% of the total aboveground residues other than grain are removed and the frequency of removal is no more than 2 years out of 3 (Lafond et al., 2009) *End of life*: Depending on composition and how they have been finished, many straw building products are decomposable at end of life (up to 100 years). The use of cement and reinforcing can be problematic in this regard (Owens, 2015)
Provision of energy *Medium priority*	*At harvest*: There is potential competition between straw as a building material and straw as a base for ethanol production (Blanco-Canqui, 2012) *In use*: The benefits of using straw products in terms of insulation and energy conservation are well documented (Ashour et al., 2011)
Provision of food *Medium priority*	*After harvest*: Provision of food is likely to be affected by the impacts straw removal from fields has on soil health and productivity. See nutrient cycling row

Table 2.4 Potential impacts on ecosystem services of common building materials that are grown: Timber

	Timber
Climate regulation *High priority*	*At harvest*: Demand for northern hemisphere timber means more forest planted so greater carbon sequestration occurs. Demand for tropical hardwoods leads to less forest and therefore less carbon sequestration and more carbon emissions, however (Van der Lugt et al., 2012) *In use*: Total timber used in the built environment is increasing as population does leading to greater carbon sequestration in the built environment (while the timber is in use). However only 25% of timber from a forest cut for building materials ends up in buildings due to wastage (Van der Lugt et al., 2012)
Purification *High priority*	*At harvest*: The effects on water quality of timber harvesting may include increased concentrations of dissolved salts, suspended solids, and nutrients (Campbell and Doeg, 1989). This will likely affect the ability of the ecosystem to filter impurities out of fresh water
Provision of water *High priority*	*During growth*: The use of herbicides and pesticides in some managed forests negatively affects water quality. This also impacts on habitat provision and nutrient cycling (Campbell and Doeg, 1989) *At harvest*: Timber harvesting often has significant effects on water quantity and quality (Campbell and Doeg, 1989). Typically, removing forest decreases evapotranspiration and temporarily increases stream runoff. The increase in stream flow diminishes over time (Campbell and Doeg, 1989). Buffer zones may protect some fresh water ecosystems but this is controversial. Buffers may need to be at least 30 m wide to be effective (Campbell and Doeg, 1989)
Habitat provision *Medium priority*	*During growth*: Plantations can have both positive and negative consequences for biodiversity, at a range of scales as detailed by: Kanowski et al. (2005). This is controversial and highly dependent on the location and history of the plantation. Some techniques exist to increase biodiversity in timber plantations. See Lamb (1998) *At harvest*: 50%–100% of biodiversity in a forest is disturbed or lost in harvesting of old growth forest (Vogtländer et al., 2010). Some evidence suggests that less environmental impact occurs when using directional felling and cut-to-length rather than tree-length methods of logging (Tenyah, 2009). Major short-term effects of timber harvesting may occur on local aquatic biota resulting from increased sediment input into streams and through increased light penetration into certain habitats due to damage to, or removal of, riparian vegetation.

(*Continued*)

Table 2.4 **(Continued)**

	Timber
	Longer-term effects may include changes to stream habitat due to a reduction in the number of large logs available in a forest that fall into streams. Large fallen logs play a major role as habitat and retention structures in streams (Campbell and Doeg, 1989). Waters may also become more acidic, which can have severe negative impacts on aquatic biota (Likens et al., 1978). Selective logging (taking out only the largest trees) can allow light and wind into forests altering ecosystems negatively
Nutrient cycling *Medium priority*	*At harvest*: Removal of vegetation and compaction of the soil that is associated with timber felling reduces soil permeability to water which increases erosive surface runoff (Campbell and Doeg, 1989). This has impacts on nutrient cycling. In many cases, streams going through undisturbed forests have low nutrient levels because of the ability of forests to retain nutrients. Catchments which have been disturbed by logging often "leak" nutrients which are transported out by streams (Campbell and Doeg, 1989). Other impacts may include an increase in soil temperature and moisture, an increase in decomposition rates, increase in nitrification and mineralization, a decrease in canopy and reflection of solar radiation, and water that is more acidic draining from the area. Eutrophication may occur in streams and lakes due to increased nutrient (particularly phosphorus and nitrogen) loading after logging (Likens et al., 1978)
Provision of energy *Medium priority*	*During growth and at harvest*: Logging residues and first thinning waste in plantations can be used to generate bioenergy (Malinen et al., 2001)
Provision of food *Medium priority*	*At harvest*: Logging generally negatively affects fresh water fish populations (Campbell and Doeg, 1989) which can be a food source. Numbers of forest-dwelling animals are also likely to decline in the event of logging. Some of these may also be food sources

Localized habitat damage is often a major concern with extracted materials. Two quick reference tables are provided in this section: stone and cement/concrete.

There are many different kinds of stone used in buildings so ecological impacts will vary depending on the type of quarry operation, its size, and its location. More than 89% of the international quarrying activity is concentrated in nine countries: China, Italy, India, Iran, Spain, Turkey, Brazil, Greece, and Portugal. Each of these countries produces more than 2 million tonnes of natural ornamental stone per annum (Founti et al., 2010).

Table 2.5 Potential impacts on ecosystem services of common building materials that are grown: Bamboo

	Bamboo
Climate regulation *High priority*	*Growing stage*: Bamboo produces a lot of leaf litter. The phytoliths produced in the epidermal cells of bamboo are prolific and form as microscopic silica grains in the leaves and stems. These become incorporated into the soil during decomposition of leaf matter and are highly resistant to decomposition. The growth of bamboo therefore is a more permanent form of carbon sequestration than some other plant types (Sabto, 2010) *At harvest*: use of waste bamboo for energy generation can be carbon-negative and be positive therefore for climate regulation (Van der Lugt et al., 2012). Demand for bamboo leads to more bamboo grown and therefore more carbon sequestration and potentially less carbon emissions. Research indicates that even if high-quality bamboo products have come from China to Europe the net carbon effect is typically neutral or better (Van der Lugt et al., 2012). Bamboo can replace tropical hardwood, so it can potentially mitigate the decrease of tropical forest area (Vogtländer et al., 2010) *In use*: Bamboo stores carbon during its life as a building material
Purification *High priority*	*At harvest*: Bamboo can be harvested without disturbing the soil, meaning less disruption to soil and water quality and therefore need of purification ecosystem services (Ben-zhi et al., 2005). Bamboo is a prolific generator of biomass (leaf litter), and is used to remediate depleted soil in some cases (Ben-zhi et al., 2005) *End of life*: Bamboo charcoal, potentially made as a byproduct from bamboo waste at the end of a building product's life or during manufacture, has the ability to absorb certain heavy metal pollutants from drinking water (Wang et al., 2008)
Provision of water *High priority*	*At harvest*: Bamboo is planted or occurs along river and lake banks in some countries to prevent soil erosion and to protect or improve fresh water quality. Plantations of bamboo for use as building materials may have similar benefits depending on where they are planted (Friederich, 2014). Bamboo also retains soil moisture in some situations meaning less water is needed to grow it than some other crops (Ben-zhi et al., 2005)
Habitat Provision *Medium priority*	*At harvest*: Bamboo is not harvested from natural forests so habitat in indigenous forest is not usually disturbed. Plantation bamboo is typically grown on areas that used to be industrial or grassland (Van der Lugt et al., 2012). In mountainous and eroded areas, where tree growing or cultivation is not possible, and where the biodiversity is low, expanding bamboo plantations can result in increasing biodiversity and carbon sequestration (Vogtländer et al., 2010), and water and soil

(*Continued*)

Table 2.5 **(Continued)**

	Bamboo
	retention (Ben-zhi et al., 2005). Bamboo is used to restore degraded land and increase biodiversity in some countries (Friederich, 2014)
Nutrient cycling *Medium priority*	*At harvest*: Studies on nutrient uptake and their accumulation show that essential nutrients (N, P, K, Mg, Ca) are removed from the soil at the rate of 62%–74% with each harvest of bamboo. Because of the fast rotation times, application of fertilizers after the harvest of the bamboo may therefore be needed for further growth (Shanmughavel and Francis, 2001). Certain mixed crop rotation systems use bamboo to increase nutrient cycling and productivity of land without fertilizer use however (Christanty et al., 1996)
Provision of energy *Medium priority*	*At harvest*: waste from productions of bamboo products can be used as a potentially carbon-negative electricity generation source (Van der Lugt et al., 2012)
Provision of food *Medium priority*	*During growth*: Parts of the bamboo plant are edible and have high nutrition value. These parts are not the same as parts harvested for building materials. In situations where certain kinds of bamboo are planted in previously degraded land areas, food provision could be increased (Satya et al., 2010). Certain mixed food crop rotation systems use bamboo to increase productivity of land for food crops without fertilizer use (Christanty et al., 1996)

The tables examining stone and cement/concrete (Tables 2.6 and 2.7) suggest that when investigating extracted or mined materials for use in buildings, it is important to know specifically where materials come from due to what the local impacts might be (an open cast mine in the middle of a rainforest will have different impacts compared to an underground mine in farm land, for example), but also knowing the location is important because of the differences between environmental laws and regulation of them in different parts of the world. The specific method of the extraction and processing will also result in differences to ecosystem services impacts. Again, this is likely to vary considerably between brands or types of similar materials. The most important issues to be aware of with extracted or mined materials include impacts on local habitat and biodiversity, interaction with the water cycle, relative scarcity of the material, and associated GHG emissions and embodied energy estimates.

Table 2.6 **Potential impacts on ecosystem services of common building materials that are extracted or mined: Stone**

	Stone
Climate regulation *High priority*	*Extraction*: In temperate areas removing vegetation and soil which occurs with large mining operations reduces evapotranspiration and can increase effective rainfall over the short term (Langer, 2001) *Processing*: The largest amount of CO_2 emissions associated with stone (particularly slate) is from the processing stage (when stone is cut and polished, for example). It is likely that the more unfinished stone is the lower the embodied carbon. Stone typically has lower embodied carbon than other common building materials over whole life cycles. The CO_2 impact of imported stone is likely to be very high due to transport-related emissions. Some companies ship rough stone from the country of origin (particularly in Europe) to India or China for processing to take advantage of lower labor costs, before exporting it back to Europe for on-sale. This practice results in significantly higher CO_2 emissions (Crishna et al., 2011)
Purification *High priority*	*Extraction*: Air pollution damage from stone quarrying is very high (higher possibly that the value it adds) (Muller et al., 2011). Quarries often affect the ability of ecosystems to filter water. In some karst areas the soil overlying the rock is usually a zone of filtration and water purification, so removing this affects the ability of the ecosystem to purify water. Soil systems and plants can be negatively affected by being overloaded with quarry dust (Langer, 2001)
Provision of water *High priority*	*Extraction*: Groundwater quality is commonly affected by quarrying through increased fine-sediment concentrations and accidental spillages. This depends on the type and location of mine. Impacts can be long term even after quarrying stops (Hobbs and Gunn, 1998). Increased rainfall can escalate runoff and sedimentation, and result in deterioration of ground water (Langer, 2001). Quarrying can also substantially modify the routing of recharge, and degrade water quality. Surface water nearby and at a distance can also be affected by changes in flows, silting, and sedimentation. Many instances of dramatic declines of the water table are caused by underground mines, rather than surface quarries (Langer, 2001)
Habitat provision *Medium priority*	*Extraction*: Major impacts of quarrying include changes in geomorphology and conversion of land use, accompanied by potential loss of habitat (for both land and aquatic creatures depending on the site of the quarry or mine), noise, dust, vibrations, chemical spills, erosion, sedimentation, changes in drainage, soil conditions, and dereliction of the mined site. Karst (cave) biota can be severely affected by mining in some areas (Langer, 2001). Chemical modification of ecosystems can occur through quarrying/mining activities, which impacts negatively on biodiversity (Ratcliffe, 1974)

(*Continued*)

Table 2.6 (Continued)

	Stone
Nutrient cycling *Medium priority*	*Extraction*: Nutrient cycling (in land and water) is severely affected by some quarrying operations, particularly in karst country. In a UK study, quarry operators estimated that waste in slate quarries was 85% compared to 29% for sandstone and 47% for granite. This large amount of solid waste (low-quality or irregular or damaged sized blocks) affects the ecologies of the site and nutrient cycles (Crishna et al., 2011)
Provision of energy *Medium priority*	*Extraction and processing*: Crishna et al. (2011) and Gazi et al. (2012) detail embodied energy information for various types of stone. Actions to reduce the amount of fuel used to produce stone (and to reduce other environmental impacts caused by quarrying) are given by Gazi et al. (2012)
Provision of food *Medium priority*	*Extraction*: Impacts of quarrying on water quantity and quality available for agricultural uses may affect food production

To summarize, the following aspects should be investigated:

1. Where has the material been extracted from? Are there environmental laws in this place that indicate good ecological practice? Has the material been processed in the same place?
2. How exactly has the material been extracted or mined and processed? Which chemicals have been used and what are the impacts of those on ecosystems over time? Are certain practices better from an ecosystem impact point of view?
3. How has the extraction and processing of the material contributed to GHG emissions? What are the implications in use?
4. How has the extraction of the material affected local biodiversity in the short and long terms? How has this been addressed by the company? Are certain practices less damaging? What remediation of land strategies are in place?
5. How does the extraction, processing, and use of the material interact with fresh water cycles over time?
6. How has the extraction and processing of the material affected local air, water, and soil quality?
7. How does the extraction of the material interact with nutrient cycles? Are there issues with end-of-life disposal or reuse of the material that need to be considered?
8. Does the extraction and processing of the material compete with food growing or energy production potentials?
9. Are there groups of people (particularly indigenous) that oppose the extraction of the material or the location of the operation? What are the main issues? Are there ecological or cultural issues that need to be considered?

Table 2.7 Potential impacts on ecosystem services of common building materials that are extracted or mined: Cement/concrete

	Cement/concrete
Climate regulation *High priority*	*Processing*: Annual global production of cement accounts for at least 7.5% of global anthropogenic CO_2 emissions (Mehta, 2001)
Purification *High priority*	*Extraction*: Air and land pollution is often associated with aggregate mining (Sandberg and Wallace, 2013)
Provision of water *High priority*	*Extraction*: Aggregate mining can negatively affect ground water in some mining conditions (Sandberg and Wallace, 2013) *Construction*: The concrete industry is one of the largest industrial users of fresh water. Just the mixing water requirement is approximately 1 trillion L every year. Vast amounts of water are also used as wash-water by the ready-mixed concrete industry, and for curing concrete (Mehta, 2001). This can compete with fresh water needs of ecosystems and people. Recycled industrial waters or brackish natural waters may be suitable substitutes for potable water for making concrete and certainly for wash-water and curing water (Mehta, 2001)
Habitat provision *Medium priority*	*Extraction*: Large quantities of raw materials, such as limestone and clay (and coal for processing), are mined for the production of concrete. This often results in extensive deforestation and topsoil loss and negatively affects riparian ecosystems (Mehta, 2001). Environmental management of aggregate mines in developing countries may be poor. Specifiers should enquire about revegetation of land after mining (Sandberg and Wallace, 2013)
Nutrient cycling *Medium priority*	*Construction*: There is growth in the use of Portland cement blends containing cementitious or pozzolanic waste byproducts of other industrial processes, such as ground granulated blast-furnace slag and fly ash. Concrete made using these byproducts often ends up in low-value applications, such as landfills and road subbases however, due to perceived reduced quality and longer setting times (Mehta, 2001). Portland cement blends containing 50% or more granulated blast-furnace slag or fly ash can actually yield much more durable concrete products than neat Portland cement (Mehta, 2001) *End of life*: In many industrialized countries, approximately two-thirds of construction and demolition waste consists of masonry and concrete rubble. Some companies and developments make an effort to reuse this as coarse aggregate in new concrete structures to reduce inputs of new aggregates (which are finite sources) (Mehta, 2001). Much of this is an example of down-cycling however

(Continued)

Table 2.7 **(Continued)**

	Cement/concrete
Provision of energy *Medium priority*	*Processing*: Producing a ton of Portland cement is energy-intensive (requiring about 4 GJ energy) (Mehta, 2001). See Section 3.3.4 *In use*: Concrete can be used as a vital component of passive solar design and can contribute to the creation of buildings that use less energy and therefore are responsible for fewer greenhouse gas emissions over their life time (Bainbridge and Haggard, 2011)
Provision of food *Medium priority*	*Construction*: The large amounts of water used to construct concrete structures may compete with irrigation needs for food growing during construction periods

2.5.3 Materials that are made or processed

Almost all building materials are processed to some degree, however some do not exist in nature (or are extremely rare in a natural state, such as glass) and must be created by humans through complex and often energy-intensive processes. Common building materials that are made include glass, steel, plastics, and a large range of composite materials. Typical issues associated with made or highly processed materials include high embodied energy figures, non-renewability of raw materials, undesirable emissions to air, water, and soil, and recyclability. Two quick reference tables are provided in this section: glass and steel.

Plastic is a more and more ubiquitous material found in buildings as materials, glues and sealants, finishes, and furnishings and fittings. Most plastics are made from petrochemicals where the major associated ecosystem services issues are related to the GHG and other toxic emissions that occur during extraction, processing, and end-of-life stages. A small but growing number of plastics for use in the building industry are made of plant-based or other starches, or even CO_2 (Patel-Predd, 2007) and have an ability to biodegrade (at least in part) making them less impactful in terms of end-of-life pollution. The range of plastics is extensive, and so therefore are the specific associated ecosystem services impacts. For this reason, a quick reference table related to plastic is not included in this section. Information about environmental impacts of some petrochemical-based plastics can be found in Section 3.4.3.

The tables examining glass and steel (Tables 2.8 and 2.9) suggest that when investigating processed/made materials for use in buildings, it is again important to know specifically where materials come from due to what the local impacts might be, but also because of the differences between environmental laws and regulation of them in different parts of the world. Specific methods used will also result in differences to ecosystem services impacts. The most important ecosystem services issues to be

Table 2.8 Potential impacts on ecosystem services of common building materials that are made or processed: Glass

	Glass
Climate regulation *High priority*	*Extraction*: Mining the silica for use in glass making may have an impact on climate regulation. The global biogeochemical silica cycle has impacts on global CO_2 concentrations through the combined processes of weathering of silicate minerals and transfer of CO_2 from the atmosphere to the lithosphere (Conley, 2002). Silicon is also a key element linking land and ocean systems (often through "dust events") and can affect carbon sequestration (Millennium Ecosystem Assessment, 2005, p. 337) *Processing*: Glass making is energy-intensive. CO_2 is released during the process of making glass due to the combustion of natural gas/fuel oil and the decomposition of raw materials (AGC Glass Europe, 2016). Ninety percent of architectural glass is made using the float method (AGC Glass Europe, 2016). Approximately three-quarters of glass in the United States is made using natural gas, but a small number of electricity-fired plants are operational. Some of these may be powered by renewable energy (EIA, 2013) *In use*: Glass is a vital component of passive solar design and can contribute to the creation of buildings that use less energy and therefore are responsible for fewer greenhouse gas emissions over their life time (Bainbridge and Haggard, 2011)
Purification *High priority*	*Extraction*: Sand mining can affect local water and soil quality. Silica sands can blow from sand mines altering the quality of both soil and air. Silica dust is a carcinogen and can contribute to lung disease (Loeb, 2016) *Processing*: Sulfur dioxide (SO_2) can be generated from the combustion of fuel used to make glass and from decomposition of sulfate in the batch materials. This contributes to acidification and decreased air quality. Nitrogen oxides (NO_x) can be generated due to high melting temperatures and in some cases due to decomposition of nitrogen compounds in the batch materials. NO_x also contributes to acidification. Evaporation from the molten glass and raw materials can cause release of particles into the atmosphere (AGC Glass Europe, 2016). Glass-making factories that use natural gas rather than heavy fuels may have reduced CO_2 and SO_2 emissions, but increased NO_x emissions. Some technologies that reduce NO_x can cause an increase in CO_2 when extra energy input is needed (AGC Glass Europe, 2016)
Provision of water *High priority*	*Extraction*: Silica mining can have negative effects on water quality. This can affect fresh water systems, ground water, and marine ecosystems depending on the source of the silica. Some sand-mining operations are extremely water-intensive *Processing*: Glass production can contribute to water pollution (AGC Glass Europe, 2016)

(*Continued*)

Table 2.8 **(Continued)**

	Glass
Habitat provision *Medium priority*	*Extraction*: Silica sand mining can have devastating impacts on local ecosystems. This can affect both terrestrial ecosystems due to open pit mining, and aquatic ones indirectly due to increased run off from open pit mining into fresh water or ocean systems, or directly through the dredging of river or sea beds (Struyf and Conley, 2012, Conley, 2002, Crawford, 2011)
Nutrient cycling *Medium priority*	*Extraction*: Silicon is the second most abundant element in the Earth's crust but is not renewable (Conley, 2002). Human-induced alteration of silicon cycles is less well understood, but it is believed, with medium certainty, to be a significant factor in altering the productivity of the ocean (Millennium Ecosystem Assessment, 2005, p. 334). See climate regulation row *End of life*: Glass is a fully recyclable material, however most building glass is not recycled due to difficulty of dismantling, sorting, and collecting waste glass from buildings and potential additives or films used in building glass that may affect recyclability. Glass is only 1% of construction and demolition waste and is generally inert (Glass for Europe, 2013)
Provision of energy *Medium priority*	*Extraction*: Depending on the location of sand mines, some of these may remove topsoil, thus affecting potential for biofuel crop production (Loeb, 2016) *In use*: Glass can be used to generate renewable energy through solar-thermal and photovoltaic applications
Provision of food *Medium priority*	*Extraction*: Depending on the location of sand mines, some of these may destroy arable farm land by removing topsoil, thus affecting food growing potential (Loeb, 2016)

Table 2.9 **Potential impacts on ecosystem services of common building materials that are made or processed: Steel**

	Steel
Climate regulation *High priority*	*Extraction*: The extraction of higher-grade iron ores results in the emission of less CO_2 than extraction of lower-grade ores (Magnetite Mines, 2016) *Processing*: Steel making is both CO_2- and energy-intensive. 0.28–1.8 tonnes of CO_2 are emitted for every tonne of steel produced (World Steel Association, 2015, Bonenfant et al., 2008). The iron and steel industry accounts for approximately 9% of total global anthropogenic CO_2 emissions (Pauliuk et al., 2013). The World Steel Association (2015) details various technologies that reduce CO_2 emissions in steel production. The production of steel results in the generation of byproducts

(*Continued*)

Table 2.9 **(Continued)**

	Steel
	that can reduce CO_2 emissions by substituting resources in other industries. For example, blast furnace slag is used by the cement industry allowing it to reduce its CO_2 emissions significantly (World Steel Association, 2015). See however discussion about the use of such wastes in Section 3.4.2
Purification *High priority*	*Extraction*: Iron ore mining can negatively affect local water and soil quality
Provision of water *High priority*	*Extraction*: Iron ore mining significantly affects fresh water sources although this depends on the country of origin and the nature of the mine. Typically, iron ore mining is very water-intensive, requiring vast amounts of water to transport slurries. The use of this water competes with other fresh water uses (Krishnaswamy et al., 2006, NZ Steel, 2016) *Processing*: Some steel producers reduce the large amounts of fresh water needed to transport slurries during extraction and during processing by recycling water and/or capturing storm water on site. Other operations may use sea water (sometimes desalinated) which may relieve pressure on fresh water resources (NZ Steel, 2016). Increased salinity of tailings and possible seepage into fresh ground water reserves should be investigated in operations using sea water for iron ore processing (Magnetite Mines, 2016)
Habitat provision *Medium priority*	*Extraction*: Open-cast iron ore mining operations are known to have devastating effects on downstream riparian ecosystems due to sediment increases, but the impacts in humid tropical areas are particularly severe due to monsoon weather patterns (Krishnaswamy et al., 2006). Typically, the impacts of iron ore mining on riparian ecosystems are much worse than other land use changes such as deforestation (Krishnaswamy et al., 2006). Many operations attempt to revegetate the landscapes affected after mining, although soil fertility is often reduced. Revegetated landscapes reduce the negative impacts of erosion and dust problems (NZ Steel, 2016, Jasper et al., 1988)
Nutrient cycling *Medium priority*	*Extraction*: Human-induced alteration of iron cycles is believed, with medium certainty, to be a significant factor in altering the productivity of the ocean. Biological nitrogen fixation has increased during the modern era as a result of increased iron fertilization by windborne dust (Millennium Ecosystem Assessment, 2005, p. 334). Mining activities are likely to be one cause of this *End of life*: Steel is infinitely recyclable without loss of quality and its byproducts and waste energies are valuable resources (World Steel Association, 2015)

(*Continued*)

Table 2.9 (Continued)

	Steel
Provision of energy *Medium priority*	*Processing*: The average energy intensity for steel production is 20 GJ/tonnes crude steel with a potential for improvement of 15%–20% (World Steel Association, 2015)
Provision of food *Medium priority*	*Extraction*: The large amount of water required in some iron ore operations may compete with water for irrigation uses in nearby food-growing areas. This may be more of a problem when ground water is used for operations (NZ Steel, 2016). There is also competition in land use between iron ore mining and high-yield crop-growing land in some places (such as in Australia) (Magnetite Mines, 2016)

aware of with processed/made materials include impacts on local habitat and biodiversity, interaction with the water cycle and marine systems, associated GHG and other toxic emissions, and embodied energy estimates.

To summarize, the following aspects should be investigated:

1. Where have the constituents of the material been extracted from? Are there environmental laws in that place that indicate good ecological practice?
2. How exactly has the material been processed? Which chemicals have been used and what are the impacts of those on ecosystem services during the process? Are certain practices better from an ecosystem impact point of view?
3. How has the making of the material contributed to GHG emissions or resulted in other pollutants or toxins over time? What are the implications in use?
4. How has the extraction and making of the material and its constituents affected local biodiversity in the short and long terms? How has this been addressed by the company? Are certain practices less damaging? What remediation of land/waterway strategies are in place?
5. How does the creation and use of the material interact with fresh water cycles over time?
6. How has the creation of the material affected local air, water, and soil quality over time?
7. How does the extraction and making of the material and its constituents interact with nutrient cycles over time?
8. Does the making of the material compete with food growing or energy production potentials?
9. Are there groups of people (particularly indigenous) that oppose the extraction of the constituents of the material or the locations of the operations? What are the main issues? Are there ecological or cultural issues that need to be considered?

What the quick reference tables in this entire section show, is that it is impossible to say whether a specific material is sustainable or not, or a better choice than others, without knowing details about where and how a specific material is grown, how it is harvested, what is likely to happen to it at end-of-life and the implications these variables will have on ecosystem services over time. The most important thing perhaps when considering if a material is sustainable or not, is to understand how it will be used in a building within the context of a specific design. It must be understood as being part of a larger system that interacts with: a specific local climate and global climate; local ecology and biodiversity; other materials in the space; and crucially the behavior of people over time. This demonstrates the importance of considering materials very carefully during the design process and allowing time for the selection of materials to in turn influence the design of a building as research is conducted. The design process, including materials specification, cannot be a simple linear process, but must be a series of back-and-forth adjustments, calculations, and estimates based on a specific context and set of conditions.

2.6 Conclusion: Materials selection and ecosystem services. A shift in thinking

Ecological regeneration goals for building developments can be provided by ecosystem services analysis for a particular place. Similarly, understanding the functions of ecosystems and the importance of them to human communities can be a criterion in materials selection processes. The processes presented here provide a starting point for creating an ecosystem services materials selection methodology that is measurable.

The application of ecosystem services analysis to materials selection has significant philosophical implications because it asks design teams to judge their environmental performance goals in terms of impact on ecosystem services, rather than on the more easily understandable norms of energy and water use, GHG emissions, and waste generation. Employing ecosystem services analysis in the pursuit of sustainable or regenerative built environments may therefore require a rethinking of key performance indicators. Rather than a "one size fits all approach," materials selection criteria can initially be of a general nature but should then also be specific to a particular site, locality, or region in order to make accurate comparisons between materials and products. Practically, the use of ecosystem services analysis in design processes will mean working much more closely with ecologists and allowing time for conducting research on both ecosystems relative to the site of the proposed development, and on specific brands of materials.

In addition to the potential of new or retrofitted developments using an understanding of ecosystem services to become even partly regenerative by acting as filters (purifying air and water), providers (of food and fresh water), and generators (of energy) for the rest of the surrounding existing built environment, selecting materials that have positive effects on ecosystem services even at just the harvesting or extraction phases of their lifecycles adds to the possibility that built environments could support the health of ecosystems rather than be agents of degradation as is currently the case.

Most buildings undergo multiple retrofits, upgrades, and additions during their life time, all of which require the use of building materials (Brand, 1994). This means each change made to a building is an opportunity to both select materials that are more sustainable and to allow even small changes to one part of a building to positively affect the environmental performance of the whole building through good design. Existing urban environments and the individual buildings within them are likely to endure for considerable time, during which they will still be degrading ecosystems and climate. If buildings could start to contribute to or perform even small aspects of ecosystem services beyond their own needs or boundaries, the negative environmental impacts of the built environment may in part be mitigated, and resilience or adaptability may be increased (Pedersen Zari, 2012). When considering how to practically apply these ideas in terms of just materials selection, using the seven ecosystem services described (see Table 1.2) as a guide to decision making could aid in the specification of materials where the impact on each ecosystem service has been taken into account.

Using an ecosystem services approach to materials selection should be considered as one part of a set of criteria that includes: the impact of materials on human health; the embodied energy, carbon, and water in material; how the use of materials in specific design configurations impact on water, energy use and GHG emissions during the life of the building; waste and pollution implications; the scarcity of the material or its constituents; and other lifecycle issues. The ecosystem services concept simply clarifies how people can understand what the impacts of different building materials are, not just on ecosystems themselves, but also in terms of how ecosystems function and relate to overall planetary systems (such as climate and hydrology) as well as human communities. Using ecosystem services analysis in the context of materials selection also challenges design teams to consider the wider design opportunities available to them to conceive of and build buildings and urban environments that may be able to integrate with or provide ecosystem services themselves.

There is a need for a reconsideration of the built environment's overall purpose and expectations and evaluation of its performance. The fact that many green, sustainable, or otherwise environmentally better than conventional design concepts, methodologies, techniques, and materials exist, but have not to date made significant positive impacts on the environmental performance of the built environment as a whole (or even to individual buildings they may be used on or in) is testimony to this (Gerhardt and Vale, 2010). The change in thinking required will not necessarily come through the highly problematic task of determining a list of perfectly sustainable materials that will fit every climate and function, or by developing new materials or technologies (Mitchell, 2012), but instead requires the adoption of new mindsets and goals for how built environments can and should function as systems in relationship to biological ecosystems and the ecosystem services we derive from them, as well as social ecologies.

References

AGC Glass Europe, 2016. Environmental Report 2016. AGC Glass Europe, Louvain-la-Neuve. Available at: http://www.agc-glass.eu/English/Homepage/Our-Values/Environment/Environmental-Report-2016/Environmental-Impact/page.aspx/2654.

Alcorn, A., 2003. Embodied Energy and CO2 Coefficients for NZ Building Materials. Centre for Building Performance Research, Victoria University of Wellington, Wellington.

Andreae, M., 1991. Biomass burning: its history, use and distribution and its impact on environmental quality and global climate. In: Levine, J. (Ed.), Global Biomass Burning: Atmospheric, Climatic and Biospheric Implications. MIT Press, Cambridge, MA, pp. 3–21.

Ashour, T., Georg, H., Wu, W., 2011. Performance of straw bale wall: A case of study. Energy Build. 43 (8), 1960–1967.

Bainbridge, D., Haggard, K., 2011. Passive Solar Architecture. Heating, Cooling, Ventilation, and Daylighting Using Natural Flows. Chelsea Green Publishing, Vermont.

Bansal, N.R., Doremus, R.H., 1986. Handbook of Glass Properties. Academic Publishers Ltd, Orlando.

Ben-zhi, Z., Mao-yi, F., Jin-zhong, X., Xiao-sheng, Y., Zhen-cai, L., 2005. Ecological functions of bamboo forest: research and application. J. Forest. Res. 16 (2), 143–147.

Blanco-Canqui, H., 2012. Crop residue removal for bioenergy reduces soil carbon pools: how can we offset carbon losses? BioEnergy Res. 6 (1), 358–371.

Bonenfant, D., Kharoune, L., Sauvé, S., Hausler, R., Niquette, P., Mimeault, M., Kharoune, M., 2008. CO_2 sequestration potential of steel slags at ambient pressure and temperature. Ind. Eng. Chem. Res. 47 (20), 7610–7616.

Brand, S., 1994. How Buildings Learn: What Happens After They're Built. Viking Press, New York, NY.

Campbell, I.C., Doeg, T.J., 1989. Impact of timber harvesting and production on streams: a review. Aust. J. Marine Freshwater Res. 40, 519–539.

Carrillo Usbeck, V., Pflieger, J., Sun, T., 2011. Life Cycle Assessment of Float Glass. PE International, Echterdingen.

Christanty, L., Mailly, D., Kimmins, J.P., 1996. "Without bamboo, the land dies": biomass, litterfall, and soil organic matter dynamics of a Javanese bamboo talun-kebun system. Forest Ecol. Manage. 87 (1–3), 75–88.

Conley, D.J., 2002. Terrestrial ecosystems and the global biogeochemical silica cycle. Global Biogeochem. Cycles. 16 (4), 1121–1129.

Crawford, R.H. 2011. Life Cycle Assessment in the Built Environment. Oxon: Spon Press.

Crishna, N., Banfill, P.F.G., Goodsir, S., 2011. Embodied energy and CO_2 in UK dimension stone. Resour. Conserv. Recycling. 55 (2011), 1265–1273.

de Groot, R. S., R. Alkemade, R., Braat, L., Hein, L. and Willemen, L. (2010) 'Challenges in integrating the concept of ecosystem services and values in landscape planning, management and decision making', *Ecological Complexity,* 7(3), pp. 260–272.

EIA, 2013 Glass manufacturing is an energy-intensive industry mainly fueled by natural gas - Today in Energy - U.S. Energy Information Administration (EIA). U.S Energy Information Administration. Available from: http://www.eia.gov/todayinenergy/detail.cfm?id=12631 (accessed 31.07.16.).

Erftmeijer, P., Robin Lewis III, R., 2006. Environmental impacts of dredging on seagrasses: a review. Marine Poll. Bull. 52 (12), 1553–1572.

Founti, M.A., Giannopoulos, D., Laskaridis, K., 2010. Environmental management aspects for energy saving in natural stone quarries. In: Global Stone Congress, Alicante, March 2–5.

Friederich, H., 2014. Bamboo for Land Restoration. Ecosystems, economy and society: how large scale restoration can stimulate sustainable development. Washington, DC, 29–30 May, Institut Veolia.

Gazi, A., Skevis, G., Founti, M.A., 2012. Energy efficiency and environmental assessment of a typical marble quarry and processing plant. J. Clean. Prod. 32, 10–21.

GEPVP (Groupement Européen des Producteurs de Verre Plat), 2005. Energy and environmental benefits from advanced double glazing in EU buildings. Glass Technol. Inter. 2005 (3), 99–102.

Gerhardt, C., Vale, B., 2010. Comparison of resource use and environmental performance of green walls with façade greenings and extensive green roofs. In: Sustainable Building Conference (SB10), Wellington, New Zealand.

Glass for Europe, 2013. Recycling of End of Life Building Glass. Glass for Europe, Brussells. Available from: http://www.glassforeurope.com/images/cont/38_83208_file.pdf.

Grbeš, A., 2015. A life cycle assessment of silica sand: comparing the beneficiation processes. Sustainability. 8 (1), 9.

Greenspec, 2016. Green Building Design. Metals: Production and Environmental Impact. Copper. Greenspec. Available from: http://www.greenspec.co.uk/building-design/copper-production-environmental-impact/ (accessed 29.07.16.).

Harmsen, J.R., Roes, A.L., Patel, M.K., 2013. The impact of copper scarcity on the efficiency of 2050 global renewable energy scenarios. Energy. 50 (1), 62–73.

Heinberg, R., 2013. Peak Everything: Waking Up to the Century of Declines. New Society Publishers, Gabriola Island, Canada.

Hobbs, S.L., Gunn, J., 1998. The hydrogeological effect of quarrying karstified limestone: options for prediction and mitigation. Quart. J. Eng. Geol. 31, 147–157.

Huang, Y., Chen, L., Fu, B., Huang, Z., Gong, J., 2005. The wheat yields and water-use efficiency in the Loess Plateau: straw mulch and irrigation effects. Agri. Water Manage. 72 (3), 209–222.

Jasper, D.A., Robson, A.D., Abbott, L.K., 1988. Revegetation in an iron ore mine - nutrient requirements for plant growth and the potential role of vesicular arbuscular (VA) mycorrhizal fungi. Aust. J. Soil Res. 26 (3), 497–507.

Joseph, G., 1999. Copper: Its Trade, Manufacture, Use, and Environmental Status. ASM International, Materials Park, OH.

Kanowski, J., Catterall, C.P., Wardell-Johnson, G.W., 2005. Consequences of broadscale timber plantations for biodiversity in cleared rainforest landscapes of tropical and subtropical Australia. Forest Ecol. Manage. 208 (1), 359–372.

Klinglmair, M., Sala, S., Brandão, M., 2014. Assessing resource depletion in LCA: a review of methods and methodolo. Inter. J. Life Cycle Assess. 19 (3), 580–592.

Koellner, T., Geyer, R., 2013. Global land use impact assessment on biodiversity and ecosystem servic. Inter. J. Life Cycle Assess. 18 (6), 1185–1187.

Korjenic, A., Petránek, V., Zach, J., Hroudová, J., 2011. Development and performance evaluation of natural thermal-insulation materials composed of renewable resources. Energy Build. 43 (9), 2518–2523.

Krishnaswamy, J., Bunyan, M., Mehta, V.K., Jain, N., Karanth, K.U., 2006. Impact of iron ore mining on suspended sediment response in a tropical catchment in Kudremukh, Western Ghats, India. Forest Ecol. Manage. 224 (1–2), 187–198.

Lafond, G.P., Stumborg, M., Lemke, R., May, W.E., Holzapfel, C.B., Campbell, C.A., 2009. Quantifying Straw Removal through Baling and Measuring the Long-Term Impact on Soil Quality and Wheat Production All rights reserved. No part of this periodical may be reproduced or transmitted in any form or by any means, electronic or mechanical, including photocopying, recording, or any information storage and retrieval system, without permission in writing from the publisher. Agron. J. 101 (3).

Lamb, D., 1998. Large-scale ecological restoration of degraded tropical forest lands: the potential role of timber plantations. Restor. Ecol. 6 (3), 271–279.

Landis, D.A., Wratten, S.D., Gurr, G.M., 2000. Habitat management to conserve natural enemies of arthropod pests in agriculture. Ann. Rev. Entomol. 45 (1), 175–201.

Langer, W.H., 2001. Potential Environmental Impacts of Quarrying Stone in Karst— A Literature Review. U.S. Department of the Interior and the U.S. Geological Survey.

Li, L., Wang, Y., Zhang, Q., Li, J., Yang, X., Jin, J., 2008. Wheat straw burning and its associated impacts on Beijing air quality. Sci. China Series D: Earth Sci. 51 (3), 403–414.

Likens, G.E., Borman, F.H., Pierce, R.S., Reiners, W.A., 1978. Recovery of a deforested ecosystem. Science. 199 (4328), 492–496.

Loeb, N.C., 2016. The Sand Mines That Ruin Farmland. The New York Times, 20160523. Available from: http://www.nytimes.com/2016/05/23/opinion/the-sand-mines-that-ruin-farmland.html.

Magnetite Mines, 2016. Social & Environmental Impact. Magnetite Mines. Available from: http://magnetitemines.com/social-environmental-impact (accessed 31.07.16.).

Malhi, S.S., Lemke, R., Wang, Z.H., Chhabra, B.S., 2006. Tillage, nitrogen and crop residue effects on crop yield, nutrient uptake, soil quality, and greenhouse gas emissions. Soil Tillage Res. 90 (1–2), 171–183.

Malinen, J., Pesonen, M., Määttä, T., Kajanus, M., 2001. Potential harvest for wood fuels (energy wood) from logging residues and first thinnings in Southern Finland. Biomass Bioenergy. 20 (3), 189–196.

Mattila, T., Grönroos, J., Judl, J., Korhonen, M.-R., 2012. Is biochar or straw-bale construction a better carbon storage from a life cycle perspective?. Proc. Safe. Environ. Prot. 90 (6), 452–458.

McDonough, W., Braungart, M., 2002. Cradle to Cradle-Remaking the Way We Make Things. North Point Press, New York.

Mehta, K., 2001. Reducing the environmental impact of concrete. Concrete Inter. 61–66.

Millennium Ecosystem Assessment, 2005. Ecosystems and Human Well-being: Current State and Trends, 5 vols. Island Press, Washington, DC.

Mitchell, R.B., 2012. Technology is not enough. J. Environ. Dev. 21 (1), 24–27.

Muller, N.Z., Mendelsohn, R., Nordhaus, W., 2011. Environmental accounting for pollution in the United States economy. Am. Econ. Rev. 101 (5), 1649–1675.

Northey, S., Haque, N., Mudd, G., 2013. Using sustainability reporting to assess the environmental footprint of copper mining. J. Clean. Prod. 40 (1), 118–128.

NZ Steel, 2016. Sustainabillity. NZ Steel. Available from: http://www.nzsteel.co.nz/sustainability/ (accessed 31.07.16.).

Olgyay, V., Herdt, J., 2004. The application of ecosystems services criteria for green building assessment. Solar Energy. 77 (4), 389–398.

Owens, T., 2015. The Pros and Cons of Straw Bale Wall Construction in Green Building. Building With Awareness. Available from: http://buildingwithawareness.com/the-pros-and-cons-of-straw-bale-wall-construction-in-green-building/ (accessed March 2016.).

Patel-Predd, P., 2007. Carbon-Dioxide Plastic Gets Funding. A startup is moving ahead with an efficient method to make biodegradable plastic. Technol. Rev. (November 14) Available at: https://www.technologyreview.com/s/409047/carbon-dioxide-plastic-gets-funding/. Date accessed: January 2017.

Pauliuk, S., Milford, R., Müller, D., Allwood, J., 2013. The steel scrap age. Environ. Sci. Technol. 47 (7), 3448–3454.

Pedersen Zari, M., Jenkin, S., 2009. Rethinking Our Built Environments: Towards a Sustainable Future. Ministry for the Environment, New Zealand Government, Wellington.

Pedersen Zari, M., 2012. Ecosystem services analysis for the design of regenerative built environments. Build. Res. Inf. 40 (1), 54–64.

Ratcliffe, D.A., 1974. Ecological effects of mineral exploitation in the United Kingdom and their significance to nature conservation. Proc. Royal Soc. A. 339 (1618), 355–372.

Reed, B., 2007. Shifting from ‘sustainability’ to regeneration. Build. Res. Inf. 35 (6), 674–680.

Sabto, M., 2010. Australian bamboo takes a stand. ECOS. 2010 (156), 22–23.
Sandberg, L.A., Wallace, L., 2013. Leave the sand in the land, let the stone alone: pits, quarries and climate change. ACME Inter E-J. Crit. Geograp. 12 (1), 65–87.
Satya, S., Bal, L.M., Singhal, P., Naik, S.N., 2010. Bamboo shoot processing: food quality and safety aspect (a review). Trends Food Sci. Technol. 21 (4), 181–189.
Shanmughavel, P., Francis, K., 2001. Bioproductivity and nutrient cycling in bamboo and acacia plantation forests. Bioresour. Technol. 80 (1), 45–48.
Struyf, E.C., Conley, D.J., 2012. Emerging understanding of the ecosystem silica filter. Biogeochemistry. 107 (1-3), 9–18.
Sverdrup, H.U., Ragnarsdottir, K.V., Koca, D., 2014. On modelling the global copper mining rates, market supply, copper price and the end of copper reserves. Resour. Conserv. Recycling. 87 (1), 158–174.
TEEB, 2011. TEEB Manual for Cities: Ecosystem Services in Urban Management. TEEB.
ten Brink, P., Berghöfer, A., Neuville, A., Schröter-Schlaack, C., Vakrou, A., White, S., et al., 2011. Chapter 10: Responding to the value of nature. In: Ten Brink, P. (Ed.), The Economics of Ecosystems and Biodiversity for National and International Policy Makers. Earthscan, London.
Tenyah, T.M., 2009. Environmental Impact of Different Logging Methods in the Bankhead National Forest, Alabama: A Comparative Analysis. Master of Science, Alabama A&M University, Normal, Alabama. Available from: http://www.aamu.edu/Academics/alns/bes/Centers/CFEA%20Publications/Thesis_ThomasTenyah2009.pdf.
TNO Built Environment and Geosciences, 2008. Potential impact of low-Emissivity glazing on energy and CO2 savings in Europe, Netherlands.
Turner, R., Daily, G., 2008. The ecosystem services framework and natural capital conservation. Environ. Resour. Econ. 39 (1), 25–35.
U.S. Geological Survey, 2016. Mineral Commodity Summaries 2016. Reston, Virginia. Available from: http://minerals.usgs.gov/minerals/pubs/mcs/2016/mcs2016.pdf.
US Congress Office of Technology Assessment, 1988. Copper: Technology and Competitiveness. Washington, DC. Available from: https://www.princeton.edu/~ota/ns20/year_f.html.
Van der Lugt, P., Vogtländer, J., Van Der Vegte, J., Brezet, J., 2012. Life cycle assessment and carbon sequestration; the environmental impact of industrial bamboo products. In: World Bamboo Congress Proceedings [9, 10-15 April, University of Antwerp, Belgium].
Vogtländer, J., van der Lugt, P., Brezet, H., 2010. The sustainability of bamboo products for local and Western European applications. LCAs and land-use. J. Clean. Prod. 18 (13), 1260–1269.
Wang, S.-Y., Tsai, M.-H., Lo, S.-F., Tsai, M.-J., 2008. Effects of manufacturing conditions on the adsorption capacity of heavy metal ions by Makino bamboo charcoal. Bioresour. Technol. 99 (15), 7027–7033.
World Steel Association, 2015. Steel's contribution to a low carbon future and climate resilient societies. World Steel Association. Available from: http://www.worldsteel.org/publications/position-papers/Steel-s-contribution-to-a-low-carbon-future.html.
Zhang, Y., Baral, A., Bakshi, B.R., 2010a. Accounting for ecosystem services in life cycle assessment, Part II: toward an ecologically based LCA. Environ. Sci. Technol. 44 (7), 2624–2631.
Zhang, Y., Singh, S., Bakshi, B.R., 2010b. Accounting for ecosystem services in life cycle assessment, Part I: a critical review. Environ. Sci. Technol. 44 (7), 2232–2242.

Part II

Choosing Sustainable Materials

Building materials

3

Brenda Vale
Victoria University of Wellington, Wellington, New Zealand

3.1 Introduction

It is perhaps unhelpful to think of materials as being sustainable, since the aim has to be to make sustainable buildings, a term which can mean many things. You might think of a sustainable building as one that has a mutually beneficial interdependent relationship with the local or global ecosystem (see Part I), or one that has a low environmental impact by not using too much energy or water or causing air pollution during construction, or one that is zero fossil fuel in use. It might even be a building that encourages its users to live a low-impact lifestyle by traveling much less, eating food based on plants, and generally trying to be recyclers rather than consumers (Vale and Vale, 2009), and perhaps a sustainable building is one that does all of these things. Such a building would depend as much on the users and their behavior as it might do on the design and choice of materials. That said, the aim in this section is to consider how to minimize the adverse impacts that building materials have on the environment, including all the materials that are used for maintenance and repair over the full useful life of the building, by choosing and using materials wisely. This part of the book is thus written from the viewpoint of the person designing the building and specifying the materials. It suggests thinking beyond the initial finished building and considering not only how it looks, and possibly how it performs, at the start but also how it will age and change over its life. The purpose of this chapter is to provide information for a practitioner or designer in a hurry who needs simple advice on what to specify to reduce environmental impact. The aim therefore is not to be comprehensive but to try to highlight the big issues to consider. For a true choice of sustainable materials Chapter 2, gives clear advice on how to approach this. Here the aim is first to think about the materials and then in Chapter 4, to think about the building and how the materials it is made of might influence the overall environmental impact.

In the past the sustainable choice of materials was easy, since these were probably the only available materials for all but the most prestigious buildings. In a medieval city in southern England, for instance, the houses might have walls of chalk and flint, since these could be extracted by digging into the underlying subsoil and the roofs might be thatched with straw, the stalks that remain from growing wheat once the grain has been harvested, and the chaff from winnowing might also be used as insulation between ceiling joists. Windows and doors would be of

Materials for a Healthy, Ecological and Sustainable Built Environment: Principles for Evaluation.
DOI: http://dx.doi.org/10.1016/B978-0-08-100707-5.00003-4

hardwoods, like oak, from local forests and floors could be of beaten earth, or finished with fired earth tiles. Paints would be made from lime and casein (from milk) with plant and earth pigments. Although details might differ the same palette of materials would be used for all ordinary buildings. Only the cathedral, palace, or houses of the very wealthy would be made from imported stone, and roofed with lead covered timber because these materials were much more expensive and had to be brought from further afield. Even so, as much use was made of water transport as possible because bringing heavy materials by land in carts on dirt tracks was a very slow and difficult process. Using local materials for the majority of buildings gave rise to vernacular traditions as the materials available from the immediate environment varied with both climate and geology. Although these vernacular traditions are well known and even admired in a world where globalization means the same kind of building can be found in both very hot and very cold cities, what are less well recognized are the vernacular traditions for limiting the taking of materials from the immediate environment to avoid its overexploitation. As an example, in the United Kingdom local forestry practices ensured that harvesting of building timbers was matched by replanting, although royal land could always be deforested to raise money for wars, from Richard I going to the crusades to Henry VIII building a navy (Nail, 2008, pp. 12–13).

The key word here is local, since if materials are taken from the local area then their taking is easy to police. If you made a house with earth walls traditionally the clay was sourced from the excavations for the foundations, but often now suitable soil may be brought in from elsewhere (Easton, 2007, p. 110). The modern problem is thus not only knowing what the materials are but also how and where they have been extracted and made, since only then can their impact be truly known. Timber from a distant tropical rainforest that is not replanted has a very different environmental effect from timber from a local replanted forest. However, some knowledge of how materials are produced at least allows them to be graded for their environmental impact, and this is explored in Section 3.2. Chapter 4, then deals with the more complicated issue of how materials are used in buildings, through taking a lifecycle approach, including briefly considering the impact of building users as part of the strategy of choosing appropriate materials for sustainable buildings. Finally, there is a brief introduction to the relationship between the choice of materials for making sustainable buildings and the link with human health, which is the focus of Part III of this book.

3.2 Materials that are grown

It seems obvious that the less processed a material is the fewer will be the resources that go into making it, and hence it should be the more sustainable choice (Vale and Vale, 1990). This has led to the not unreasonable idea that to make sustainable materials we should be using the so-called "natural" materials that are plant-based, such as straw, palm, bamboo, and wood (Pearson, 1991). Wool, which can be used as both insulation and as carpet and fabric in building interiors, would also come into this category, though as discussed in Section 3.4.3.2 wool has issues of its own when it comes to environmental impact. This section first considers the natural materials which can be grown and then looks at materials which are extracted from the earth's crust and their processing (earth, stone, brick). It ends by considering made and synthetic materials, such as steel and plastics. Obviously there are overlaps between categories since ore has to be extracted to make steel, and some plastics, like cellulose, are based on plant fibers.

The first thing to realize is that although plants are a way of harvesting sunlight they need land to grow and if that land is growing building materials it cannot generally also be used to grow food, energy, or be left as habitat for fauna and flora. That said, some land may be far better suited to growing timber than food, and some food systems will need trees in the form of windbreaks for better cropping, so symbiotic relationships are possible. The nearer the site of production and consumption the easier it is to see these relationships, however they are generally obscured with distance. The case study on palm thatching in Mexico (see Chapter 11) illustrates this point well. In the Mexican highlands the palm trees grow on land that is also used for crops and grazing but they are harvested using human labor and in a way that ensures their long-term survival by never taking too much from the tree. The harvested palms are then used for local buildings. As another example, Finland is a country almost 80% covered in boreal forest because of the cold climate (Tran, 2013, p. 58). The vernacular there is one of building in timber, often using whole logs notched at the corners. In summer the forest yields berries and a tradition has grown up of families going into the forest to harvest these. These are examples of local people understanding how much the land can yield on a sustainable basis.

3.2.1 Grasses

In the past plant materials in the form of grasses (straw, reeds) and other small plants like heather and leaves (palm thatching uses the leaves of the palm tree) were harvested and used for roofing as in most climates the rain has to be kept out of the building interior. This is done by making a reasonably thick layer of overlapping materials which had the added benefit of providing insulation, either to keep the heat in or the sun out. In cold climates walls were also often made of the same materials, such as the New Zealand Māori house (whare) with roof and walls made from bundles of raupō (reeds), with just a small entrance door (Isaacs, 2015, pp. 60–62). However, these buildings needed regular maintenance and repair and the interior conditions were far from what would be acceptable today.

Modern use of small plant materials tends to focuses on two types—straw and bamboo. Straw, traditionally used for thatching, is seen as a waste product (MacDougall, 2008). However, what needs to be remembered is that in the past straw was never grown for thatching, nor was straw a "waste" product. In traditional mixed farming straw was used for animal bedding where, mixed with animal dung, it became a balanced fertilizer to be returned to the fields, with the high carbon content in the straw being mixed with the high nitrogen content in the animal dung (Augustin and Rahman, n.d., p. 4). Taking a small quantity of these vital nutrients away from the soil to thatch a roof was something that happened occasionally. A long straw roof should last 25 years with a ridge of sedge that is replaced more often (10–15 years), and even then the old straw can still be used for making manure and returned to the soil. Straw in the form of combed wheat reed should last longer, 25–35 years (Thatch Advice Centre, n.d.). Straw only became a waste product as farming became industrialized and specialized, so farms were no longer mixed, with grain and livestock on the same holding. At the same time the long straw thatching tradition of the United Kingdom was lost in the search for greater yields of grain from the same land, leading to the use of short-stemmed varieties of wheat so the plant put more energy into the ear and less into the stalk. This produced a waste product no longer useful for thatching and no longer needed for livestock bedding, leading to other problems, such as straw burning on arable farms in an effort to get rid of it, which adds to CO_2 production, and to the modern idea that straw is a waste product.

In 19th century Nebraska this apparent waste product in the form of straw bales was seen as a building material (Swan et al., 2011). Because of the potentially high insulating value a straw bale wall can have if built correctly, using straw bales has become a fashionable way of making a sustainable building (Swentzell Steen et al., 1994; Hollis, 2005; King 2006). It has the advantages of being relatively cheap and manageable by self-builders. It also has "organic" qualities in that the walls are thick and can be somewhat uneven when plastered, thus showing that the building is not just the product of a modern industrialized society. However, these ideas are worth probing in this discussion of what might be a sustainable building material. Firstly, putting the straw in the walls is taking nutrients from the soil and unlike a thatched roof where the straw is pinned to the roof structure with wooden pegs (usually hazel) to make a waterproof interior, the straw is often plastered with a cement-based render, so composting the straw at the end of the building life may be more problematic. Modern expectations are that a building should be totally water and vermin proof, not something that was necessarily true of buildings in the past, and a cement render can aid in this. Secondly, only the walls can be of straw and other materials are required for the foundations and the roof structure. Thirdly, straw walls are sometimes reinforced with materials that might be regarded as not sustainable, such as metal straps, in order to ensure that they meet modern building codes (Pritchard and Pitts, 2006). It is also one thing to build straw bale walls in rural locations but if the straw is moved by lorry to suburban and urban locations then the environmental impact of this will add to the overall impact of the straw walls. Finally, straw walls are thick, so more land is required and the footprint of the building is increased. Despite this a three-bedroom single-storey 200 m^2 NZ straw bale house is described as having "...a small footprint" but this is hardly a small house (The Little Pig Building Company Ltd., 2016). There seems to be a tendency, in New Zealand at least, to make large houses of straw bale, possibly because the material is cheap. Like all materials, there is a place for straw bale, for modest buildings satisfying local needs, but it is not the sustainable panacea it often appears (for more discussion of building with straw bales see Chapter 4).

3.2.2 Hemp

Another popular plant-based sustainable material is hemp, traditionally a source of fiber and oilseed. Modern industrial hemp grown for the same purposes is, perhaps fortunately, low in the active ingredients in marijuana. In the recent search for more sustainable buildings, researchers and designers have also looked to this modern source of hemp. For instance, hemp hurd or shiv (the woody interior of the hemp plant stalk) can be mixed with lime to make hemp concrete and hemp concrete blocks (Ip and Miller, 2012). Hemp fibers (the outer part of the plant's stalk) have also been mixed to make a hemp fiber reinforced concrete, but the fiber has other values so the hurd-based product is more common (Ingrao et al, 2015). Concrete made with hemp and hydrated lime rather than cement may not be suitable for structural work, like bridges, but can be used for medium and low rise commercial and residential buildings (Kidalova et al., 2012). House walls can be made by mixing the hurd with hydrated lime (1 part hemp to 1.5–2 parts hydrated lime) and tamping the mixture into a thick wall layer on the outside of a structural timber frame. The lime hurd mix is built up in layers against and between the timber uprights using formwork. When dry it can be finished with a lime render or a rain-screen cladding (Sutton et al., 2011). Like other lime-based products depending on the porosity of the surface the lime absorbs carbon dioxide in a process called carbonation (see Section 4.6.2), so this type of construction can theoretically help in greenhouse gas reduction.

3.2.3 Bamboo

Bamboo, a grass, has been promoted as a modern natural building material because of the short time of 5–7 years it takes from planting to maturity (Banik, 2015). Because of this, its apparent sustainability credentials have seen it used in expanding markets for anything from skateboards to clothing, including laminate flooring (Tice, 2007), thus increasing demand for the raw material. However, this produces other problems. An issue associated with use of natural materials in modern buildings, because these materials are seen as sustainable, is mono cropping rather than the traditional symbiotic relationships such as using some straw for thatching while returning most to the soil as manure, as described above. Bamboo is an example of this change. In Penglipuran, a traditional village in Bali, bamboo is harvested for building in the form of doors, structural members, and roof tiles but its cutting from the forest surrounding the village is carefully controlled. This means only cutting at certain phases of the moon, so that the resource is maintained on a sustainable basis. Bamboo needs to be regularly harvested to avoid congestion in its growth and to make room for the emerging shoots, but this harvesting has to be controlled and if done traditionally by selective harvesting could be labor-intensive, and hence costly (Banik, 2015, Section 3.3.8).

The move now is to grow bamboo in plantations on a sustainable basis as, apart from traditional uses, it also serves as a commodity that can be sold for flooring, paper, and fiber, via the cellulose industry (East African Bamboo project, n.d., p. 4). It also has many traditional uses (e.g., hats, brooms, and fences). However, this is not the natural forest with a near natural ecosystem and associated fauna but a plantation growing bamboo not to satisfy local need for materials but rather an international desire for sustainable building products such as bamboo plywood. Bamboo has almost been given the status of a sustainable silver bullet just because it regenerates so quickly and grows in areas that are underdeveloped (UN Habitat, n.d.). As soon as land is cropped for bamboo it is unavailable for other uses, such as food growing, or as natural habitat. This is the problem with the so-called natural materials. The tradition of using these materials is that they are extracted from the environment in small quantities as they are needed to satisfy local demands, but whether their use remains "sustainable" when their production is "industrialized" is questionable.

3.2.4 Vines

Before moving to wood, a material that has formed the basis of many traditional vernacular buildings as structure, cladding, and finishing, another group of plants should at least be mentioned, even if they are unlikely to be considered in the search for modern sustainable building materials. Building with natural grown materials has to make use of some type of fixing in order to keep the materials in place. Binding elements together using vines, or other grown fibers twisted into rope, has a long tradition. Before the widespread use of metal fixings, the only alternative was to use wooden pegs to join two structural members together, or joint the timber in such a way that it could hold together without fixing, the classic example being the Chinese bucket arch or Dou Gong, which is a complex series of mortise and tenon joints, which gives lateral stiffness to timber post and beam structures, and allows for movement in earthquakes (Chen et al., 2014). However, in many tropical areas lashing was the common way of jointing, as discussed in the case study on Mexican palm thatching (see Chapter 11). This raises an issue for natural materials and sustainable buildings. Often what is picked out is a single material and not the whole system in which that material was traditionally used, whether this is careful management of the local resources through restricted harvesting, or understanding why buildings in certain locations, such as those subject to tropical storms, would be lashed together rather than fixed by nails or other modern means. In looking to natural materials to make our buildings sustainable we are not normally looking towards traditional ways of building and using them. Expectations, especially in the developed world, have changed. We prefer to see natural materials made into products that are familiar, such as bamboo laminate flooring or bamboo plywood, that can be fixed or glued in conventional "modern" ways, rather than making houses of lashed bamboo beams and poles. However, it is important that we specify these new plant-based materials that are more acceptable to modern tastes knowingly.

3.2.5 Wood

Timber, however, remains one natural material that is both cropped and is widely used in the building industry in many forms, both as dressed timbers and as engineered wood products (see Section 3.4.4.3). Some 60% of global timber in the form of wood from trees both from inside and outside forests is still used around the world for fuel, and wood has an important future as a renewable energy resource (Trossero, 2003). Forests are also seen as key in a low-carbon economy, both for the production of fuel wood and as a means of climate change mitigation through managed carbon sequestration (Kallio et al., 2016). Given all these aspects and the shrinkage in global forest coverage, whether it is possible to have more buildings of timber to replace the apparently unsustainable materials of steel and concrete needs to be carefully examined.

Only certain treeless areas of the world, such as hot deserts, have never had traditions of building with wood. In some places, like the United Kingdom where the forests disappeared through overharvesting, the timber traditions have been replaced by newer masonry ones. Urbanization was also a spur to masonry construction (see Chapter 10) as it was a more fireproof way of building at increased building footprint densities. Some countries which are earthquake-prone, like Chile and New Zealand, maintain a timber building tradition, as do countries like Canada and Finland which are heavily forested. Most of these continuing traditions have, however, seen a change in the size of available timber, which has tended to get smaller, and also to be from fast-growing trees rather than slow-growing hardwoods. As discussed above, fixings have also changed from natural materials and jointing to metal nails and plates. All of this change has a bearing on the sustainability of timber as a material.

One other important tree product needs to be mentioned before a further discussion of wood—leaves. Here the many species of palm in different parts of the world have given rise to traditions of thatching, as discussed in the case study of Mexico (see Chapter 11). Taking a few leaves helps the tree to generate more but obviously cutting all the leaves for thatching is not possible if the plant is to survive. There is a danger that a developed world desire for natural materials could put undue pressure on palm habitats that hitherto have sustained local needs. A classic example of this is the person from the Middle East who saw palm thatching in Mexico and immediately arranged to have a similar structure built in his home country, importing the materials and skilled labor to achieve this (Rios Calleja, 2015). This may be building using natural materials but it is not a sustainable situation.

Unlike palm or straw for thatching, which are first harvested and then prepared for use by hand, timber for building is a natural material that requires more processing, and even more so in modern times than in the past. Gone is the time when two suitably curved trees were selected and felled to form two crucks that when leant against each other made a characteristic pitched roof cottage (Brunskill, 1971, 1978, pp. 54–55). Even this type of construction required some processing, to fell and prepare the timber for the various joints required to join the crucks and splice in head and cill plates that carried the vertical wall members. Most importantly these huge members had to be set on foundation stones to separate them from the damp earth as like all natural materials timber is trying to decompose from the moment it is harvested. For some hardwoods (deciduous trees) this process is very slow and for some softwoods (coniferous trees) relatively quick. One basic method of preventing rot is to ensure the timber is dry before it is put into a building, meaning in the past there was often a considerable time between felling and using timber beams and posts, as these had to be stacked and left to dry in the air, often for long periods. Times vary for where the timber is dried as wood is always trying to be in equilibrium with the moisture content of the air (Simpson and Hart, 2000), meaning that in a building timber will still be absorbing and losing moisture with changes in the seasons. Although air drying is still used in the developed world building industry for specialized timber, much structural timber, such as the notional 50 × 100 mm stud for framing is kiln dried to achieve the required moisture content, as it is no longer acceptable for a building to be constructed over many years, but rather the expectation is it will be finished over many weeks. This means that traditional vernacular methods of dealing with timber have been industrialized, entailing more use of fossil fuel energy in the place of human labor and less energy-intensive methods. Much timber is now kiln dried to speed up the process. This runs counter to traditional views of timber where times of felling are controlled by local lore, winter usually being the preferred time. However, the nature of timber was recognized in that whether trees were felled in summer or winter "...both approaches share the common imperative that the timber be handled with the utmost care, thus putting it in a class above all other building materials" (Phelps, 1982, p. 44).

What emerges is a change in the way we view timber. In the past the skill was with the forester in selecting, felling, and drying the timber and with the carpenter and joiner in assembling it in the best way possible that would preserve it for the longest time, even if that meant ensuring plenty of airflow around the timbers. We also now expect buildings to be draught-free and infiltration-controlled rather than random, making such airflows more difficult to achieve. Now timber has to meet certain specifications with the loads it can carry and the expectation is that every piece of processed timber that arrives on site will be the same size and dimensionally stable, whereas we are still dealing with a natural material with all its inherent variability. The consequence of this, at least in the timber building tradition of New Zealand, is that rather than exposing structural timber as in the past, it is now concealed within other materials, many of which are not timber. Again a sustainable method for using a sustainable material has been lost in the search for cheap and quick building.

Kiln drying of timber is one of the ways of speeding up the whole process, but it is also energy-intensive. A study by Anderson and Westerlund (2014) for using a heat pump to make more use of waste timber in the drying process at the sawmill, states that typically 47% by wet weight of wood entering Swedish sawmills is converted to timber while a further 12% by wet weight of the waste is normally used in the kiln drying process, with the rest of the energy coming from other local energy sources. Techniques have been developed to maximize the yield and ensure that more useful squared timber has been extracted from each log, taking into account what the timber will be used for downstream (Gil and Frayret, 2016).

However, buildings have always been made of green unseasoned wood and there are still advocates for doing this in the making of sustainable buildings. The wood will continue to dry until it reaches equilibrium moisture content with the ambient air, so the real problem is fixing it so it can move as it dries. Green wood is also cheaper just because energy has not gone into drying it (Hackenberg, 1976). Some hardwoods, like oak, were used green because seasoned oak is so hard it is difficult to work with tools (Watts, 2013, p. 396). Green oak-framed housing is promoted in the United Kingdom as a more sustainable way of building (Border Oak, 2015), but is more associated with larger, detached houses in the country than urban dwellings, especially since it is not the cheapest way to build.

As mentioned above timber reacts with the air surrounding it. If a piece of timber is thoroughly immersed in water it will not rot and if exposed to the air so it can dry out if it gets wet it will last. When it comes to timber post foundations another condition occurs where part of the timber is always in contact with the damp ground and the part above ground level with the air. This junction is the point where fungal attack and rotting start (Phelps, 1982, p. 48). Traditional builders understood the need to keep timber dry and devised appropriate details. For example, in a traditional Norwegian log building with a sod roof several layers of birch bark were placed above the poles that formed the slightly sloping ceiling to form an impervious interface (Phelps, 1982, p. 86) (see also Chapter 10). The modern equivalent would use a layer of plastic (see Chapter 14).

Because of the natural if unfortunate tendency of timber, especially softwoods, to rot, modern varieties that are grown for building, such as radiata pine, are treated with chemicals so that the rotting process will be delayed. This can be viewed as a sustainable step since preserving timber means conserving forests by not perpetually demanding more wood (Richardson, 1978, p. 9), although other "green" minded people might shudder at the idea of using chemical wood preservatives because of their potential harm to both people and the environment (Coffin, 2003). However, using fungicidal preservatives to protect wood has a long history. Alexander the Great ordered timber to be dipped in olive oil for bridge building, a practice copied by the Romans, and Noah was commanded to finish the Ark with pitch inside and out (Richardson, 1978, pp. 11–12). Charring has also been recommended as a way to extend the life of fence posts (Willis, 1910). It was partly the shortage of timber in WWII that led to the development of building framing that used smaller section timbers, such as the gang nail roof truss, and the consequent extensive use of copper chrome arsenic (CCA) pressure impregnation of timber as a means of preserving it, especially in wet conditions. This chemical impregnation removes treated plantation-grown softwoods from the "natural" materials group but at the same time means they can be used in vulnerable situations which would otherwise require durable hardwoods. The latter are often taken from unmanaged tropical forests, a problem discussed in more detail below. Concerns over chemical treatments have led to the search for less environmentally harmful preservatives, one example being pressure impregnation with furfuryl alcohol (Lander et al., 2004). Effort has also been put into understanding how chemicals leach into the environment from CCA-treated timber when it is used in wet ground situations (Hingston et al., 2001; Kartal et al., 2007). Alternative preservatives that are readily available are boron, which is usually used for treating framing timber, and LOSP (light organic solvent preservative), which will off-gas, but neither are perhaps ideal, and the advice to avoid putting timber in the ground and use concrete piles may be the best (BRANZ, 2014).

Processing to turn round logs into square timbers gives rise to waste materials in the form of wood chips and sawdust, which in turn have been combined with modern chemical products to produce a range of materials, like chipboard and oriented strand board that are used for flooring and linings in modern interiors. Glue laminated timber and modern methods such as laminated veneer lumber make use of small pieces of wood for structural materials (see Section 3.4.4.3.). The purist would say these are no longer natural materials because they are contaminated with plastic glues but they are probably useful in the search for sustainable materials because they are trying to get as much use as possible out of the whole tree. In conventional logging about 50% of the tree is discarded in the form of roots, crown, and small branches (Okai, 2003), while chips and sawdust are produced at the sawmill during the conversion process. Despite the use of oil-based adhesives timber-based structural products tend to have a lower impact than the more industrialized materials of steel or concrete, though this does have to be assessed on a lifecycle basis to take account of maintenance and renewal (see Chapter 4).

Overall timber is promoted as a material that can contribute to making green and sustainable buildings. A UK study of 36 experts endorsed the increasing use of timber in promoting green buildings (Wan et al., 2014) even though the United Kingdom has limited access to home-grown building timber. The only caveat for those specifying timber is to ensure that it comes from a sustainably managed source. This has led to systems of labeling timber and timber products like plywood such as that of the Forestry Stewardship Council (FSC, n.d.). A problem with such eco-labeling programs is that they may harm small companies in developing nations, and that not all labels for the same product, e.g., plywood, cover the same thing (Hock, 2001). However, they can be helpful in the quest for making sustainable buildings.

3.2.6 Rewards for using materials that are grown

Having looked at the advantages and disadvantages of materials that might automatically be thought sustainable because they can be grown, this section probes how institutions that rate building sustainability view these same materials. Building rating systems like LEED, which is the product of the US Green Building Council, have a checklist approach whereby points are gained for various aspects of the design stage of the building. These points are then totaled and an overall rating is given for the building design, such as a silver or gold rating (US Green Building Council, 2016). Early on the LEED system encouraged the use of natural materials but over time it seems that how natural materials are acquired and used has been given more weight. LEED now requires that materials are certified as coming from a sustainable source or have a lifecycle label in order for LEED points to be awarded. However, the lifecycle of the material is only a relatively small part of the lifecycle of the building in use as will be shown in Chapter 4, One version of LEED for new buildings also gave points for use of materials that had a 10-year regeneration cycle (specifically 1 point if 5% of materials by cost were classed as rapidly renewable (Building Green Inc., 2015)). This would encourage use of materials from crops like bamboo and cotton, the latter as an alternative but much more expensive insulation material, and other natural interior finishes such as wool carpet. In this system animal products only earned points if the animal was not harmed, so leather was excluded but wool was allowed (Building Green Inc., 2015). The latter suggests a lack of understanding of farming processes, given that producing leather is a way of valuing a potential waste product from eating meat. An issue with animal products, like wool and leather, is their carbon footprint since sheep and cows are methane producers and as a greenhouse gas, methane is 25 times worse than carbon dioxide over 100 years in terms of its global warming potential (US Environmental Protection Agency, 2016). This raises the issue of whether having a rating system in which a plant-based material can be immediately equated with an animal-based one without taking global warming potential into account is appropriate.

Both LEED and the earlier UK BREEAM system in later versions place emphasis on ensuring materials, especially natural materials like wood and bamboo, come from sources that are certified as being sustainable (Ari, 2010). However, certification of the sources of materials may be more problematic for developing countries, particularly because of the cost of setting up and maintaining the process, though there is evidence of benefits when it is in place (Carlson and Palmer, 2016). This is an area that is still under development.

3.3 Materials that are extracted

This section considers materials that are dug up from the earth, including using earth itself as a building material, and that are somewhat processed. These materials could be thought of as fitting in the mid-range of environmental impact. It does include materials like stone and brick that used to be viewed as the main materials for making buildings that would last. It also includes concrete and a discussion of the process of making it. The main difference from the grown materials discussed above is that the extracted materials generally last much longer without replacement, so that in the history of building in some places, such as Germany and the

United Kingdom, their use superseded natural materials, in both cases timber, in the quest for making culturally significant structures, such as churches. This is a generalization as in some countries with a culture of building in timber, this material continued to be used, examples being the 12th century stave churches of Norway and the timber temples of Japan and China. The need for repair and replacement of plant-based buildings has been enshrined in some traditions, such as the 20-year cycle for rebuilding Shinto shrines in Japan (Lannom, 2014).

3.3.1 Earth

Earth, or more properly subsoil, is a widespread extracted material, occurring in many places on the planet in a form suitable for building, and consequently it has a long history, and has produced different techniques in different climates. Like grown natural materials earth is only serviceable as a building material if it is kept dry. This is easier to do in hot dry climates, like that of Egypt and other parts of Africa, than in temperate wet ones, like the United Kingdom where it was used in the past where there was no alternative material available, making it the choice of last resort. However, its very use in hot climates has led to some mistaken assumptions about the role of earth buildings. Earth is not a good insulating material, and given it is basically clay, like brick, this should be no surprise. It does have the ability to store heat, so if an earth (or brick) wall is insulated on the outside and this insulation layer is kept weather-tight the wall will lose heat much more slowly, and can effectively store heat. In a desert climate, which is always hot during the day and cold at night, this ability to store heat can be used to temper the interior of the house on a daily basis in buildings made of earth with no insulating layer. The sun warms up the exterior of the earth wall and the heat starts to migrate into the building interior. By the time the sun has set and the cold of night has arrived this heat will have migrated to the inside surface of the walls making for a comfortable indoor environment in the evening. The heat then starts to migrate back through the wall to the colder exterior and the next day the process repeats. This ability to act as a heat store has led to earth being considered a sustainable material in all climates. True, digging it up and processing it on site is a very low-impact approach to building, but in colder climates earth will at best make a wall that will need insulating if it is to provide a comfortable interior without the use of a lot of heating energy. Using earth walls (and possibly floors) inside buildings that are designed using passive solar principles so that the sun heats up the earth during the day is an approach that works best in climates where the sun shines on most cold days, such as in mountain and desert areas (see Chapter 13).

In the past the earth for building was dug up after removing the top layer of soil containing the organic matter, which is too valuable to throw away and unsuitable for building. The various techniques of using earth, whether rammed dry (pise), mixed with straw and water and placed in layers on the walls (cob), or made into blocks (adobe) which are first dried and then laid in courses with an earth mortar to make walls are well documented (Williams-Ellis et al., 1947; Keefe 2005; Schroeder, 2016). In the developed world, however, people can now buy adobe blocks ready-made and ship them to site. These blocks often contain cement as a stabilizer, and thus only differ from the so-called concrete block in terms of the matrix. The adobe is cement-stabilized earth, while a concrete block is cement-stabilized sand, perhaps mixed with some gravel. Moreover, adobe blocks are generally solid and concrete blocks can be made hollow, saving on resources. In some circles in the developed world adobe is viewed as a sustainable building material (Minke, 2012; Schroeder, 2016), whereas in developing and poorer societies where adobe is the only available material it is disliked, especially because of its poor performance in earthquakes (Hawkey, 2002). In the developed world earth walls have been used because the material is cheaper than the equivalent in brick or timber, or because such construction lends itself to DIY. Alternatively, the use of earth in wealthy societies may be to show that you are "someone who cares about the planet."

Unless mixed with cement, walls made of earth have to be kept dry, something that is easy to achieve in a dry climate. In wetter climates earth walls have to have a foundation isolated from the wet earth and a good roof overhang to protect the wall from driving rain. An external earth render is also a useful barrier, but will need replacing over the life of the building. However, the main problem with building in earth is the fact that with the possible exception of rammed earth (pise) it is relatively weak and no more than two-storey construction is generally the norm. This, together with waterproofing and maintenance problems, was one reason why with increasing wealth stone became the material for buildings that were large, culturally significant, or belonged to the wealthy, although in some areas where stone was prolific it was also the material of choice for smaller structures.

3.3.2 Stone

Building stone is the result of compression and change over long time of nonstone materials, such as sand (sandstone) and shell (limestone), or results from volcanic activity (scoria). Like earth, its sustainability credentials relate to the fact that it occurs naturally and can be quarried and processed for use with little energy input, in the past using mostly human labor but more recently using powered machinery. Unlike earth, stone has apparent permanence, although it does weather, and this made it attractive to those seeking to leave a great monument behind, such as the pyramids of Egypt, which are perhaps the first megalomaniac structures. These were of limestone, quarried from nearby bedrock, with special facing stones brought in from elsewhere, as transporting stone was always difficult (Fonte, 2007). However, if you were a ruler with grandiose ambitions and plenty of human labor available creating such edifices was less problematic. Like earth, however, once removed the stone is only regenerated on a geological time scale, so although it may have a relatively low environmental impact it is definitely not a renewable natural material. In terms of the energy for extracting and processing stone from a modern quarry a life-cycle study by the University of Tennessee (2007, p. 7) found it took 1 GJ of energy to produce 1 tonne of stone (0.8 million Btu/ton). As a quick comparison a meta study of the embodied energy of materials (Hammond and

Jones, 2006) found that the energy in making bricks was 3 MJ/kg, which equates to 3 GJ/tonne. However, a stone wall in a modern house would be thicker and heavier than the equivalent brick wall (a brick cavity wall works because the two thin leaves of brick are tied together to act as one), so more concrete would be required for the foundations and the building footprint would be bigger for the same floor area (see Chapter 4). This shows the danger of only looking at the material without considering the whole building when it comes to environmental impact. As a result it is easy to see why brick might be preferred over more costly stone for making small sustainable buildings.

The big advantage of stone over natural materials like timber in sustainable buildings is its longevity, although as noted in Section 3.1, it is not maintenance-free. Builders in the past were well aware of this and the stones prepared at the quarry to go into great buildings, like cathedrals and churches, were given three marks. The first was that of the man doing the work so he could be paid, the second was the mark to indicate where the stone was to go in the building, and the third a quarry mark to show where matching stone could be found when it came to repair and replacement (Alexander, 2007).

Poor-quality stones like chalk have also been used for building when the local environment has no other materials to hand that are better. In the UK chalk has been used much like earth, being made into "cob" walls in layers which are then left to dry out. Times of crises, such as WWI and WWII, when materials were short saw a revived interest in these materials (Vale, 1973), and some writers are still citing chalk as a sustainable material (Pearson, 1998, p. 144).

Although we normally think of stones for making masonry walls or in foundations under timber posts to isolate them from the ground, some stone can be split and used for roofing. The big urban expansion of 19th century England was roofed with thin slates, split from Welsh rock using steam power (Schunck and Oster, 2003, p, 12).

The main reason that stone is not used for sustainable buildings is the cost, much of which relates to transporting it from quarry to site. Because it is heavy this will add to the embodied energy (see Chapter 4) that goes into the stone building, which is why in the past large buildings, like the medieval cathedrals, were often situated so they had easy water access for delivery of the stone. There is a cheaper long-lasting and relatively maintenance-free alternative that is also extracted from the earth—brick.

3.3.3 Brick

Bricks cannot be produced without fire and this means energy. When left to dry in the sun bricks are called adobe, as discussed above. Brick is another material with a very long history, partly because brick buildings survive. They are much less likely to be damaged by water and weather than earth buildings and the fact a fired brick is stronger than an earth brick means that they can be used to build higher structures. Fired bricks were made in Mesopotamia in the third millennium BC and there were brick buildings from the 4th century BC onwards in Greece, a country where there was plenty of limestone available for building (Stefanidou et al., 2014). The Roman Empire could almost claim to have been built on brick and Latin had two terms for brickwork, *opus testaceum*, which meant just of brick, and *opus mixtum*, where brickwork was mixed with stonework (Adam, 1994, pp. 58–64). Bricks have formed the backbone of many large buildings that have survived, from the 6th century Hagia Sophia, much of which apart from the main stone piers is of brick with lime mortar, including the dome (Moropoulou et al., 2002), to the mills of the industrial revolution, such as the 1798 six-storey steam-powered cotton mill, Murray's Old Mill, on the Rochdale Canal in Manchester (Nevell, 2005, p. 199), to the 1889 16-storey Monadnock Building in Chicago by Burnham and Root. This was described as "...the highest that burden-bearing masonry walls ever carried," and had footings that virtually filled the basement level, rendering it unusable for plant (Starrett, 1928, p. 25). The fact that all these buildings are extant illustrates the longevity of brick construction.

Unlike the adobe brick which was large and heavy the fired brick fits into the human hand and because of this brick buildings have a satisfying quality of having been made. Many patterns of bonding bricks within a wall were developed, further enriching the appearance of a brick building. When fired different clays also produce different-colored bricks, in England ranging from "...yellows to deep reds, greys and purples" (Jones, 1936, p. 98), again adding to traditions of local building. From a sustainability viewpoint, the problem with brick is that the clay has to be extracted, originally by hand, often leaving very large holes behind, then the bricks have to be molded, again originally by hand, and then fired at high temperature. In the past bricks were made and stacked in a "clamp" with a wood fire started underneath. This led to uneven firing of the bricks, although those that were less burned could still be used for interior walls, and kiln methods were introduced to give more even firing, such as the Hoffman kiln of 1858. The latter automated the firing process, with the bricks stacked in chambers and the fire moving from chamber to chamber. With industrialization the excavation and molding processes were also mechanized.

The traditional clamp could make use of forest thinnings and wood that was unsuitable for building or other purposes, as still happens in parts of the developing world today just because such methods are labor rather than capital intensive (Baily, 1981), but the energy content of modern bricks depends on the fuel used. As a result research efforts have been put into alternative ways of firing bricks, such as using agricultural waste (Obeng et al., 1997) or tyres, and even solar energy via a large mirror and a reflecting concentrator (Villeda-Muñoz et al., 2011). The extent to which such endeavors are worthwhile depends on what else goes into making a sustainable building. If the building is zero energy in use then it might be worth looking at reducing the energy that goes into making the materials (see Chapter 4). However, bricks already possess good qualities in terms of longevity and low maintenance, and in terms of overall environmental impact it might be better to focus efforts on designing to reduce repair and maintenance over the life of the building. Nevertheless, there are sometimes symbiotic relationships, as in the example where a UK brickworks leased the holes they created while extracting the clay to the local authority to use for landfill, and then put down pipes as the holes were filled to take off the methane produced by the decomposition of the landfill. The methane was then mixed with natural gas to fire the bricks, reducing use of fossil fuels by up to 50% (Vale and Vale, 2000, p. 144).

3.3.4 Concrete

Concrete is the most used building material in the world (Crow, 2008). It also has poor sustainability credentials because of the carbon dioxide emitted during the manufacture of cement although it has found favor as a material for making sustainable buildings (it lasts a long time, withstands severe winds and floods, and has thermal mass) (Georgopolous and Minson, 2014, pp. 11–17). Because of the energy that goes into the production of cement, which is the basis of modern concrete, arguably it is a made rather than an extracted material. The reason it has been included in this section is because concrete has a very long history in buildings and it has properties that are similar to those of stone and brick—it is very durable and it is thermally massive, so it can be used to store heat if appropriately designed. Additionally, the main aggregate constituents of concrete, sand, and gravel, are extracted from the land, as well as the limestone and clay or shale that are mixed and heated to make cement.

The Romans used concrete in the dome in the Pantheon and in Roman times a natural cementitious agent, pozzolana—a type of volcanic ash—was used in place of the modern Portland cement. Although this made a concrete that was less strong, its durability is greater and its environmental impact is much lower than modern concrete (Hill, 2013). Portland cement is made by heating limestone or shales in a kiln, where clinker is formed which is then ground to a powder. Because these raw materials are widely available, cement is made in many locations. However, there are environmental issues, not only because energy is used in the process, but because carbon dioxide and other pollutants are emitted, with the manufacture of Portland cement accounting for 7.5% of human produced greenhouse gas emissions (Anon., 2010). Consequently a lot of research effort has gone into lowering the environmental impact of concrete, concentrating on the process of making Portland cement and on both the cement binder and the matrix.

Portland cement is made in a kiln and to reduce the impact of the product both the process and the fuel used to power it have to change. Alternative fuels have been proposed for the latter, such as using dried waste sewage sludge (Rodriguez et al., 2013), which is used currently as fuel in cogeneration plants that produce heat and electricity. Using sludge as fuel for cogeneration generates an ash that still presents a disposal problem. However, if the dried sludge is used as fuel in the making of Portland cement the ash gets bound up in the clinker, eventually finding its way into the concrete. Other waste products such as old tyres are used and agricultural and municipal wastes are also proposed as fuels (Prokopy, 2008). However, for the specifier at present it is hard to know where the Portland cement for the concrete is coming from, so if the cement and concrete industry are serious about alternative processes a labeling scheme that gives the origins and history of the product needs to be introduced.

What is clear is that the search is on for cement production with lower greenhouse gas emissions, both through energy efficiency measures in the process (Brunke and Blesl, 2014) and the use of alternative materials. One study in Malaysia (Mehmannavaz et al., 2014) investigated using palm oil residue and pulverized fuel ash as partial replacements for limestone. The palm oil was ground waste material from a palm oil mill and the fuel ash came from a coal power plant dust collection system. Fuel ash is a known pozzolanic material. The Malaysian experiment showed that using 30% ground palm oil residue and 20% pulverized fuel ash as replacements was optimal. More mainstream, perhaps, are proposals to use replacement binders such as calcium aluminate, calcium sulfoaluminate, alkali-activated binders, and supersulfated cements (Juenger et al., 2011). All four are known technologies, but cost has often been a factor in the preference for Portland cement. Alkali-activated binders refer to the use of fuel ash, which is a true waste product in the form of fly ash from coal-fired power stations and blast furnace slag, which would seem to have sustainability advantages, but in a market-led economy whereby lowest cost will always score above less environmentally damaging alternatives it is easy to see why Portland cement remains so popular.

Fly ash as a waste product of coal-fired generation has long been used as a pozzolanic additive in making Portland cement (Thomas, 2007), but herein lies another issue for the selection of materials to make sustainable buildings; should a waste product of a process that also has a significant global warming impact be encouraged? To achieve a 2°C or less rise in global temperature it has become apparent that coal-generated electricity is no longer a viable option both now and in the future and the remaining coal needs to be left in the ground. "...over 80 per cent of current coal reserves should remain unused from 2010 to 2050 in order to meet the target of 2°C" (McGlade and Ekins, 2015). Fly ash concrete blocks are lightweight and have insulating properties and have been extensively used in UK houses (Sear, 2008, pp. 211–234). However, as suggested this waste product should not be generated in the first place, and using agricultural waste, as in the Malaysian study of palm residue above, may be equally harmful in another way, given that such waste should be returned to the soil to maintain long-term fertility. The United Nations have recently reported that soil deterioration largely from intensive chemical-based farming means the world only has 60 more harvests remaining (Arsenault, 2014). As a result research that suggests using rice husks and fly ash to make low-cost concrete blocks (Salas et al., 1987; Mehmannavaz et al., 2014) may be misplaced.

One form of naturally occurring lightweight concrete that is discussed as a sustainable material is pumice (Jeppson, n.d.). Like fly ash blocks this is a way of creating foundations that have some inherent insulating qualities. Used as a slab, pumice concrete (pumicecrete) has the advantage of thermal mass and insulation, without the need for oil-based insulation materials with appropriate properties, such as expanded polystyrene. However, this is a long way from the use of high-strength reinforced concrete for high rise and other large buildings that make up so much of the current use of concrete. Such structures cannot be replaced by using naturally occurring pumice. (Further discussion of reinforced concrete is found in Section 3.4.4.1).

Concrete is not just ubiquitous but has many uses apart from structure, such as for floors, foundations, and blocks, among which foundations might be the most important, as concrete can be left in damp ground for centuries and still carry loads. However, concrete has been used for other building components. Concrete roof tiles first appeared in the 1840s in southern Bavaria using a local quick-setting naturally occurring cement (Dobson, 1959, pp. 3–10). In New Zealand concrete has also been used for sanitary fittings, such as baths (Siers, 2007, p. 106) and laundry tubs (BRANZ, n.d.). It is made into paving slabs and paths. Cement is also combined with other fibers to make sheet products (see Section 3.4.4.2). Concrete also finds its way in passive solar design as thermal mass, often in an insulated floor slab, which when polished can also serve as an internal finish. However, the acknowledged advantages of concrete come with environmental caveats. Finding alternatives to materials based on Portland cement, with all its inherent environmental issues will be neither easy nor cheap. What is clear is that making cement and using naturally occurring pozzolanic materials leaves holes in the ground, which are again an interruption to local eco-systems (see Chapter 2). The hope is that such holes will make some retrospective contribution to these and not just be used for landfill in order to dispose of waste (the stuff societies throw away) in a "waste" or leftover hole.

3.4 Materials that are made

3.4.1 Glass

Separating the discussion of concrete as an extracted material from glass as a man-made material may seem odd. The reason for so doing is that concrete has been made in the past from naturally occurring cementitious materials but glass, unless appearing as part of volcanic activity, always needs a process to produce it, and this process involves energy. That said, modern processes for producing concrete also need energy as discussed above. However, boundaries have to be drawn somewhere.

In a cold or temperate climate making buildings that can make use of solar energy means using glass, and often a lot of it. Because glass admits light, or short-wave radiation, which is then absorbed by the surfaces inside the glass and readmitted as long-wave radiation that cannot pass back out through the glass, the interior heats up. The Romans, for whom glass was a very expensive product, knew about this principle and made use of it in the caldarium (hot room) of the Baths of Caracalla, which had openings closed by sheets of mica that behaved in the same way (Butti and Perlin, 1980, pp. 20–25). This property of glass again illustrates the need not to think about which material is sustainable but rather about how to make a sustainable building through the optimal use of available materials. To paraphrase

Professor Ng's 2012 comment in a Passive and Low Energy Architecture (PLEA) conference in Lima, if you want to make a low-energy building in a tropical climate you don't start with an all-glass tower. However, the iconic all-glass tower has come to represent the ultimate in corporate architecture and has been reproduced in most climates, however unsuitable. When it comes to sustainability, an all-glass building is probably best left for plants.

Glass is a mixture of sand, soda ash, and limestone and other residual materials, often mixed with cullet (broken glass). The mixture is heated and the resultant molten substance drawn off in a continuous process. Apart from the familiar flat glass sheets the molten mixture can also be extruded and blown to make fibers, which can be used as an insulation material in the form of batts in dry situations, and also foamed when it becomes a closed cell insulation material that is impervious to water. The fiber batts can be treated with resin to render them suitable for use in masonry cavity walls. The use of cullet in glass manufacture shows how glass can include recycled content. Nor is glass a particularly high-energy material (Alcorn, 2003), even though energy is required in its manufacture.

The other problem with glass is not only does it let light in, so buildings heat up, especially in direct sunlight, but at night it is a poor insulator and easily lets out heat. The classic passive solar house has glass oriented towards the sun in winter and shaded in summer, and this glass may well be double or even triple units with various noble gases in the cavities between the panes (argon, krypton) to reduce convection currents and hence heat transfer between the glass layers. Low-emissivity coatings also reduce radiant emissions through the glass but these complexities add to its cost and to its embodied energy (Asif et al., 2001). However, the use of glass for collecting solar energy also needs an interior thermal mass, perhaps in the form of the floor slab and internal partitions, insulated from the cold outside, to store solar energy from sunny days for use on overcast days and at night. The principles of designing for climate were well known 6000 years ago to the Chinese, even without the use of glass (Perlin, 2013) and later to the Greeks and Romans (Vitruvius, c30BC, VI). The same principles are part of the vernacular of different societies, such as rural houses in southern China which have windows to the south but present a solid wall to the north. Passive solar principles have been periodically rediscovered by architects and designers, as in the 1940s solar houses produced by the Keck brothers in the United States (Shreve, 2013, pp. 59–115). Glass is a material that is essential in making sustainable buildings but using glass is not a guarantee that the building will be sustainable, the latter will depend far more on the design of the building and its performance in use (for further discussion see Chapter 2).

3.4.2 Metals

Metals are another of the so-called modern materials (along with glass and concrete) that have a history almost as old as that of people. In building construction metals are used in their elemental form, such as copper, and as alloys, such as brass.

Metals are hard and shiny and, apart from mercury, solid at normal room temperatures. They are impervious to moisture but most are active elements seeking to combine with oxygen and other elements in the environment, and this makes them prone to corrosion. They can also react in an electrolytic way when dissimilar metals are in contact with each other in a damp situation. Metals are malleable, so thin sheets can be made, and they can also be cast. Some, like aluminum, are ductile and can be drawn through dies to make complex shapes, such as glazing bars.

When it comes to environmental impact, metals take energy to manufacture at all stages from the extracted ore to the useable material, but their environmental impact to an extent depends on their recycled content, gaining a reduction in embodied energy with an increase in the latter (see Table 4.1). However, because metals need energy and skill to manufacture and work they have always been more expensive in the past than materials that could be grown, such as wood, and although metals were used for fixings and as reinforcement (the Parthenon in Athens used iron bars sheathed in lead as protection from rust (Bates, 1917), they were not used for structural purposes until the industrial revolution and the development of ways of producing metals in much larger quantities.

3.4.2.1 Steel

Initially the large-scale development of metals for the construction industry was in the form of cast iron, and later wrought iron and steel, which is a refinement of wrought iron. The basic difference between the three materials is their carbon content. The properties of cast iron (brittle and resistant to rust but good at resisting compression) and wrought iron (good at resisting tension) meant that iron frame structures were possible, and these were developed in the bridges that supported the new transport systems, first canals and then railways, and in the large mill buildings that housed the manufacturing enterprises of the industrial revolution (Richards, 1958, pp. 16–17). Following the development of the Bessemer process to produce low-carbon steel without the slag content found in wrought iron, engineers in 1880s America first used this new product for building bridges. Experiments with bridge design enabled them to work out the stresses the material could take, paving the way for its use in tall buildings (Starrett, 1928, pp. 36–38). The fact that steel led to the making of a new building type, the skyscraper, has led to it being seen as a new material as opposed to an evolving iron-based building material.

Steel is an important building material, with the annual growth rate in steel production averaging 3.8% from 2010 to 2104, with demand expected to continue with global urbanization (World Steel Association, 2015). World steel production produces approximately 25% of industrial greenhouse gases, or about 9% of all anthropomorphic greenhouse gas emissions (Pauliuk et al., 2013) compared with 7.5% for cement (Anon., 2010). Steel takes energy to make but has advantages for making sustainable buildings in that it can be recycled. As long ago as the 1950s a third of world steel was made from scrap with consequent savings in ore and coking coal using either the open hearth process or the electric furnace (Pounds, 1959). The scrap steel has already had impurities removed during its refining, and thus less energy is required to remove the impurities from a batch with recycled steel content. Scrap is usually classed in three forms: "home" scrap which is made within the steel processing plant from defective materials and trimmings and is reused within weeks; "new" or "prompt" scrap comes from the process of making the steel into products, and is usually available within months; and "obsolete" scrap is formed at the end of the useful life of steel products and will require sorting and preparation before reuse, such as removing tin or zinc coatings (Yellishetty et al., 2011). Currently scrap is mostly used in the manufacture of construction steel, which can take steel of lower quality, although it is anticipated that steel demand will exceed supply of raw materials by 2030, meaning recycled steel will have to be used in more demanding applications, such as transportation and machinery. Scrap steel can contain traces of copper and tin, which could be a problem for the car industry which, for aesthetic reasons, needs a pristine surface finish on its products (Pauliuk et al., 2013). For steel to be 100% recycled such defects would have to be accepted by consumers, or maybe we just find an alternative to cars.

Stainless steel, which contains chromium, nickel, and molybdenum, can also be recycled and does contain recycled content (25% obsolete scrap and 35% new scrap, with 45% being new raw materials (British Stainless steel Association, n.d.)). In exposed situations it also has a longer life than ordinary steel.

The good news for designers is that the steel offered to the construction industry will probably have recycled content. There is a carbon footprint label for steel under the Hong Kong Carbon Labelling Scheme, but at the moment it may not be comparable with other carbon footprints for other building materials, such as the Singapore Green Labelling Scheme, because of the different ways of defining the boundaries of the lifecycle assessment methods (Wu et al., 2014).

3.4.2.2 Copper

Copper has been used in building as a sheet roof covering and for pipes and wires, the latter two being its main current uses. Copper use expanded during WWII both in its natural form (especially in communications) and in its alloys (in shipbuilding) (Copper Development Association, 1960, pp. 56–57). Copper is both ductile and malleable and resistant to water corrosion. As a roofing material it reacts with air and water to form a patina that then prevents further corrosion, giving it a long life, even of hundreds of years (Copper Development Association, 1959, pp. 7–8). The fact it can be easily soldered to form stable joints has led to its extensive use as pipes and tubes, and gutters, downpipes and flashings. It can be recycled easily, which is the basis of its sustainability credentials. In the United States almost half of all copper on the market is recycled, with most virgin copper used for wiring. Of the recycled material more than half is new scrap, the remainder being obsolete and salvaged from old electric cables and discarded pipes (Copper Development Association Inc., 2015). Gordon et al. (2006) suggest that in the United States the total quantity of copper is about equally divided between three areas, these being the copper still in ores yet to be mined, the copper in products that are currently in use, and the copper dumped in waste deposits, and they stress the need to move from extraction of raw materials to use of recycled products. Extraction of copper from waste dumps could also become essential.

The fact that copper has always been recycled has contributed to it being suggested as a sustainable material of choice for roofing and cladding, given a life of 200 years is claimed (Anon., 2006). The downside of copper as a sustainable material is that it is a raw material in short supply, with average ore concentrations under 1%. This means that more than 100 tonnes of ore have to extracted, which uses energy and toxic chemicals that have to be disposed of, to obtain 1 tonne of refined metal (Messner, 2007, p. 113). Although environmental legislation in the developed world has led to a lessening in the environmental impact of copper over its lifecycle, this is offset by what is happening in terms of inefficient copper mining in developing countries (Messner, 2007, p. 128).

Copper is a material that needs to be used appropriately and sparingly, since it is a vital resource in supplying people with heat and electricity, both now and in the future, as there is no truly suitable cost-equivalent alternative (Messner, 2007, p. 116). Aluminum could be used to conduct electricity, but being a less good conductor more metal would be needed, raising the cost (Gordon et al., 2006). The problem is the copper-producing industry wants to sell as much as possible now, and so will stress the sustainability aspects of copper, whereas true sustainability should be planning so that the material is available both to this and to future generations (for further discussion see Chapter 2).

3.4.2.3 Aluminum

Like copper, aluminum has found a place in the building industry because it is resistant to corrosion. The major user of aluminum at 36% of total is the transport industry, with the building industry accounting for 25% (Council for Aluminium in Building, 2008, p. 14). In the latter aluminum is used in window joinery and in curtain walls, as well as in small items such as hardware. Aluminum is also processed into sheets and tubes, thus finding a place in HVAC applications. It can also be cast and drawn, leading to use in door and window fixings and handles. After WWII there was surplus aluminum because of the wartime expansion of the aircraft industry, and aluminum schools and houses were made and marketed, as well as many smaller products like saucepans and furniture (Vale, 1995, p. 171) but such moves did not lead to any significant building-sized developments. Although aluminum takes energy to produce it is easily recycled and unless exposed to salty atmospheres, is usually long-lasting. Currently 50% of aluminum is produced using hydroelectricity, which lowers its carbon footprint, and the global recycling of aluminum is increasing, currently standing at 35% of the total, which is important given it takes only 5% as much energy to recycle aluminum as to produce it new (Council for Aluminium in Building, 2008, p. 5).

However, when it comes to sustainability things are not straightforward. Building dams to make hydroelectricity is not without negative environmental and social impacts and when this electricity is used for smelting aluminum the money generated may not benefit the local people who are those most affected. This electricity may also be subsidized and charged at a much lower rate than electricity to local consumers to attract the aluminum smelters to come (Fearnside, 2016). In a country like New Zealand, where aluminum is still smelted using hydroelectricity, amounting to 13% of total national energy use, and where aluminum joinery is the norm, maybe this does not matter, though taxpayers are still supporting the smelter (The [NZ] Treasury, 2013). However, in Brazil, as Fearnside (1999) pointed out, the concession was that the cost of electricity to the smelter was never to be more than 20% of the sale price of aluminum, meaning that the smelter paid only a third to half the cost of producing the power. Residential customers paid six times more for their electricity than the smelter paid. So to be confident about specifying aluminum in a sustainable building it is vital to know where it was produced and who "paid" for its production.

3.4.2.4 Zinc

In the building industry zinc coatings have been used to prevent other materials, like mild steel, from corroding, although perhaps the more important use of zinc is as a constituent of brass (Craddock, 1987) where it finds many uses outside the building industry, such as for brake linings. In the building industry it is commonly used for galvanizing, particularly steel sheet used for roofing and other purposes. However, these uses of zinc are dissipative, meaning it is much harder to recover and recycle the metal because in use it dissipates into the atmosphere. Because of this problem of recovery Gordon et al. (2006) suggest that it would be difficult to have enough zinc to go round if everyone made use of it at the same rate as those living in the developed world, suggesting it is hardly a sustainable resource.

3.4.2.5 Lead

Lead, like copper, is an elemental metal with a long history of use in buildings. It is soft, gray, and malleable, so in Roman times was used for pipes, tank linings, and in the aqueducts constructed to bring water into urban areas (the word plumber comes from the Latin for lead). It was used for roofing many churches and cathedrals and was made into guttering and down pipes, often with decorated hoppers as patterns were easily beaten into the soft metal (Weaver, 1909). It found use as a flashing material, again because it could be easily worked. Unlike copper, which is still used for roofing and pipework, lead is no longer used in modern construction in the same way (see Part III for discussion of the health issues associated with lead). Lead now is more likely not to be found in construction work but where shielding from radiation is needed, or in battery cells. The current car industry still depends on lead for its starter batteries, as do emergency battery power supplies. The use of lead in batteries is partly because it is so extensively recycled, and is the most recycled of the common metals (International Lead Association, 2014).

Because of health issues attached with lead plumbing its use in sustainable buildings is minimal with, perhaps one exception, that of lead flashings. There are substitute materials but none are as good as lead at giving a long-lasting, trouble-free way of flashing. More important is designing to avoid the use of extensive flashings, and this means avoiding dormers and changes in roof level and keeping to a simple roof profile (Vale and Vale, 2000, pp. 182–183). Again, the making of sustainable buildings is as much about sensible design choices as it is about the choice of materials.

3.4.2.6 Other metals and issues

Nickel and tin have both been used in building in the form of alloys. Nickel is a component of stainless steel (discussed above) and tin when alloyed with copper forms bronze.

The claimed sustainability credentials of metals are that they can be recycled, which is true, though levels of recycling differ, as discussed. However, some metals are in limited supply and it would be wise in trying to make more sustainable buildings to use each metal only where no easy or common substitute can be found. Copper is useful for conducting electricity and for making pipes. Structural steel is essential for making tall buildings, but not for framing up small ones, where wood is potentially a less environmentally damaging alternative. Wood has been used for window joinery for centuries while aluminum joinery is a relative newcomer. Metals used in combination with other materials, such as the composite cladding material Alucobond, are briefly discussed in Section 3.4.4.

3.4.3 Plastics

It was the increasing expense of one resource and the fact the supply for a perceived essential use was dwindling that gave the big impetus to the development of plastics in the 19th century. The dwindling resource was ivory and the essential product billiard balls (Kaufman, 1963, pp. 26–27).

After World War II the built environment, particularly the home environment, was the target of the chemical and plastics industries that had expanded in support of the war effort and were now looking for a new market (Freinkel, 2011, pp. 25–27). The history of plastics in the home is much older than this, having begun in the late 19th century. Alexander Parkes was the inventor of Parkesite, the first real plastic that could be formed into useful products and it won him a medal in the 1862 International Exhibition (Kaufman, 1963, p. 21). Parkesine is not dissimilar from the modern celluloid. However, at the start these materials were used for small trinkets, such as combs, buttons, and collars rather than for building components. This was fortunate, as these plastics are unstable and shrink with age and in breaking down release corrosive gases, which is why many celluloid knife handles from this period, masquerading as mother of pearl, are attached to some very rusty knives. The inflammable nature of celluloid was also always in the background and Bakelite became the first plastic that was both totally synthesized within the laboratory and that was safe enough to be used for electrical insulation (Fielding, 1948, p. 13). It then also found its way into buildings as light switches, power sockets, and door furniture.

The 1950s mark the real date of the invasion of plastics into buildings. Plastic, or more correctly the family of polymer chemicals, is cheap, and so fits with the Western, postwar, socialist modernist aim of a more equitable society. It is also long-lasting and flexible in what it can do, from paper-based melamine laminates for kitchen worktops to the molded acrylic bath. However, that very longevity is part of plastic's problem since, unlike natural materials, plastics do not biodegrade very readily. This is bad for manufacturing since if a product does not break or wear out it is hard to sell any more once everyone has one. It is also potentially bad for the environment since the elements that go to make up plastic, most of which come from the waste products of oil refining, are not available for reuse in natural ecosystems, unlike the biodegradable elements that make up natural materials.

Plastics come in two types—thermosetting (Bakelite is thermosetting) and thermoplastic (the vinyl and styrene group)—with the former being, as the term suggests, set by heat but the latter able to be recycled by being melted with heat, providing the heat is never sufficient to destroy the chemical composition. Thermosetting plastics can be destroyed by heat but cannot be melted and reformed. Whether plastics like PVC should be recycled is another issue (Daniels, 2016). However, there is another problem with some of the raw materials used in plastics (cellulose, phenol, urea, casein, and acetylene), and the family of chemicals, or intermediate raw materials, that are used to convert them. Although some of the basic materials for plastics could be grown, such as cellulose from wood or other fibers and casein from milk, other oil-based chemicals are still needed for the process of conversion to plastic. Phenol resins, which are most important in the thermo-setting plastics are extracted from coal tar or prepared from benzene. Formaldehyde has to be produced synthetically from methane, itself the main component of natural gas. Acetylene is made from coal and lime and is the main raw material for the vinyl group of plastics. Hence the sources of the many plastics currently used in the building industry are, at least in part, fossil fuels.

Perhaps a better way of thinking about plastics and sustainable building is to consider which products are perhaps the only option (polythene dampproof membranes), which products provide benefits that might outweigh deficits (insulation) and which products can have substitutions (linoleum instead of vinyl flooring).

3.4.3.1 Necessary plastics

The purpose of building is to create a defense against unwanted weather, and this will normally mean keeping out water in the form of rain, sleet, or snow. Because many materials deteriorate if they get wet, especially the natural materials which might be favored when making sustainable buildings, then buildings must be detailed so that certain materials stay dry. Some plastics, like polyethylene, are very useful materials for keeping things dry. Just as concrete foundations may be unavoidable in making a sustainable building (see Section 3.3.4), so keeping water out of belowground spaces will also rely on polypropylene (PP) geo-textile membranes or polyethylene (PE) damp-proof membranes.

Suppose you wanted to design a passive solar house. You need a concrete slab to store the solar energy that enters through the glass windows. This slab will need insulation under and around it to ensure the heat from the sun is not lost to the ground. To keep the insulation dry a membrane is needed between the under-slab insulation and the ground and this would normally be low-density polythene (LDPE) for a small building like a house. Should you specify reused or recycled material? Addis (2006, p. 143) suggests the risk of using salvaged plastic membranes is too high, as once they fail in situ the replacement will be both difficult and costly as you will have to take up the floor slab. However, polyethylene is a recyclable product and some membranes will have been made with a proportion of pellets from recycled material. Polythene is more easily recycled if it does not include plasticizers (for discussion of these see Part III). At present there is no way of knowing what you are specifying in terms of recycled content.

Another way of looking at the membrane problem is to quantify the energy that goes into making it against the energy it might save, in this case by protecting the slab insulation from getting damp (this type of tradeoff is discussed further in Section 4.4.3). Because house sizes differ between countries (Khajehzadeh and Vale, 2016) an assumption is made of a two-storey three-bedroom house with a total floor area of 100 m^2 (large for new housing in the United Kingdom and small for new housing in New Zealand). Allowing extra for the depth of slab and insulation might need a polythene damp-proof membrane of 80 m^2 (the membrane needs to pass under and up the sides of the slab and insulation and overlap the damp-proof course in the wall). The energy that goes into making such a membrane in LDPE (low-density polyethylene) is 103 MJ/kg (Mithraratne et al., 2007, p. 218). A typical 1200 gauge polythene membrane weighs 265 g/m^2 (ICOPAL, 2013) so, allowing for overlaps, the energy in the membrane is 2.7 GJ. A house in the United Kingdom that meets the German Passiv Haus standard will use no more than 15 kWh/m^2 for space heating and achieve a 75% reduction in space heating compared to a normal UK house (The Passivhaus Trust, n.d.). Over one winter in London a Passivhaus of 101 m^2 used 1220 kWh for space heating (Ridley et al., 2013). Thus a nonpassive house would use 4880 kWh, or 17.6 GJ. This means the energy going into the polythene membrane is 15% of the energy used for heating that house in 1 year. The damp-proof membrane under the floor slab will last as long as the house foundations. This puts the use of a plastic damp-proof membrane into perspective.

In the past there was no plastic and floors were either separated from the wet by building a suspended timber floor above ground, or the slab was finished with an impervious material, like a fired clay quarry tile. While it is possible to insulate between the joists in a suspended timber floor this is lightweight construction and cannot be used for storing solar energy. The traditional tile-finished slab on ground has no insulation under it and unless the ground conditions under it are very dry, it will also not be useful for storing energy.

Another area where plastics have found a role in making low-energy and hence more sustainable buildings is as a vapor barrier. In cold climates vapor barriers are essential to prevent moisture generated inside the building (from breathing, cooking, washing, etc.) from condensing within the wall or ceiling construction and damaging the insulation. This is of particular importance in very cold climates where the condensation can build up as ice over winter to melt into the house come the spring. Plastic membranes are also used for air barriers in timber frame construction, to prevent draughts entering the interior. Another important use for membranes is to separate the building interior from radon gas, which occurs naturally in the ground in some parts of the world (Daoud and Renken, 2001). Because radon is a known carcinogen no-one would argue about the benefits of using a LDPE membrane to keep radon out of a building interior.

The issue is there is often no correct solution for making a sustainable building as it will depend on the climate, the ecology and geology, the type of construction used, and the building use. However, as the example of the damp-proof membrane shows avoiding plastics altogether because at least some of the constituents come from oil, may not be necessary. After all, avoiding all plastics within a building (biros, files, computers, telephones, etc.) is another issue and something almost impossible in modern life (Freinkel, 2011, pp. 2–3). The only question that remains is whether there is enough oil to continue to maintain a large plastics industry, and if the answer is yes then that industry needs to be based on a recycling agenda to avoid the problem of too many plastics entering the waste stream. (The issue of oil is complex since to avoid climate change we need to stop using it now, but without oil it is impossible to make plastics.) The real problem with plastics is that many are treated as short-life products. In 2004, 9% of New Zealand waste was plastics but plastic packaging also had the lowest rate of recycling, at 23% of total compared to 88% for aluminum packaging (Statistics New Zealand, 2009). This low rate is despite the fact most plastic packaging is thermoplastics, so could be recycled (Curlee, 1986). This is an issue of relative costs, in that the virgin plastic is so cheap, recycling is not cost-effective, something that will have to change if we are to move to a sustainable society.

3.4.3.2 Benefits of plastics

For any building to be sustainable it ideally has to get its energy, whether for the materials of which it is made or that used in its operation, from renewable sources. When it comes to operating energy, which is usually the larger component over the life of the building (see Chapter 4) there are two possible approaches. The first is to insulate the fabric of the building, whether in a hot or cold climate, so that the energy used for conditioning the space is kept to a minimum so it can be supplied by renewables, which currently, with the exception of nuclear power, are more costly in terms of creating and operating plant than conventional energy sources (US Energy Information Association, 2016). This will also entail design being appropriate for the particular climate, such as not making an all-glass building in the tropics. The second is to accept that we have to use buildings as in the past and that rather than conditioning the whole building we wear suitable clothes, have spots internally which are heated or cooled (much like the communal fire in the calefactorium or warming room in a medieval monastery (Kinder, 2002, p. 277)), and put up with any discomfort resulting from not heating or cooling the whole space. Regarding the latter approach, an academic in charge of a solar energy research center in a university in Thailand commented that it would make much more sense to let people do office work out of doors in the shade rather than try to air condition offices in the tropics using solar power.

Accepting that the second approach to making sustainable buildings will probably not be acceptable in the 21st century then insulation materials are important, and some of the important water-resistant ones, like extruded polystyrene (such as Styrofoam), are plastics. Obviously where construction can be kept dry, which may entail using a plastic-based vapor barrier, then there are alternatives in the form of wool, glass fiber, and mineral fiber. However, none come without an environmental impact. Fiberglass batts have been associated with health issues for those involved in installing them, though a vapor or air barrier will prevent fibers entering the building interior. When combined with a resin to waterproof them they can be used in wet conditions, such as infill for brick cavity walls, but then the resin is a plastic-based product. Mineral wool is rock that has been expanded through the application of heat, so it takes energy to process and leaves a hole in the ground. At first sight wool seems the obvious choice but wool comes from sheep and sheep emit methane so wool insulation does have a carbon footprint. A ewe and the lambs she raises form a livestock unit which gives off 11.5 kg of methane a year, which equates to 265 kg of CO_2 (Vale and Vale, 2009, p. 210). At the same time a sheep might produce a fleece weighing 3 kg. As the lambs go for meat, as will the sheep eventually, as an estimate maybe half these emissions (133 kg of CO_2) are attributable to the wool. Wool insulation can be assumed to have a density of 8.0 kg/m^3, similar to that of fiberglass batts (BRANZ, 2016). This means 1 m^2 of 100 mm thick wool insulation weighs 0.8 kg. Assuming the insulation is all wool (and most wool insulation products are mixed with polyester to make sure it has good loft and does not settle under its own weight, thereby reducing the trapped air which gives insulation its insulating properties), each sheep produces 3.75 m^2 of insulation. For the 100 m^2 two-storey house above (5×10 m), further assumptions have to be made. Assuming the house is detached and window openings are at least 10% of the floor area, this gives an insulated wall of 135 m^2 (15 m^2 of openings), together with 50 m^2 of ceiling. The 185 m^2 required would take the wool from 49 sheep, and would have a carbon footprint of 6517 kg. Alcorn gives a CO_2 emissions value of 59.9 kg/m^3 for expanded polystyrene (Mithraratne et al., 2007, p. 223), so the same 100 mm thick insulation using plastic would have a carbon footprint of 1108 kg. This is a situation where the natural material might not be the best when it comes to making sustainable buildings. However, using recycled wool insulation tells a very different story, since this has a CO_2 emissions value of 11.4 kg/m^3 (Alcorn, 2003), and if used in the same house would have a carbon footprint of 212 kg.

More important is to set these emissions against possible greenhouse gas reductions, which will depend on climate and on the fuel used for house heating. This is now moving into the discussion of lifecycle energy and lifecycle assessment, and this example will be explored further in Chapter 4. The point of entering this discussion now is to show that it is not just a question of "plastics bad" and "natural materials good" but rather one of being aware of exactly which material is being selected and being sure that it is appropriate for the situation in which it is being used (dry or wet). Some closed-cell insulating plastics, like Styrofoam, are water-repellant and can be used in damp situations, and are even sufficiently robust for salvage and reuse in some situations (Vale and Vale, 2000, p. 170).

Another area in building where plastics appear to offer huge benefits is as glues and sealants, particularly those that are waterproof and hence protect the building interior from damp. In many ways what these materials have done is to make maintenance easier and less frequent as the plastics are stable and do not degrade as quickly as many natural glues (often based on casein which comes from milk or boiled animal connective tissues) and sealants (linseed oil-based putty). That said, these natural glues have worked for centuries as descriptions from 3500 years ago of Egyptian woodworking demonstrate (Darrow, 1930, p. 9). Now wood is more likely to be glued with PVA (polyvinyl acetate), which apart from being waterproof has the advantage of drying clear. The problem with plastic glues, aside from their derivation from oil, is that some solvents, which evaporate as the glue "dries" contain volatile organic chemicals (see Part III). However, like the discussion of polythene barriers above, plastic glues are a small part of the total materials that go into buildings, though on-site applications would need to be made in appropriately ventilated conditions. (Glues used to make products like plywood are discussed further in Section 3.4.4).

The statement above about being a small component of the overall impact of a building also tends to hold true for paints. Modern paints have plastics like acrylics added, making them easier to work with, in terms of coverage per application, and last longer, thus speeding up the job of painting and saving on maintenance in terms of not having to repaint as often. This saves time and hence money. Paints have been made from natural materials for years—whitewash is made from lime (hydrated quicklime from burning limestone) and water, with the addition of salt and sometimes ingredients like molasses or milk. It has to be mixed and left and when applied is almost clear and streaky, only drying to a whiter finish. Several coats are needed but it will never give the thick covering and brilliant white of modern emulsion paint. Whether you see this modern covering effect as a benefit of adding plastics to paint is perhaps in the eye of the beholder. In one way paints that contain plastics are avoidable but they also provide benefits, certainly in terms of time saving, that more traditional paints do not. That said not all the chemicals added to paints are things you necessarily want in your home This is where choices have to be made when it comes to making sustainable buildings—do you value the health of the ecosystems on which humanity depends and the health of the building occupants over the quality of the surface finish, or do you want to make a sustainable building as cheaply and as quickly as possible so such buildings are not just for the rich? There are no easy answers when it comes to choice of materials.

3.4.3.3 Avoidable plastics

When Franklin and Associates (1991) made a lifecycle assessment of plastics in the US construction industry they provided a table of plastic resins used in buildings and possible alternatives (Table 3.1).

Table 3.1 **Uses of plastics and possible alternatives**

Building element	Plastic resin	Alternatives
Carpet underlay	Polyurethane	Hair and jute Rubber
Flooring	PVC	Ceramic tile Linoleum
Guttering and downpipes	HDPE PVC	Aluminum Steel
Insulation	Polystyrene Polyurethane	Fiberglass
Lighting components	Polystyrene PVC	Aluminum Glass Steel Wood
Mobile home skirting	PVC	Aluminum Steel Wood
Paneling and cladding	PVC Unsaturated polyester	Aluminum Steel Wood
Pipes, fittings, conduits	ABS HDPE Polystyrene PVC Unsaturated polyester	Aluminum Concrete Copper Iron Steel Vitrified clay
Showers and baths	Unsaturated polyester	Iron/porcelain Steel/porcelain
Weatherstripping	PVC	Rubber
Window units	PVC	Aluminum Steel Wood

Franklin Associates, 1991.

Missing from the list Table 3.1 are the melamine resins used in finishes for kitchen cupboards and worktops but what is important is the fact that alternatives do exist for all the plastic products cited. A number of these are discussed more fully below as use of some alternatives also has negative environmental implications. Others, like insulation, have been discussed (see Section 3.4.3.2).

Sometimes traditional ways of doing things are seemingly as good as the modern plastics-based products. In 1933 colored vinyl tiles were first exhibited at the Century of Progress exhibition in Chicago in the Vinylite House (Konrad and Kofoed, 2014, p. 209). This was not a stand-alone house but a three-roomed apartment of plastic wall panels within the Hall of Science (Bacon, 2001, p. 185). For those passionate about sustainable buildings this was the same exhibition where the Keck brothers exhibited their House of the Future that, although built around an air-conditioning system, made use of solar energy entering through the glass walls. The Kecks went on to build passive solar houses with glazed walls facing the sun from the 1940s onwards (Jandl, 1991, pp. 128–137, 201). Sadly it was not the passive solar design principle but the vinyl tile that was to become almost ubiquitous in the postwar home.

The alternative is linoleum, which like vinyl comes as sheet flooring or tiles. It dates back to the middle of the 19th century, being a mixture of oxidized linseed oil, cork dust, and gum or resin applied to a backing of cotton or jute. Until after WWII it was the choice for floors in areas of high use and in wet areas, such as kitchens and bathrooms. Battleship lino does what it says, it is a very heavy-duty grade designed for use in warships and army barracks, as well as hospitals and schools. It seems that linoleum was ousted by vinyl because of the colorful and cheap nature of the latter and the fact that lino just lasted too well and people wanted change (Marmoleun, n.d.). There is no problem with using lino in place of vinyl flooring but there is a problem in that many people, even designers, do not see them as being different floor coverings (Petrović et al. (2016)), even though all you need to do is smell them (lino smells linseed-oily and vinyl smells of chemicals).

Before WWII a kitchen might contain a built-in wooden dresser, a free-standing table with perhaps an enamel top, a sink with separate wooden drainer, and a free-standing cooker. However, melamine as a surface finish revolutionized the kitchen post WWII. Although strictly a composite product being made of layers of kraft paper interspersed with melamine thermosetting resin cured by heat under pressure, because the plastic component is so essential to the way the finished product behaves it is included here as an avoidable plastic product. Marketed as Formica or Laminex, its big advantage was its color and its seeming indestructability, combined with the way it could be applied to surfaces to make the coordinated fitted kitchen (Hine, 1987, p. 64). However, not all postwar kitchens depended on plastic finishes, as at this time in New Zealand the kitchen was more likely to be updated with wooden cupboards with turned wood handles and a linoleum bench top, which now would be considered a more sustainable alternative (Anon., 1956). This would seem a sensible approach to making a kitchen in a sustainable manner.

Another area where plastics have almost become ubiquitous is in the area of pipework and drainage, not least because being lightweight and corrosion-resistant they are fit for purpose. Their use is potentially on the increase in the replacement of failing infrastructure (Whelton and Nguyen, 2013). However, there have always been alternatives. In many vernacular traditions a roof overhang with suitable ground treatment deals with rain off the roof without the need to collect and channel it. Wooden gutters have been used in places as far apart as the United Kingdom (in Sheffield wooden gutters are still used as plastic ones can be wrecked by snow falling off roofs) and New Zealand. The modern fashionable use of copper gutters and downpipes in place of plastic raises the problem of use of copper (see Section 2.4.2). Lead, which has been used for pipes and gutters for over 20 centuries is, at least for pipe work, ruled out on health grounds, as is much more recent asbestos cement for rainwater goods. This leaves copper, cast iron, and galvanized steel. The latter both have issues around the energy it takes to make them and the fact the raw materials have to be dug up. Copper, as discussed above, needs to be used sparingly since it is an important material for wiring, so its use for guttering may not appropriate. Cast iron at least lasts a long time once installed (Trace, 2006) and can cope with having ladders rested on it for maintenance, something not recommended for plastic gutters. In favor of plastic guttering is that it can be made of recycled plastic (Anon., 2005), which in turn can be salvaged and reused—at least the lengths can, given the glued joints will normally have to be scrapped. Nevertheless it is a material for which there are substitutes, even if all of these also have drawbacks.

What can be said is having good rainwater disposal is essential for maintaining the envelope of a building in good condition, and that gutters also need to be cleaned regularly so need to be designed with easy access for maintenance, which might have a bearing on the selection of gutter materials. A lovely illustration of this principle is found at Gidea Park in the United Kingdom, which was the site of two housing competitions, one in 1911 and one in 1934 (Vale and Vale, 2007). The 1911 houses are broadly speaking Arts and Crafts examples with pitched roofs and exposed rain water goods, while the 1934 houses mostly have flat roofs, some with concealed guttering. The 1911 houses are still lived in and look as they did when first built, whereas the 1934 houses are either in disrepair or have had extensive (and at least in one case very costly) refurbishment. To keep a building in good condition and make it last as long as possible, and hence achieve important aims for sustainable buildings, the goal should always be to shed the rain water as quickly as possible.

When it comes to plastic plumbing the alternative is copper, with its problems (see Section 3.4.2). The easiest way to save the resources that go into internal plumbing is to arrange to have fewer bathrooms, many of which may be seldom used (Khajehzadeh and Vale, 2016). This will also have the advantage of saving on bathroom fittings and finishes (Vale and Vale, 2009, pp. 150–153).

Plastic is now also used for drainage with its simple fit-together joints in place of earthenware pipes that used to be socketed and joined using sand and cement mortar. There is one product by the UK firm Hepworth which uses clay pipes from 100–300 mm diameter joined with push-fit plastic couplings (Hepworth, n.d.) which uses the plastic where it is needed to make the installation simple but also benefits from having clay pipes which can be put on a bed of recycled construction waste and buried in the soil rather than having to be surrounded by granular bedding, as required for all-plastic drainage systems. The clay pipes contain 15% recycled material from manufacturing waste and themselves can be salvaged and reused or at worst broken down to form bedding and hardcore. This seems one example of substituting for plastic what can be substituted except for the vital plastic joint which makes construction a much simpler process and also, being flexible, is likely to mean reduced chance of failure over time. This underscores the issue that neither considering materials alone nor components is sufficient when it comes to making sustainable buildings. Instead the whole building (and its ecological and climate context) must be the focus, part of which is the effort that will go into making it on site.

3.4.3.4 Plastics: good or bad?

When it comes to issues of human health there is doubt over having so much plastic in the human environment, including the built environment. This is discussed in detail in Part III. Plastics turn up in many aspects of building and only a few have been considered in detail here. Door furniture and fittings can be plastic but can equally well be made from metal and wood. Paints, which can contain acrylics and vinyl, have more traditional alternatives (see Section 3.4.3). Plastic wall cladding in the form of laminates has made an appearance, especially in bathrooms, and many modern products that uses small or waste timber (MDF, plywood) will be held together with plastic glues. These are also considered under composite products (see Section 3.4.4). Like all building, to be truly sustainable the answer is not to build new but to make best use of something that already exists (see Chapter 15). Some plastics, like waterproofing barriers and insulation are, as discussed, unavoidable when it comes to making modern buildings that minimize or eliminate energy for space conditioning and that will last without needing major refurbishment. Some plastic can be avoided, but even alternative materials will make an impact, so the need is to ensure that the alternative has a lower overall impact, which can only be verified by making some type of lifecycle assessment so that maintenance and replacement cycles can be included. It is not the choice of plastic that is crucial but how it is used by the designer.

3.4.4 Composites

The issue of composite products has already come up, e.g., clay drainage pipes with plastic push fit connectors. This section looks at some commonly found composite building products in terms of their environmental impact. While this section, as with others in this part of the book, can never claim to be comprehensive the aim has been to choose a variety of different types of composite materials for discussion.

3.4.4.1 Reinforced concrete

Concrete reinforced with steel is as important in the development of large modern buildings as structural steel. In early 20th century United States the skyscraper was a steel-framed building whilst in Europe contemporary large buildings were more likely to be in reinforced concrete (Starrett, 1928, pp. 36, 51). Suffice it to say, reinforced concrete is a French invention from the middle of the 17th century, but it may owe something to the British invention of chicken wire in 1844. Joseph-Loius Lambot made a ferro-cement boat and took out a patent for the idea in 1855 (Courland, 2011, p. 218). Better known is Monier's patent for reinforced concrete flower pots and Coignet's first building in reinforced concrete, the 1853 house in north Paris. From then on the material became ubiquitous.

Fundamentally reinforced concrete combines the best of a material (concrete) that resists compression and a material (iron or steel) that resists tension. It is also considered fireproof and dimensionally stable. However, all these materials take energy and resources to make. Like concrete, the steel reinforced version can be recycled by crushing it and removing the steel bits with magnets. The crushed material can be used for granular fill and the steel recycled. Crushed waste concrete can be used for aggregate in new concrete in place of virgin materials like gravel. However, only small amounts of recycled aggregate are currently used despite the estimated generation of concrete and masonry waste of around 1 billion tonnes a year (Mehta, 2001). Mehta also suggests recycled aggregate which contains masonry will be more porous, so making concrete in this way will use more water and this in turn will affect the properties of the final product, so like much recycling of metals and other materials mixing the recycled aggregate with new material is recommended.

Because of the potential longevity of the material if correctly designed it may be better to think in terms of reusing the building or at least the reinforced concrete structure rather than the material, in the way Le Corbusier envisaged with his 1914 Domino House.

3.4.4.2 Fiber cement sheets

Because of their use of a cement matrix it seems sensible to follow a discussion of reinforced concrete with these cement-based products. In a sheet or roofing slate the cement sets around the fibers which provide the reinforcing. In the past, asbestos, which is a naturally occurring mineral, was used for this purpose. At the time, a roofing slate might contain 10% by dry weight of asbestos fibers, which were then mixed with the cement and water, although other products might contain up to 75% by dry weight of the fibers (Rosato, 1959, pp. 62–66). The sheets are light in weight and hence mean that less structural material is needed to support the lightweight sheets and roofing slates. So, from a sustainability perspective using a lightweight material based on a natural mineral could save resources, though the very adverse health effects of asbestos (see Part III) mean that it is no longer used in the manufacture of building materials and its removal from existing buildings is carefully controlled. As a replacement for asbestos other fibers, such as cellulose, are now used. In New Zealand fiber cement sheets are made from cellulose fibers from plantation-grown radiata pine, cement, sand, and water and "Small amounts of other chemical additives are used to help the process, or provide products with particular characteristics" (Hardie, 2016). This is a return to WWII products like lignocrete, used for making panels of cement with chemically treated sawdust (so as to render it inert) as the aggregate. These were bonded to a plywood frame to form the walls of the Tarran bungalow in the United Kingdom 1944 Temporary Housing Programme (Neel, 1943). Advantages of fiber cement sheets are their light weight but they have also been involved in the leaky homes issue where the sheets were used as the basis of monolithic cladding, being finished with a coat of render. However, where the product has been used with good detailing and has been well maintained it can be part of making a weather-tight building. Like all building products the issue of whether the product is a sustainable building material comes down to how it is used and maintained, which means some kind of lifecycle analysis.

3.4.4.3 Timber composites

Composite timber products, often called engineered wood, are potentially good environmentally because they make use of smaller bits of timber, so this leads to a saving in the land needed for making buildings as more products can be extracted from the same number of trees. The problem is that most of these small timbers are glued together and some of the glues are not good for people or the environment.

Glulam

Use of glue laminated timber (glulam) again dates to the early and middle 19th century when composite wood members were made with animal glue (Glue Laminated Timber Association, 2014), making this like a "grown" material. In the later 19th century Otto Hetzer patented the glulam portal frame using casein glue, a waterproof glue derived from a protein in cow's milk, which was used for buildings and railway station canopies (Glue Laminated Timber Association, 2014). The use of synthetic resins in glulam timbers happened after WWII with the introduction of completely waterproof phenolic resin-based adhesives (like the plastic Bakelite, made from phenol and formaldehyde). Just as the war had given a big impetus to the use of plywood and other types of bonded wood, so after the war glulam work became much more popular with these new glues.

The reason for looking at the brief history of the material is to show how what could be thought of as a sustainable product using small bits of wood to achieve long spans held together with glues produced from natural materials has now become dependent on oil for the phenolic resins. That said, a glulam structure will have a lower life time embodied energy than its equivalent in reinforced concrete (Robertson et al., 2012) because the timber structure will lock up carbon rather than releasing it to the atmosphere in the normal process of growth and decay of timber. When it comes to the end of the life of a glulam building it might be possible to reclaim and reuse the members where these are bolted together but extracting the timber from the glulam structure is potentially much more problematic.

A more recent development has seen timber veneers laminated together in the form of small framing members such as studs and joists. The advantages of these

LVL (laminated veneer lumber) products are that they are dimensionally stable and do not warp, and are uniform in their manufacture and stronger than the equivalent in natural wood. This product therefore has advantages in use over natural timber but it does contain glue for sticking the veneers together. However, if used in place of steel LVL can reduce environmental impact (NZ Wood, n.d.). This suggests the size of a sustainable building may well dictate the choice of materials. For a small structure at the scale of a house, use of timber members may well have the lowest impact. However, where the structure is such that steel or reinforced concrete might be chosen then glulam or LVL might be preferable. Currently the highest timber frame building in the world is in Bergen, Norway. At 14 storeys, it has been made using 1 m square glulam columns (Treet, 2015). For a skyscraper, if a skyscraper can ever be a sustainable building, steel will be the probable choice because of the ratio of the thickness of the structural members to available floor area.

Plywood

The veneering of lower-quality timber with a thin surface layer of higher-quality timber is a technique that dates back to ancient Egypt and plywood takes this idea to make a sheet product out of thin veneers that are peeled off a log using a rotary lathe. The history of plywood is similar to that of glulam, with a big impetus given to the product after its many uses in WWII coupled with a postwar shortage of timber. Prior to that, the invention of the rotary lathe by Immanuel Nobel (father of Alfred, the inventor of dynamite) gave impetus to the idea of creating a sheet material out of layers of veneers glued together. Plywood panels for doors were produced in the United States at the end of the 19th century (Jester, 1995, p. 132).

To make a sheet of ply, an uneven number of sheets (three being the thinnest ply) are glued together with the grain of each sheet at right angles to that of the preceding sheet for additional strength, the whole assembly then being either cold or hot pressed. The advantage of ply is that a good product can be made from less than perfect veneers in terms of grain quality in the inner layers with a high-quality outer layer; additionally 20 sheets of veneer can be obtained from a piece of timber 25 mm thick (Tainuo Woods, n.d.). Like gulam it is also dimensionally stable and it can be curved, a property that has been used in the making of both furniture and aeroplanes. Again, the development of plywood with reliable properties was also linked with the development of glues. Early cassava glues were vegetable-based and casein glue from milk gave the first waterproof ply sheet. In the US German synthetic resin glue was introduced in 1931 and the development of US-made phenol formaldehyde glues in 1935 was another spur to the use of plywood in the building industry (Jester, 1995, p. 134).

The problem when it comes to making sustainable buildings is knowing whether the plywood comes from sustainably grown wood, which means looking for certified products either Forest Stewardship Council (FSC) or Sustainable Forestry Initiative (SFI), both of which are nongovernment independent organizations. The modern glues are also problematic and although new soya-based glues are available, like casein, this is making a glue out of a product that has a high protein value for humans. To be waterproof it also needs to be mixed with chemicals (Lei et al.,

2014). Glues from waste products, like animal glues from bones and hides, are not as waterproof and so the nature of the plywood product changes.

A variant of plywood is blockboard, which is made from sandwiching thin strips of wood as the core between two plywood sheets. Overall, this will have more wood and less glue content and is sufficiently thick to be used for cupboard carcassing.

Particle boards, fiberboards, and hardboard

Other boards that make use of small bits of wood are the various forms of particle board (chipboard) and medium-density fiberboard (MDF). These are made from wood chips, sawmill shavings, and sawdust in a glue matrix, the whole being pressed. The board with the biggest wood particles is oriented strand board in which flakes of wood with the grain running in various directions (perhaps more correctly called disoriented strand board since it is the disorientation that gives the board its strength) are pressed together in the glue matrix to make boards ranging from 6 to 18.5 mm thick. Strand board is used for flooring, sheathing, and making structural I-beams, particularly in North America. Chipboard (which is not the same as the paper board used for making things like cereal boxes) as the name suggests is made in the same way from smaller bits of waste wood and is often produced in tongue-and-grooved flat sheets for flooring. When covered with other materials, like plastic laminate, it is also used for cupboard carcassing, especially for fitted kitchen units. For MDF (medium-density fiberboard) the soft or hard wood pieces are reduced to fibers in a machine before being combined with the resin and having heat and pressure applied. This makes a fine-grained board that is denser than plywood and particle board that can even be used for making wood cuts. In building it can be used for wall panels but is more likely to be the carcassing for cupboards and units.

The issue with these boards made from timber offcuts and wood fibers is the modern glues as these are what have been implicated in indoor air quality problems (see Part III). Consequently, the search is on for replacement glues that give the same performance, in terms of being waterproof under normal building conditions, and longevity as modern glues but at present the advice might be to use these alternative glues in dry indoor applications but not if the boards are to be used outdoors.

One other common composite wood fiber product is hardboard, known under trade names such as Masonite. Hardboard is made from pulped wood fibers that are compressed under heat into thin sheets, without the use of glue, and is also known as high-density fiber board. The fibers can come from forest and wood waste, and recycled wood materials (Packard Forest Products, 2014). Hardboard is often used in building interiors as a substrate to mask small irregularities in uneven surfaces, usually as a result of age, such as floors which are to be given a further covering of sheet material or nonceramic tiles. Tempered hardboard is treated with linseed oil and heat, which gives it increased moisture and impact resistance (Akers, 1966, p. 140), to the point where it has been used as external cladding, especially in North America. In 1933 a house clad entirely in Masonite was exhibited in the Chicago Century of Progress Exposition (Jandl, 1991, p. 128). Hardboard can also be finished with wood veneer and plastic laminate for a decorative finish.

Engineered wood products end of life

The glues in engineered wood products mean they will not biodegrade in the same way as timber. They have to be reclaimed and used again or incinerated industrially so that unwanted combustion products can be dealt with. Only waste hardboard, because it is made without glues, can be shredded and used as animal bedding or garden mulch (Packard Forest Products, 2014).

3.4.4.4 Metal composite panels

When it comes to cladding frame structures, apart from glass, the modern trend is to use metal panels as spandrels to cover the floor slab and any services at this level. Having a solid panel can also feel less vertiginous for the occupants than floor to ceiling glass, especially in the upper levels of a high-rise building. The solid panel is also a chance to put some insulation in the cladding. From the sustainable building point of view the first question might be whether making a curtain wall of glass and other lightweight claddings is a good idea, given it can lead to overheating and unnecessary use of air conditioning, especially in hot climates. Some panels, like Alucobond, which are an aluminum–polythene–aluminum sandwich (Kaneba, n.d.) in panels from 3 to 6 mm thick, have little thermal resistance, so would need to be used with insulation where appropriate.

3.4.4.5 Glass composites

Glass can be given different properties by being combined with other materials. For example, laminated glass is a glass–plastic film (polyvinyl butyral (PVB) or ethylene vinyl acetate (EVA))–glass sandwich. Some special glasses such as laminated and toughened cannot be recycled (BRANZ, 2015). Low-e glass has a molecular layer of metal on the inside surface of the double (or triple) glazing unit, which accounts for its insulating properties. This can be recycled but like the majority of glass not as window glass but as glass fiber insulation or powdered for use as filler in paints (Ecospecifier Pty Ltd, 2016). Glass, as discussed in Chapter 2, can be recycled though currently most end of building life glass finishes up in landfill (Glass for Europe, 2013).

3.4.4.6 Bituminous products

Bitumen or tar impregnated materials have been used for roofing, especially installed in layers for covering the flat roofs favored by modernism. Modern bituminous roofing felt is impregnated polyester fleece or glass fiber produced as a roll, which is then welded on site to make a continuous covering. The same type of material can also be used for damp-proof courses and roofing shingles. A lightweight, corrugated bitumen-impregnated roofing sheet is also available, known commercially as Onduline. This "is manufactured from a mono-layer of organic fibres impregnated with bitumen under intense pressure and heat. The colour pigmentation system 'stains' the colour into the sheet with a UV resistant resin" (Dachnik, 2014). Bitumen can also be mixed with aggregate to form asphalt, which is also used for roofing as well as for floors, pavements, and roads. Asphalt pavements have a history of recycling dating back to 1915, but the recycled material has to be mixed with the new in proportions of 10%–30% recycled (Mehta et al., 2013, p. 241). A Danish company separates the bitumen from roofing products for recycling as asphalt paving (Tarpaper Recycling, n.d.), but this is, as yet, uncommon.

3.5 Conclusion

There is no "silver bullet" when it comes to choosing materials for making sustainable buildings. What the history of building materials shows, as briefly touched on in this chapter, is that they have moved from being locally sourced, where their environmental impact would at least be noticeable, to a situation where materials are industrially produced and internationally traded. This may make for cheaper buildings that are put up more quickly. It will also make for more extreme buildings, in that they are larger and higher. The price of this is that the environmental impact of the built environment has grown. However, materials are only one part of making buildings. Chapter 4, looks at the impact of putting materials together to make a building and what the priorities are when it comes to making that building sustainable.

References

Adam, J.-P., 1994. *Roman Buildings Materials and Techniques*. B. T. Batsford Ltd, London.

Addis, B., 2006. Building with Reclaimed Components and Materials: A Design Handbook for Reuse and Recycling. Earthscan, London.

Akers, L.E., 1966. Particle Board and Hardboard. Pergamon Press, Oxford.

Alcorn, A., 2003. *Embodied Energy and CO_2 Coefficients of NZ Building Materials*. Victoria University of Wellington, Centre for Building Performance Research.

Alexander, J.S., 2007. The introduction and use of Masons' marks in romanesque buildings in England. Mediev. Archaeol. 51 (1), 63–81.

Anderson, J.-O., Westerlund, L., 2014. Improved energy efficiency in sawmill drying system. Appl. Energy. 113, 891–901.

Anon., 1956. Remodelling an old fashioned kitchen. Home Build. 19 (4), 30–31.

Anon., November 10, 2005. Firms unveil rain gutter made of recycled. *New Materials Japan*.

Anon., February 15, 2006. Feature: sustainability: copper and the environment. *What's New in Building*. Available from: http://search.proquest.com/docview/236636589?rfr_id=info%3Axri%2Fsid%3Aprimo (accessed 10.01.16.).

Anon., August 9, 2010. Low carbon cements; recent research and developments on alternatives to Portland Cement. *Chemistry and Industry*. Available from: http://web.b.ebscohost.com.helicon.vuw.ac.nz/ehost/pdfviewer/pdfviewer?sid=7d8ee7ea-3c93-4bc8-b941-6df21faefa4f%40sessionmgr110&vid=0&hid=110 (accessed 7.12.15.).

Ari, M., 2010. LEED Materials: A Resource Guide to Green Building. Princeton Architectural Press, New York, NY.

Arsenault, C., December 5, 2014. Only 60 years of farming left if soil degradation continues. *Scientific American*. Available from: http://www.scientificamerican.com/article/only-60-years-of-farming-left-if-soil-degradation-continues/ (accessed 11.01.16.).

Asif, M., Davidson, A., Muneer, T., 2001. Embodied energy analysis of aluminium-clad windows. Build. Serv. Eng. Res. Technol. 22 (3), 195–199.

Augustin, C., Rahman, S., n.d. Composting animal manures. Available from: https://www.ag.ndsu.edu/manure/documents/nm1478.pdf (accessed 17.07.16.).

Bacon, M., 2001. Le Corbusier in America: Travels in the Land of the Timid. MIT Press, Cambridge.

Baily, M.A., 1981. Brick making in Columbia; a case study of alternative technologies. World Dev. 9, 201–213.

Banik, R.L., 2015. Morphology and growth in *Bamboo*. Volume 10 of *Tropical Forestry*, 43–89. Available from: http://link.springer.com.helicon.vuw.ac.nz/chapter/10.1007/978-3-319-14133-6_3/fulltext.html (accessed 9.12.15.).

Bates, W.N., 1917. Archaeological discussions. Am. J. Archaeol. 21 (4), 449–476.

Border Oak, 2015. Sustainability. Available from: http://www.borderoak.com/about-us/sustainability (accessed 9.12.15.).

BRANZ, n.d. Renovate: bathrooms and laundries 1940s-1960s. Available from: http://www.renovate.org.nz/1940-60s/kitchens-and-bathrooms/bathrooms-and-laundries/ (accessed 10.07.16.).

BRANZ, 2014. Working with treated timber. Available from: http://www.level.org.nz/health-and-safety/materials-handling/working-with-treated-timber/ (accessed 12.07.16.).

BRANZ, 2015. Glass. Available from: http://www.level.org.nz/fileadmin/downloads/Materials/LevelMGlass.pdf (accessed 20.07.16.).

BRANZ, 2016. BRANZ appraised: pink batts insulation. Available from: http://www.pink-batts.co.nz/assets/Technical/Pink-Batts-Insulation-238.pdf (accessed 12.07.16.).

British Stainless Steel Association, n.d. Environmental aspects of stainless steel. Available from: http://www.bssa.org.uk/sectors.php?id = 99 (accessed 20.01.16.).

Brunke, J.-C., Blesl, M., 2014. Energy conservation measures for the German cement industry and their ability to compensate for rising energy-related production costs. J. Clean. Prod. 82, 94–111.

Brunskill, R.W., 1978. Illustrated Handbook of Vernacular Architecture. 2nd edition Faber and Faber, London.

Building Green Inc., 2015. NC-v2.2 MRc6: Rapidly Renewable materials. Available from: http://www.leeduser.com/credit/nc-v2.2/mrc6 (accessed 12.12.15.).

Butti, K., Perlin, J., 1980. A Golden Thread. Marion Boyars, London, Boston.

Carlson, A., Palmer, C., 2016. A qualitative meta-synthesis of the benefits of eco-labelling in developing countries. Ecol. Econ. 127, 129–145.

Chen, Z., Zhu, E., Lam, F., Pan, J., 2014. Structural performance of Dou-Gong brackets of Yingxian Wood Pagoda under vertical load—an experimental study. Eng. Struct. 80 (1), 274–288.

Coffin, B., 2003. Coalition sues EPA to ban toxic wood preservatives. Risk Manage. 50 (2), 6.

Copper Development Association, 1959. Copper Roofing. Copper Development Association, London.

Copper Development Association, 1960. Copper Through the Ages. Copper Development Association, London.

Copper Development Association Inc., 2015. Copper—the World's Most useable Resource. Available from: http://www.copper.org/environment/lifecycle/g_recycl.html (accessed 20.01.16.).

Council for Aluminium in Building, 2008. Aluminium and sustainability. Available from: http://www.c-a-b.org.uk/wp-content/uploads/080707_sustainability_s-locked-version-Justin.Ratcliffe-v1.pdf (accessed 20.01.16.).

Courland, R., 2011. Concrete Planet. Prometheus Books, New York, NY.
Craddock, P.T., 1987. The early history of zinc. Endeavour. 11 (4), 183–191.
Crow, J.M., March 2008. The concrete conundrum. *Chemistry World*, pp. 62–66. Available from: http://www.rsc.org/images/Construction_tcm18-114530.pdf (accessed 9.12.15.).
Curlee, T.R., 1986. Plastics recycling: economic and institutional issues. Conserv. Recycl. 9 (4), 335–350.
Dachnik, A., 2014. Onduline corrugated roofing. Available from: http://www.doit.house/onduline-corrugated-roofing.shtml (accessed 10.07.16.).
Daoud, W.Z., Renken, K.J., 2001. Laboratory assessment of flexible thin-film membranes as a passive barrier to radon gas diffusion. Sci. Total Environ. 272 (1-3), 127–135.
Daniels, N., 2016. Recycled PVC: an uncertain future? Mater. World. 24 (1), 11.
Darrow, F.L., 1930. The Story of an Ancient Art: From the Earliest Adhesives to Vegetable Glue. Perkins Glue Company, Landsdale, Pennsylvania.
Dobson, C., 1959. The History of the Concrete Roofing Tile. B. T. Batsford Ltd, London.
East African Bamboo project, n.d. Bamboo cultivation manual. Available from: https://www.unido.org/fileadmin/user_media/Publications/Pub_free/Guidelines_for_cultivating_Ethiopian_highland_bamboo.pdf (accessed 8.12.15.).
Easton, D., 2007. The Rammed Earth House. Chelsea Green Publishing Company, White River Junction, Vermont.
Ecospecifier Pty Ltd, 2016. Eco-priority guide: windows (glass). Available from: http://www.ecospecifier.com.au/knowledge-green/setting-priorities/eco-priority-guide-windows-glass/ (accessed 10.07.16.).
Fearnside, P.M., 1999. Social impacts of Brazil's Tucurui Dam. Environ. Manage. 24, 485–495.
Fearnside, P.M., 2016. Environmental and social impacts of hydroelectric dams in Brazilian Amazonia: implications for the aluminum industry. World Dev. 77, 48–65.
Fielding, T.J., 1948. History of Bakelite Limited. Bakelite Limited with The Banyard Press, London.
Fonte, G.C.A., 2007. Building the Great Pyramid in a Year: An Engineer's Report. *Algora Publishing e-book*. Available from: http://site.ebrary.com.helicon.vuw.ac.nz/lib/vuw/reader.action?docID=10476780&ppg=117 (accessed 7.12.15.).
Forestry Stewardship Council (FSC), n.d. Forest for all forever. Available from: https://us.fsc.org/en-us (accessed 10.07.16.).
Franklin Associates, 1991. Comparative energy evaluation of plastic products and their alternatives for the building and construction and transportation industries. Available from: https://greenbuildingsolutions.org/Main-Menu/Resources/Comparative-Energy-Evaluation-of-Plastic-Products-1991.pdf (accessed 6.07.16.).
Freinkel, S., 2011. Plastic: A Toxic Love Story. Houghton Mifflin Harcourt, Boston, New York.
Georgopolous, C., Minson, A., 2014. Sustainable Concrete Solutions. John Wiley and Sons Ltd, Chichester.
Gil, A., Frayret, J.-M., 2016. Log classification in the hardwood timber industry: method and value analysis. Inter. J. Product. Res. 54 (15), 4669–4688.
Glass for Europe, 2013. Recycling of end-of-life building glass. Available from: http://www.glassforeurope.com/images/cont/38_83208_file.pdf (accessed 19.07.16.).
Glue laminated Timber Association, 2014. Glulam – Performance record. Available from: http://www.glulam.co.uk/performanceHistory.htm (accessed 11.06.16.).
Gordon, R.B., Bertram, M., Graedel, T.E., 2006. Metal stocks and sustainability. Proc. Natl. Acad. Sci. U S A. 103 (5), 1209–1214.
Hackenberg, L.M., May/June 1976. Build a house with green wood. *The Mother Earth News*. Available from: http://www.motherearthnews.com/green-homes/green-wood-zmaz76mjztak.aspx (accessed 9.12.15.).

Hammond, G.P., Jones, C.I., 2006. Embodied Energy and Carbon Footprint Database. University of Bath, UK.
Hawkey, S., 2002. Building more appropriately with adobe houses. Appropriate Technol. 29 (2), 58–59.
Hepworth, n.d. Clay drainage supersleve. Available from: http://hepworthclay.wavin.com/web/solutions/foul-water/gravity-sewer/clay-drainage-supersleve/clay-drainage-supersleve-1.htm (accessed 18.06.15.).
Hill, D., 2013. Ancient roman concrete could hold secrets to improving modern construction. Civil Eng. 83 (9), 44.
Hine, T., 1987. Populuxe. Bloomsbury Publishing Ltd, London.
Hingston, J.A., Collins, C.D., Murphy, R.J., Lester, J.N., 2001. Leaching of chromate copper arsenate wood preservatives: a review. Environ. Polluti. 111 (1), 53–66.
Hollis, M., 2005. Practical Straw Bale Building. Landlinks Press, Collingwood, VIC.
Hock, T., 2001. The role of eco-labels in international trade: can timber certification be implemented as a means to slowing deforestation. Colorado J. Inter. Law Policy. 12 (2), 347.
ICOPAL, 2013. Monarflex 1200 Gauge DPM Manchester. ICOPAL Ltd, England. Available from: http://www.icopal.co.uk/~/media/IcopalUK/Download/Datasheets/Structural_Waterproofing/dpm/TDS1089-Monarflex-1200ga-DPM-Datasheet.pdf (accessed 31.08.16.).
Ingrao, C., Giudice, A.L., Bacenetti, J., Tricase, C., Dotelli, G., Fiala, M., . . .Mbohwa, C., 2015. Energy and environmental assessment of industrial hemp for building applications: a review. Renew. Sustain. Energy Rev. 51, 29–42.
International Lead Association, 2014. Lead Action 21. Available from: http://www.ila-lead.org/UserFiles/File/ILA9927%20FS_Recycling_V06.pdf (accessed 20.01.15.).
Ip, K., Miller, A., 2012. Life cycle greenhouse gas emissions of hemp–lime wall constructions in the UK. Resour. Conserv. Recy. 69, 1–9.
Isaacs, N., 2015. Making the New Zealand House 1792-1982. PhD Thesis. Victoria University of Wellington. Available from: http://researcharchive.vuw.ac.nz/handle/10063/4804 (accessed 20.07.16.).
James Hardie, 2016. Fibre cement. Available from: http://www.jameshardie.com.au/fibre-cement/ (accessed 19.07.16.).
Jandl, H.W., 1991. Yesterday's Houses of Tomorrow. The Preservation Press, Washington, DC.
Jeppson, B., n.d., Pumice and LEED certification. Available from: http://www.hesspumice.com/pumice-pages/PDFs/LEED-Certification-and-Pumice-Products.pdf (accessed 11.07.16.).
Jester, T.C. (Ed.), 1995. *Twentieth-century Building Materials*. The McGraw-Hill Companies, New York.
Jones, S.R., 1936. English Village Homes. B T Batsford Ltd, London.
Juenger, M.C.G., Winnefeld, F., Provis, J.L., Ideker, J.H., 2011. Advances in alternative cementitious binders. Cement Concrete Res. 41 (12), 1232–1243.
Kallio, A.M.I., Salminen, O., Sievänen, R., 2016. Forests in the finnish low carbon scenarios. J. Forest Econ. 23, 45–62.
Kaneba, n.d. What is alucobond. Available from: http://www.kaneba.co.nz/what_is_alucobond.html (accessed 8.07.16.).
Kartal, S.N., Hwang, W.-J., Imamura, Y., 2007. Evaluation of effect of leaching medium on the release of copper, chromium, and arsenic from treated wood. Build. Environ. 42 (3), 1188–1193.
Kaufman, M., 1963. The First Century of Plastics: Celluloid and Its Sequel. The Plastics Institute and Illife Books, London.
Keefe, L., 2005. Earth Building: Methods and Materials, Repair and Conservation. Taylor and Francis, London.

Khajehzadeh, I., Vale, B., 2016. Large Housing in New Zealand: Are Bedroom and Room Standards Still Good Definitions of New Zealand House Size? In: Paper presented at the 9th Australasian Housing Researchers Conference, 17–19 February 2016, The University of Auckland. Available from: https://cdn.auckland.ac.nz/assets/creative/schools-programmes-centres/URN/documents/AHRC-Papers/Khajehzadeh_Vale-Definition-of-House-Size.pdf (accessed 18.06.16.).

Kidalova, L., Stevulova, N., Terpakova, E., Sicakova, A., 2012. Utilisation of alternative materials in lightweight composites. J. Clean. Prod. 34, 116–119.

Kinder, T.N., 2002. Cistercian Europe: Architecture of Contemplation. Wm. B. Eerdmans Publishing Co, Cambridge.

King, B., 2006. Design of Straw Bale Buildings: The State of the Art. Green Building Press, San Rafael, CA.

Konrad, K.A., Kofoed, P.D., 2014. Vinyl Tile. In: Jester, T.C. (Ed.), Twentieth-Century Building Materials: History and Conservation. Getty Conservation Institute, Los Angeles, CA, p. 320.

Lander, S., Westin, M., Schneider, M.H., 2004. Eco-efficient wood protection: furfurylated wood as alternative to traditional wood preservation. Manage. Environ. Qual. 15 (5), 529.

Lannom, G., 2014. Japan's Ise Grand Shrine: forever ancient/forever new. Faces. 31 (3), 30–33.

Lei, H., Du, G., Wu, Z., Xi, X., Dong, Z., 2014. Cross-linked soy-based wood adhesives for plywood. Inter. J. Adhesion Adhesives. 50, 199–203.

MacDougall, C., 2008. Natural building materials in mainstream construction: lessons from the U.K. J. Green Build. 3 (3), 1–14.

McGlade, C., Ekins, P., 2015. The geographical distribution of fossil fuels unused when limiting global warming to 2°C. Nature. 517 (7533), 187–190.

Marmoleun, n.d. The Marmoleum Clique. Available from: http://www.marmoleumclique.com/battleship-linoleum/battleship-linoleum/ (accessed 19.03.16.).

Mehmannavaz, T., Sumadi, S.R., Bhutta, M.A.R., Samadi, M., Sajjadi, S.M., 2014. Low carbon footprint mortar from pozzolanic waste material. Res. J. Appl. Sci. Eng. Technol. 7 (16), 3374–3379.

Mehta, M., Scarborough, W., Armpriest, D., 2013. Building Construction: Principles, Materials and Systems. second ed. Pearson, e-book, Boston.

Mehta, P.K., 2001. Reducing the environmental impact of concrete. Concrete Inter. October, 61–66.

Messner, F., 2007. Towards a sustainable copper industry? In: Ayres, R.U., von Gleich, A., Gössling-Reisemann, S. (Eds.), Sustainable Metals management Securing Our Future – Steps Towards a Closed Loop Economy. Springer e-book, Dordrecht, pp. 97–113.

Minke, G., 2012. Building with Earth: Design and Technology of a Sustainable Architecture. Birkhäuser, Basel.

Mithraratne, N., Vale, B., Vale, R., 2007. Sustainable Living: The Role of Whole Life Costs and Values. Butterworth-Heinemann, Oxford.

Moropoulou, A., Cakmak, A.S., Biscontin, G., Bakolas, A., Zendri, E., 2002. Advanced Byzantine cement based composites resisting earthquake stresses: the crushed brick/lime mortars of Justinian's Hagia Sophia. Constr. Build. Mater. 16 (8), 543.

Nail, S., 2008. Forest policies and social change in England. Available from: SpringerLink Books at http://download.springer.com/static/pdf/493/ (accessed 3.07.16.).

Neel, E., 1943. Tarran system of construction. Architect. Design Constr. October, 212–213.

Nevell, M., 2005. The arrival of the cotton mill, 1783–1800. In: Casella, E.C., Symonds, J. (Eds.), Industrial Archaeology: Future Directions. Springer, New York.

NZ Wood, n.d. The environmental and cost benefits of structural wood. Available from: http://www.nzwood.co.nz/learning-centre/the-environmental-and-cost-benefits-of-structural-wood/ (accessed 10.06.16.).

Obeng, K., Ocran, K.A.G., Anaba, D., 1997. Palm kernel shell as fuel for burning bricks. Build. Res. Inf. 25 (3), 131–136.

Okai, R., 2003. Can logging residues be used to help meet timber demands for the downstream processing sector? Available from: http://www.fao.org/docrep/ARTICLE/WFC/XII/0236-B4.HTM (accessed 10.06.16.).

Packard Forest Products, 2014. Hardboard. Available from: http://packardforestproducts.com/products/wood-composites/hardboard/ (accessed 10.07.16.).

Pauliuk, S., Milford, R.L., Müller, D.B., Allwood, J.M., 2013. The steel scrap age. Environ. Sci. Technol. 47 (7), 3448–3454.

Pearson, D., 1991. The Natural House Book. Conran Octopus, London.

Pearson, D., 1998. The New Natural House Book. Simon and Schuster, New York, NY.

Perlin, J., 2013. Let It Shine: The 6,000-year Story of Solar Energy. New World Library, Novato, CA.

Petrović, E., Vale, B., Wilson, M., 2016. Vinyl and linoleum flooring: health issues as perceived by lay people and architects. J. Green Build. 11 (2), 159–177.

Phelps, H., 1982. The Craft of Log Building. Lee Valley Tools Ltd, Ottawa.

Pounds, N.J.G., 1959. World production and use of steel scrap. Econ. Geogr. 35 (3), 247–258.

Prokopy, J., 2008. Growing benefits while shrinking footprints. Environ. Design Constr. 11 (9), S4, S6.

Pritchard, M.B., Pitts, A., 2006. Evaluation of strawbale building: benefits and risks. Architect. Sci. Rev. 49 (4), 372–384.

Richards, J.M., 1958. The Functional Tradition in Early Industrial Buildings. The Architectural Press, London.

Richardson, B.A., 1978. Wood Preservation. The Construction Press, Lancaster.

Ridley, I., Clarke, A., Bere, J., Altamirano, H., Lewis, S., Durdev, M., Farr, A., 2013. The monitored performance of the first new London dwelling certified to the Passive House standard. Energy Build. 63, 67–78.

Rios Calleja, J., 2015. Environmental Impact Assessment of Vernacular Thatch Building Tradition in Mexico: Case Studies of Three Palm Species and Related Technology along the Trans-Mexican Volcanic Belt. PhD Thesis. Victoria University of Wellington.

Robertson, A.B., Lam, F.C.F., Cole, R.J., 2012. A comparative cradle-to-gate life cycle assessment of mid-rise office building construction alternatives: laminated timber or reinforced concrete. Buildings. 2 (3), 245–270.

Rodriguez, N.H., Martinez-Ramirez, S., Blanco-Varela, M.T., Donatello, S., Guillem, M., Puig, J., . . .Flores, J., 2013. The effect of using thermally dried sewage sludge as an alternative fuel on Portland cement clinker production. J. Clean. Prod. 52 (August), 94–102.

Rosato, D.V., 1959. Asbestos: its industrial applications. Reinhold Publishing Corp, New York, NY.

Salas, J., Alvarez, M., Veras, J., 1987. Rice husk and fly ash concrete blocks. Inter. J. Cement Compos. Lightweight Concrete. 9 (3), 177–182.

Starrett, W.A., 1928. Skyscrapers and the Men Who Build Them. Charles Scribnenr's Sons, New York, NY, London.

Schroeder, H., 2016. Sustainable Building with Earth. Springer, Switzerland, eBooks.

Schunck, E., Oster, H.J., 2003. Roof Construction Manual: Pitched Roofs. De Grulter, Basel.

Sear, L.K.A., 2008. Properties and Use of Coal Fly Ash: A Valuable Industrial By-product. Thomas Telford Publishing, London.

Shreve, S.D., 2013. Futures so Bright: Solar Homes in Mid-twentieth Century America, PhD Thesis. University of Iowa.

Siers, J., 2007. The Life and Times of James Walter Chapman-Taylor. Millwood Heritage Publications, Ahuriri, Hawkes Bay.

Simpson, W.T., Hart, C.A., 2000. Estimates of Air Drying Times for Several Hardwoods and Softwoods. United States Department of Agriculture. Available from: http://www.fpl.fs.fed.us/documnts/fplgtr/fplgtr121.pdf (accessed 9.12.15.).

Statistics New Zealand (2009). *Measuring New Zealand's progress using a sustainable development approach: 2008*. Available at: <http://www.stats.govt.nz/browse_for_stats/snapshots-of-nz/Measuring-NZ-progress-sustainable-dev-%20approach/sustainable-development/waste.aspx> (accessed 06.16).

Stefanidou, M., Papayianni, I., Pachta, V., 2014. Analysis and characterisation of Roman and Byzantine fired bricks from Greece. Mater. Struct. 48 (7), 2251–2260.

Sutton, A., Black, D., Walker, P., 2011. Hemp Lime. Information Paper IP 14/11, Watford, IHS BRE Press.

Swan, A.J., Rteil, A., Lovegrove, G., 2011. Sustainable earthen and straw bale construction in North American buildings: codes and practice. J. Mater. Civil Eng. 23 (6), 866–872.

Swentzell Steen, A., Steen, B., Bainbridge, D., 1994. The Straw Bale House. Vermont, the Chelsea Green Publishing Company, White River Junction.

Tainuo Woods, n.d. Technical—plywood advantages. Available from: http://www.tnwood.com/T_PlyAdvantage.html (accessed 20.06.16.).

Tarpaper Recycling, n.d. From roof to road. Available from: http://tarpaper.se/ (accessed 10.07.16.).

Thatch Advice Centre, n.d. Life expectancy of thatched roofs. Available from: http://www.thatchadvicecentre.co.uk/thatch-information/care-thatch/longevity (accessed 6.07.16.).

The Little Pig Building Company Ltd., 2016. *Albury*. Available at: <http://www.littlepigbuilding.co.nz/plans/> (accessed 07.16).

The [NZ] Treasury, 2013. New Zealand Aluminium Smelters (NZAS) Information Release. Available from: http://www.treasury.govt.nz/publications/informationreleases/nzas/pdfs/nzas-2707299.pdf (accessed 20.01.16.).

The Passivhaus Trust, n.d. What is Pasivhaus? Available from: http://www.passivhaustrust.org.uk/what_is_passivhaus.php (accessed 24.02.16.).

Thomas, M., 2007. Optimising the use of fly ash in concrete. Available from: http://www.cement.org/docs/default-source/fc_concrete_technology/is548-optimizing-the-use-of-fly-ash-concrete.pdf (accessed 21.01.16.).

Tice, C., 2007. Growing up fast. Entrepreneur. 35 (2), 54–55.

Trace, P., 2006. A cast iron solution. Building Conservation Directory. Available from: http://www.buildingconservation.com/articles/allarticles.htm (accessed 20.06.16.).

Tran, T.H., 2013. Sustainable patterns of living based on an investigation of footprint in Hanoi-Vietnam, Wellington-New Zealand and Oulu-Finland, PhD Thesis, Victoria University of Wellington. Available from: http://restrictedarchive.vuw.ac.nz.helicon.vuw.ac.nz/handle/123456789/8698 (accessed 14.06.16.).

Treet, 2015. Treet. Available from: http://treetsameie.no/ (accessed 20.06.16.).

Trossero, M.A., 2003. Wood energy: the way ahead. Available from: http://www.fao.org/docrep/005/y4450e/y4450e02.htm (accessed 16.07.16.).

UN Habitat, n.d. Green materials: bamboo. Available from: http://unhabitat.org/green-building-materials-fact-sheet-bamboo/ (accessed 16.07.16.).

University of Tennessee Centre for Clean Products, 2007. Limestone quarrying and processing: a life-cycle inventory, natural stone council. Available from: http://www.naturalstonecouncil.org/content/file/LCI%20Reports/Limestone_LCIv1_October2008.pdf (accessed 10.12.15.).

US Energy Information Association, 2016. Levelized cost and levelized avoided cost of new generation resources in the annual energy outlook 2016. Available from: http://www.eia.gov/forecasts/aeo/pdf/electricity_generation.pdf (accessed 14.09.16.).

US Environmental Protection Agency, 2016. Overview of greenhouse gases. Available from: https://www3.epa.gov/climatechange/ghgemissions/gases/ch4.html (accessed 10.07.16.).

US Green Building Council, 2016. Better buildings are our legacy. Available from: http://www.usgbc.org/leed (accessed 10.07.16.).

Vale, B., 1973. A Review of the Ministry of Agriculture's Earth Houses, Working paper 17, Technical Research Division, University of Cambridge.

Vale, B., 1995. Prefabs: As History of the UK Temporary Housing Programme. E and FN Spon, London.

Vale, B., Vale, R., 1990. Green Architecture. Thames and Hudson, London.

Vale, B., Vale, R., 2000. The New Autonomous House: Design and Planning for Sustainability. Thames and Hudson, London.

Vale, B., Vale, R., 2007. Cultural aspects of making sustainable buildings. In: Proceedings SB07 Conference Transforming our Built Environment, November, Auckland.

Vale, R., Vale, B., 2009. Time to Eat the Dog? The Real Guide to Sustainable Living. Thames and Hudson, London.

Villeda-Muñoz, G., Castañeda-Miranda, A., Pless, R.C., Vega-Durán, J.T., Pineda-Piñón, J., 2011. Clay-Brick firing in a high-temperature solar furnace. Ingeniería Investigación y Tecnologia. 12 (4), 395–408.

Vitruvius (c30BC) Ten books on architecture. Available from: https://ia902604.us.archive.org/24/items/vitruviustenbook00vitruoft/vitruviustenbook00vitruoft.pdf (accessed 28.07.16.).

Wan, L., Toppinen, A., Juslin, H., 2014. Use of wood in green building: a study of expert perspectives from the UK. J. Clean. Prod. 65, 350–361.

Watts, A., 2013. Modern Construction Handbook. third ed. Ambra V, Vienna.

Weaver, L., 1909. English Leadwork: Its Art and History. B T Batsford, London.

Whelton, A.J., Nguyen, T., 2013. Contaminant migration from polymeric pipes used in buried potable water distribution systems: a review. Crit. Rev. Environ. Sci. Technol. 43 (7), 679–751.

Williams-Ellis, C., Eastwick-Field, J., Eastwick-Field, E., 1947. Building in Cob, Pise and Stabilised Earth. Country Life, London.

Willis, C.P., 1910. The Preservative Treatment of Farm Timbers. UNT Digital Library, Washington, DC. Available from: http://digital.library.unt.edu/ark:/67531/metadc87486/. (accessed 9.09.16.).

World Steel Association, 2015. World steel in figures 2015. Available from: https://www.worldsteel.org/dms/internetDocumentList/bookshop/2015/World-Steel-in-Figures-2015/document/World%20Steel%20in%20Figures%202015.pdf (accessed 20.01.15.).

Wu, P., Low, S.P., Xia, B., Zuo, J., 2014. Achieving transparency in carbon labelling for construction materials – lessons from current assessment standards and carbon labels. Environ. Sci. Policy. 44, 11–25.

Yellishetty, M., Mudd, G.M., Ranjith, P.G., Tharumarajah, A., 2011. Environmental life-cycle comparisons of steel production and recycling: sustainability issues, problems and prospects. Environ. Sci. Policy. 14 (6), 650–663.

Materials and buildings

4

Brenda Vale
Victoria University of Wellington, Wellington, New Zealand

4.1 Choosing materials

The discussion of separate materials in Chapter 3, is not intended to be exhaustive but rather to show the problem of choosing materials with which to make sustainable buildings. The general trend is that in the past choices were more sustainable because not only the materials, but also the glues that joined them and the paints that finished them, were based on natural materials. This often meant that materials were less waterproof, so more maintenance was required for the building upkeep, although in the past labor was cheap so this was less of an issue. Although the materials of the past were "natural" they were not necessarily "healthy," lead paints being a good example. The modern situation is a preference for reducing the labor involved in building, both in first construction and in maintenance, and using materials which have sufficient synthetic components to render them more stable and durable. If materials are also produced by machines in a factory for assembly on site, rather than building from scratch, this will also save labor (and potentially materials as there could be less wastage from poor handling and site storage) but at the expense of using more energy, usually derived from fossil fuels. The argument is this approach also saves money but if we really want to make sustainable buildings then we have to have a sustainable economic system within which such buildings are produced and maintained. Although arguably outside the scope of this book it is important to be aware that choice of materials for buildings will not alone solve the problems of resource shortages, and environmental pollution and degradation.

The other issue to emerge from the discussion above is that there are no easy answers when it comes to choosing materials but rather many shades of gray. When it comes to designing buildings it is important to think holistically and consider not what the building is on the day it is handed over but how it will perform over its whole life and what happens at the end of its life. Thus any move to making sustainable materials choices will have to deal with lifecycle analysis (LCA), which is the subject of Section 4.1.1.

4.1.1 Lifecycle analysis

LCA is a much simplified model of reality and as such it should never be used to predict final outcomes. It involves accounting for the energy that goes into making the materials of a building and those involved in its maintenance, and the energy that is used in running it over the building's life. Ideally it also includes the energy used on site for putting up a building, the energy used to bring all the materials to site, and the energy used in demolishing and disposing of wastes at the end of the

Materials for a Healthy, Ecological and Sustainable Built Environment: Principles for Evaluation.
DOI: http://dx.doi.org/10.1016/B978-0-08-100707-5.00004-6

building's useful life. This long list suggests that a number of assumptions will have to be made in order to make such an analysis as some things will be hard to predict. The real use of LCA, however, is in comparing alternatives, whether these be choosing between different materials for a wall, or choosing between making a building that can gain at least some of its energy for running from renewable resources, such as the sun. This type of analysis can be productive during the design stage of a sustainable building. However, most designers are experts in building design not in calculating the life-cycle impact of building materials and products. Consequently designers need to make use of data prepared by other people, or rely on labels prepared by manufacturers, who have a vested interest in selling their products. This information is usually given in the form of an embodied energy coefficient, which is the energy involved in making a set weight or set volume of a material such as wood or aluminum.

4.1.2 Embodied energy coefficients

As shown in Table 4.1 the impact of a particular material will vary according to where it was extracted and processed, and even to how it was transported to the production site. This is partly because the energy make up of each nation's electricity supply will vary and electricity is important in the processing of some familiar materials, like aluminum. In a study of published EE coefficients Hammond and Jones (2008) compared the range of embodied energy values in MJ/kg for steel, concrete, and timber from different sources showing the wide range that results from studies in different countries and over time in the same country. Hammond and Jones' values from a study of 250 sources have been included in Table 4.1.

The variability of the data in Table 4.1 suggests that any LCA of energy in buildings is hardly going to be reliable and all published data on embodied energy do not show clearly the relative impact of the use of materials, because measurement in terms of weight cannot capture this. What is needed for meaningful results is a comparison of different forms of construction. Table 4.2 shows a comparison of the energy LCA

Table 4.1 Comparison of embodied energy coefficients

Steel[a] (MJ/kg)	Timber (MJ/kg)	Concrete (MJ/kg)	Country	References
10.1–32	0.3–2.5	1–2	New Zealand	Alcorn (1996, pp. 22–24)
8.6–31.3	2.8–9.7	1.4–19	New Zealand	Alcorn (2003, pp. 24–25)
26.3	9.2	0.80–1.14	United Kingdom	Eaton and Amato (1998, p. 57)
9–32	5.7–10	–	United Kingdom	West et al. (1994)
6–12	3–4	0.6–1	Northern Europe	Berge (2000, pp. 20–23)
6–81.8	0.3–61.3	0.07–23.9	Various	250 database sources (Hammond and Jones, 2008)

[a]Range of values for steel reflects the effect of recycled steel content.

Table 4.2 Total embodied energy in GJ for three exhibition buildings and percentage variation

Building	UK EE coefficients (GJ)	Australian EE coefficients (GJ)	German EE coefficients (GJ)	Variation (%)
Crystal Palace (1851)	348,189	346,379	379,027	4.2
Shanghai Exhibition Centre (1955)	859,082	758,772	858,924	5.7
Dutch Pavilion (2000)	93,054	93,590	89,186	2.1

Shen, 2012, p. 247.

for three exhibition buildings, including all the materials used for construction and maintenance using coefficients from three different countries (Shen, 2012, p. 247).

This comparison shows a lot less variation than for the individual coefficients in Table 4.1. Providing LCA is always used for comparison rather than thinking that it will give a true picture of the building in use, then it can be a good tool for the designer. However, as Tables 4.5 and 4.6 show it can only be used when there is a design from which quantities of materials can be extracted, which means it cannot be used until well beyond the concept stage of design. However, with a design established then it is easy to test the effect of choosing different materials for elements of the building, such as wall cladding or floor covering.

Embodied energy coefficients also change as methods of manufacture change, especially as they become more energy and resource efficient. Alcorn's sets of New Zealand EE coefficients have been updated regularly as processing methods and the national electricity fuel mix have changed. Other changes have occurred because of alterations to the way coefficients have been calculated (Khajehzadeh and Vale, 2015). What this suggests is the need to pick the set of coefficients that seem most relevant to where the building is situated and to use the same coefficients throughout the analysis. What this advice ignores is the fact we now live in a global society, which includes trade in building materials. The case study of New Zealand materials by Nigel Isaacs (see Chapter 12) demonstrates that whereas a house could be built of NZ sourced and manufactured materials until the 1970s, this is no longer the case. Studies like that of Franklin Associates (1991) assume all products are produced in the home country (USA) and claim a margin of error for their calculation of impact of $\pm 10\%$. In the end, building LCA is a "rough-and-ready" approach to assessing environmental impact. It is, however, somewhat better than guesswork and the hope that your choice of materials is the sustainable one.

4.1.3 Other environmental impacts

Another issue is how the energy and environmental impact of a material are calculated. Energy has tended to be the main means of assessing the impact of a material, and embodied energy coefficients have been produced for different countries, such as Alcorn's sets for New Zealand. However, energy is not the only impact. As discussed above, natural materials will have a low embodied energy coefficient because most take little energy to process but they will need land to grow if grown specifically for the modern building industry, so they will have an "ecological footprint." This is a measure of the land it takes to produce a particular product (Wackernagel and Rees, 1996). Some materials take a lot of water to produce and have a large water footprint. A study of steel, including all steel-making processes from raw materials to unalloyed steel, revealed over 700 kg of water were involved in producing 1 kg of steel (Kluender, 2013). The water footprint is difficult to identify in buildings, although a Chinese study found that rural residential buildings had lower water footprints that their urban equivalents, the latter being 55%–130% greater (Chang et al., 2016). An earlier Australian study also found that "non-residential buildings generally have a large amount of embodied water," with the structural elements containing the most (McCormack et al., 2007). As the world urbanizes and cities are made more compact in the name of being more sustainable the fact that materials with high embodied water tend to be used is usually forgotten.

Many materials make greenhouse gases as a result of their manufacture, both from the energy used and from the chemical processes involved (Table 4.3). As a result coefficients of greenhouse gases are also calculated for building materials (Alcorn, 2003).

Ideally, all these additional impacts should be included in an LCA, but then unless energy production and water capture are also expressed in a common unit, such as the land area used in ecological footprint calculations, it is hard to compare the impacts. Other impacts may appear in the form of pollution and the production of unwanted byproducts, such as clinker, a glass like substance, and the hard residue left from burning coal and slag, both of which can be produced when extracting metals. Coal clinker can be used for paving and blast furnace slag has many potential uses (National Slag Association, 2013), but waste heaps are also found, which again occupy land.

This would mean an LCA that only looked at the energy component of for example concrete would miss the CO_2 emitted as part of the process of manufacture and also whether any slag byproducts were dumped or used. Looking just at CO_2 production, another problem is that values for this also vary, depending on how the

Table 4.3 Greenhouse gas emissions in production of 1 tonne of various materials

Material	CO_2 tonnes/tonne	PFCs[a] tonnes/tonne
Cement	0.40	–
Lime	0.73	–
Iron and steel	1.94	–
Aluminum	1.58	0.2304

[a]PFCs (perfluorinated compounds) have long lifetimes in the atmosphere and will potentially influence the climate in the future (IPCC, 2013).
Mithraratne et al., 2007, p. 100.

cement is made. Kajaste and Hurme (2016) found that cement in Brazil emitted 652 kg CO_2/t while that from the Middle East was 920 kg CO_2/t. However, this is not the full story. Assuming a timber frame house of 10 × 10 m has a 100-mm thick concrete slab this will require 10 m^3 concrete which is made from water, aggregate, and cement. Using a concrete calculator this will need 3000 kg (3 t) of cement (Concrete Services, 2015). The emissions associated with the concrete slab are, therefore, 1875 kg CO_2 (1.9 t) in Brazil and 2760 kg CO_2 (2.8 t) in the Middle East. In comparison, in 2003 the average USA household produced 12.4 t CO_2 each year from operating its systems (heating, lighting, appliances, hot water, and refrigeration) (Hinkle Charitable Foundation, n.d.). Additionally, when MIT produced an LCA study of equivalent concrete and wood houses and concrete and steelframed office buildings over a 60-year life they found the concrete buildings used less energy for their operation and this outweighed the higher global warming potential of the same buildings when first constructed. They also found that the embodied global warming potential was the same as the global warming effect of operating the building for 2–8 years, or alternatively over the 60-year building life 88%–98% of CO_2 emissions came from the energy to run the buildings (Ochsendorf et al., 2011, p. 79). From this it seems important to discuss the components of an LCA building energy analysis to understand which are relatively large and which are relatively small.

4.1.4 LCA components

Apart from how a product is produced and the energy and environmental impact of the various stages, other things need to be considered in using a material in a building and in making an LCA. The first is transport, as the material has to be brought to site and how it travels will vary its overall impact. Then there is the energy used by tools and equipment on site to make the building. The other big issue is what happens to the materials at the end of the building's life. They will either have to be recycled, which again takes energy and space, incinerated, which could involve unwanted emissions, or sent to landfill, again taking up land that could be used for other purposes. Table 4.4 sets out the stages of the impact of three buildings with percentage values for the impact of each stage. The big impacts are the energy involved in the operation of the building and that embodied in its fabric. The energy involved in the construction and demolition phases is small by comparison.

An LCA is never a picture of the whole impact of a building. It normally ignores things like the impact of the share of making the tools used on site and the share of the insurance industry that insures the buildings, or the finance industry that pays for it, or the share of roads and shipping that bring the materials to site but for the purposes of those choosing materials from which to make sustainable buildings this is, perhaps, a step too far. However, although these represent smaller impacts in the overall LCA if they are discounted this should be done knowingly.

Other small impacts that are often ignored in an LCA that is undertaken at the stage of choosing materials are transport of materials to site and disposal and

Table 4.4 Breakdown of the impact of a building as a percentage of its 50-year life time impact

	NZ lightweight single-family house (Mithraratne et al., 2007, p. 164)	**NZ super-insulated single-family house (Mithraratne et al., 2007, p. 164)**	**Low-energy building, Spain, five floors, 60 dwellings (Janez et al., 2013)**
Stage in LCA	% of total primary energy	% of total primary energy	% of total primary energy
Building materials/ products	32.0	60.0	43.0
Operation	66.0	36.0	52.0
Construction phase	2.0	4.0	2.7
End of life	–	–	2.3

demolition at the end of the building life. These issues are important in working out the true environmental impact of the building but they are not the big issues in a building energy LCA—the big issues are the energy embodied in the materials for making and maintaining the building and the energy involved in operating it.

Table 4.4 also shows that even for low-energy buildings the operating energy is still a large proportion of total lifecycle energy. The advice remains when it comes to choosing materials for making sustainable buildings that the most important goal is to reduce operating energy, and this is far more important than trying to source materials that are made or procured close to the building site. This fact has long been recognized, although much academic debate has occurred over the relative values: an LCA study by Adalberth (1994) of three well-insulated Swedish family houses found that over 50 years the energy embodied in the building materials was equivalent to running the houses (heating, electricity, and hot water) for 5–7 years (10%–14%), and the energy for producing the insulation materials (mineral wool and polystyrene) was 1–2 years (2%–4%) of operating energy demand. These figures also show that climate plays a role, since the demand for heating in New Zealand and Spain will be less than in Sweden, which will affect the relative values of embodied and operating energies. The end-of-life figure is similar to another study in Italy by Blengini (2009) where a well-managed end-of-life phase in terms of trying to reclaim materials gave an environmental benefit of 0.2%–2.6% of the full lifecycle impact of a conventional building.

Even here assumptions have to be made when it comes to estimating how long materials will last and how often routine maintenance, like painting, will have to be undertaken (Mithraratne et al., 2007, pp. 131–132). Harder to predict is the use of materials that might occur due to changes in building use or changes in fashion that lead to major internal refurbishments. We can go back and measure the impact of these (Leah, 2015) but we cannot predict them.

4.1.5 Lifecycle energy impact

Because of these complexities, in building design a simplified lifecycle energy analysis, which uses the embodied energy coefficients to calculate the energy it takes to make the initial building and to maintain and repair it over its life is often the norm. This energy can then be compared to the energy it is expected to take to operate the building over the same life. It is very unlikely either value will be a true reflection of what will happen over a building's life but provided comparisons are made on the same basis and using the same data sets, these can be useful. As shown by the case study on the Hockerton Housing Project (see Chapter 14), this can lead to some unexpected results, whereby a building using concrete, which is a high embodied energy material, has a lower overall energy lifecycle impact because that concrete is very well insulated and stores heat to create a zero-energy building. This is the balance that designers can use to explore their choices of materials at the design stage, through making comparisons. As seen in Parts I and III, this approach can also reveal key issues in making sustainable buildings that go beyond the choice of materials but instead involve how materials are used in the design of the performance of the building.

4.2 Designing to minimize building lifecycle impact

Perhaps the first step in designing the ideal sustainable building is to ponder whether the building is truly needed. The most sustainable move is not to build, which might mean using an existing building. However, if it is radically altered then the impact of doing this could be close to the cost of building new, but this is where the energy LCA can be used. Chapter 15, describes the reuse of an existing house where the choices of what to do and the materials used were dictated by trying to reduce environmental impact while creating a house that met the wishes of the occupants.

Relating to the need for having the building is how long it will be needed. A building that lasts 1000 years and is still being used for the purpose for which it was designed (such as an early medieval cathedral or church) is a sustainable building, even if it makes no use of fashionable renewable energy technologies such as PV panels on the roof. The obvious corollary to this is designing a building that is easy to maintain and where the detailing will not lead to failure of important parts of the building, such as the structure. Mention has already been made of leaky building syndrome (see Section 3.4.4) where a mixture of bad detailing (design) and bad workmanship, together with use of unsuitable materials (kiln dried timber rather than treated timber) led to the need for extensive rebuilding in both Canada and New Zealand. The need to build well also cropped up in the discussion of the buildings as a result of the two competitions at Gidea Park in the United Kingdom (see Section 3.4.3) where the more traditional Arts and Crafts houses have worn much better and have been better maintained than the later modernist versions.

Sustainable building is essentially about building well and building to last, and probably has much more in common with the vernacular approach than with the idea of architecture as an art form.

4.2.1 Materials recycling

Another lesson learned from the vernacular is the longer a building lasts the less need there is to recycle it and the lower will be its overall lifecycle impact. Nevertheless, recycling of building materials has a history as long as building because there is always change and some buildings will always make way for new ones. In societies where money is still short any useful materials that emerge when a building is past repair and has to be demolished will be reused. Stones from the Coliseum in Rome were used to make roads (Aquino, 1995, preface) and parts of buildings, like the roof and timber framing of Shelwick Court in Herefordshire, have been moved from somewhere else (Landmark Trust, 2016). The problem with recycling building components and materials is not that they may not necessarily meet standards, though in the developed world this is an issue, but simply that it is cheaper to knock down and dispose of the rubbish rather than carefully disassemble, sort, and prepare for reuse, just because labor is expensive relative to resources. Where labor is cheap and resources are expensive, as in the past, recycling will happen.

One of the problems faced in the developed world is that skill has been replaced by legal requirements that are bound up in building standards and regulations. Timbers could be rescued from old buildings and those that were sound would be known through experience not through measuring their ability to withstand loads. Rotted and worm-eaten parts would be rejected (and used for firewood) and the good salvaged material used for new buildings. Some joinery fittings were so valued that they traveled from building to building with their owners (Harris, 2007, p. 12). Bricks could also be salvaged and reused because in the past they were laid in lime mortar, which is much easier to clean off without damaging the brick. Modern cement mortar is much more difficult to remove, though it can be done, but again this is a labor-intensive process, and new bricks may as a consequence be much cheaper than recycled bricks. Stone, just because it is more expensive, tends to be recycled. Many Victorian slate roofs in the UK failed after 100 years because the nails failed not the slates, which were often sorted on site with the best used as part of the reroofing. Where products can be given an added value, such as in the use of recycled hardwoods for interior fit-outs, then these will also find their way on to the market. However, this may well be using timber that is potentially structurally sound for a lesser purpose. Other more costly items, such as internal joinery (doors, skirting boards, and architraves) may also find a market, especially in the refurbishment of older properties when people are seeking to match original details (Storey et al., 2003).

Part of this approach includes making whole buildings not from waste building products but from other waste products. The Morrow Royal Pavilion outside Las Vegas uses recycled bottle glass as the aggregate and flyash to create a concrete named GreenStone that is made from "99.8% renewable materials" (Hojgrova, 2012). However, just because beer bottles are thrown away, rather than being recycled and refilled with beer as happened in the past, and given that flyash comes from burning coal, the product seems less than sustainable. Currently these are waste products but they are waste products that would not exist in a more sustainable society. Similar ideas have happened in the past, such as the 1972 patent for Envirite, a building material made from foamed waste glass and animal manure as "...a means of utilising the million tons of livestock dung produced annually in the state" (Pawley, 1975, p. 64). There have also been designs for bottles on the assumption that they will make a one-way trip and rather than being dumped as waste they could become a walling material. In the Wobo project Heineken wanted "...a brick that would hold beer" (Pawley, 1975, p. 26). Although this was an idea that never became reality, in a sustainable future there will be no avoidable waste. Much can be learned from the poorest in the world who already make their homes from the discarded products of those richer than themselves. When you have nothing, anything has a value, when you are rich, not much does.

This suggests that in the developed world the first rule of recycling materials from old buildings should be not to do it but rather question whether the building should be demolished in the first place. In some parts of the world land can dramatically increase in price so that relatively new buildings are demolished to make way for taller or more expensive ones long before the end of their useful life (Song, 2016). This is again an example of the market preferring profit over planet, putting making money over doing what is good for the environment.

4.2.2 Relative values

This section explores how the choice of materials impacts on different aspects of building lifecycle. To do this it uses a hypothetical room and looks at how the LCA of the room changes with different design decisions.

As this simulation is being done in New Zealand using NZ embodied energy coefficients the door and window orientations will be reversed compared to northern hemisphere designs because the sunshine will be coming from the north. Think of this room as being a stand-alone structure for a home office or a separate studio, so it will have a roof structure and will need to be weatherproof. The dimensions given are all finished internal dimensions (Table 4.5).

The first simulation will look at making the walls of straw bales, a "sustainable" material. It will compare these with a normal and better insulated timber-framed wall. Table 4.5 sets out the dimensions of the other components as a result of choosing straw bale as the walling material.

Table 4.6 looks at the embodied energy of these elements when the building is first constructed and after 50 and 100 years. This means assuming replacement cycles for the various elements, such as the exterior and interior render. Volumes and weights of materials are calculated as necessary using manufacturers' information, which means having the design at a stage that has sufficient detail to make

Table 4.5 Room with straw bale walls

Component	Material	Dimension
Side wall (two the same)	Straw bale	$0.5 \times 2.3 \times 5.0 = 5.75$ m^3/wall
	Wood framing[a]	50 m 100 × 50 timber; 5 sheets ply
	Cement, lime, sand render to both sides of wall	25 mm thick 16 m^2
Window wall	Straw bale	$(0.5 \times 2.3 \times 2.5) - (1 \times 1 \times 0.5) = 2.4$ m^3
	Cement, lime, sand render	7.0 m^2
	Window	1 m^2
Door wall	Straw bale	$(0.5 \times 2.3 \times 2.5) - (1 \times 2 \times 0.5) = 1.9$ m^3
	Cement, lime, sand render	6.0 m^2
	Door	2 m^2
Foundations (750 mm deep)	Concrete	7.1 m^3
Slab (100 mm thick)	Concrete	1.75 m^3
	Under slab insulation	17.5 m^2
	DPM	27 m^2
	Hardcore with sand blinding	17.5 m^2
Flat roof	Plasterboard ceiling	17.5 m^2
	Timber structure	60 m 100 × 50 timber
	Insulation	17.5 m^2
	Decking	27 m^2 10 sheets ply
	Butyl membrane	27 m^2
Window	Aluminum and timber Double-glazed argon-filled	1 m^2
Door	Aluminum and timber Double-glazed argon-filled	2 m^2

[a]Straw bale construction with a timber and plywood surround (Downton, 2013).

this possible. Sheet materials are assessed on the volume of the number of sheets needed, so there will be wastage, but this follows normal building practices. All values for embodied energy coefficients are for New Zealand and taken are from Alcorn unless stated otherwise and the most recent values are used.

Before considering the operating energy several points are worth noting. Because the straw bales are wide the concrete foundations are the largest component of the ground works. The choice of a flat roof with a butyl membrane which needs replacing every 25 years increases the value of this component to the extent that by 100 years it becomes the largest single part of the embodied energy. As an alternative we could use a metal standing seam roof on a 4 degree pitch across the shorter dimension, which would not change the volume of materials used in the

Table 4.6 Straw bale room embodied energy at point of construction

Element	EE coefficient	EE Year 0 (GJ)	Replacement cycle	EE Year 50 (GJ)	EE Year 100 (GJ)
Foundations	2019 MJ/m^3	14.3	–	14.3	14.3
Slab	2019 MJ/m^3	3.5	–	3.5	3.5
Hardcore	65 MJ/m^3	0.1		0.1	0.1
Insulation polystyrene	1401 MJ/m^3	2.5	–	2.5	2.5
DPM (0.27 kg/m^2)	51 MJ/kg	0.4	–	0.4	0.4
Subtotal ground works		**20.8**		**20.8**	**20.8**
Straw bale walls	30.5 MJ/m^3	0.5	–	0.5	0.5
Cement, lime sand render[a] 2.25 m^3 550 kg/m^3	1.09 MJ/kg	1.3	30 years	2.6	3.9
Wood framing					
Softwood	2204 MJ/m^3	0.6	–	0.6	0.6
plywood	5200 MJ/m^3	0.9		0.9	0.9
Subtotal walls		**3.3**		**4.6**	**5.9**
Roof timber	2240 MJ/m^3	0.7	–	0.7	0.7
Ply decking	5200 MJ/m^3	1.8	–	1.8	1.8
Insulation fiberglass	1026 MJ/m^3	1.8		1.8	1.8
Butyl membrane[b]	149 MJ/m^2	4.0	25 years	8.0	16.0
Ceiling (5 sheets)	7080 MJ/m^3	2.4	–	2.4	2.4
Subtotal roof		**8.9**		**14.7**	**22.7**
Window[c]	1013 MJ/m^2	1.0	45 years	2.0	3.0
Door	1013 MJ/m^2	2.0	45 years	4.0	6.0
Subtotal elements		**3.0**		**6.0**	**9.0**
Total EE		**36.0**		**46.1**	**58.4**

[a]Hammond and Jones (2008).
[b]Fung (2010).
[c]Asif et al. (2001), pro rate area value.

walls, with one now slightly higher than the other (shorter dimension now becomes 4.51 m), so this slight change is ignored and areas and roof structure remain the same. Such a roof should have a 50-year life. Table 4.7 compares the values for the two roof types.

Table 4.7 Comparison of straw bale room roof constructions

Roof component	EE coefficient	EE Year 0 (GJ)	Replacement cycle	EE Year 50 (GJ)	EE Year 100 (GJ)
Roof timber	2240 MJ/m^3	0.7	–	0.7	0.7
Ply decking	5200 MJ/m^3	1.8	–	1.8	1.8
Insulation fiberglass	1026 MJ/m^3	1.8	–	1.8	1.8
Standing seam (4.9 kg/m^2)	34.8 MJ/kg	4.6	50 years	9.2	13.8
Ceiling	7080 MJ/m^3	2.4	–	2.4	2.4
Subtotal metal roof		**11.3**		**15.9**	**20.5**
Butyl roof		**8.9**	**25 years**	**14.7**	**22.7**

What Table 4.7 demonstrates is that there is little difference between the two choices, especially given that these types of analyses are always approximations. The choice depends on how long the room is going to last, as the standing seam roof only offers a lower overall embodied energy at year 100. However, assuming a more optimistic 60-year life then the standing seam roof is better at year 50 (11.3 GJ with 10 more years of life remaining). Because of the overall similarity in impact the next comparison will use the simple flat roof with butyl membrane. Table 4.8 compares the embodied energy of using insulated timber frame in place of straw bale. This affects the foundations, which are less massive due to the much thinner walls and the walls themselves, but the roof and window and door components remain unchanged.

The immediate response to Table 4.8 is to echo Barry Commoner's comment "There is no such thing as a free lunch." Only at year 100 does the need for painting make the timber frame room marginally greater in embodied energy than the straw bale one. However, no allowance was made for painting the straw bale room either internally or externally, leaving the render in its natural state (painting of the ceilings in both rooms has also been omitted but would be the same for both), but these slight differences are more than compensated for by the approximation of the LCA method. The real issue is that the much larger foundations to support the wide straw bales add to the overall embodied energy of the room, even though the embodied energy of the straw bale walls (3.3, 4.6, and 5.9 GJ) is less than that of the timber frame walls (9.8, 11.8, and 14.3 GJ).

Embodied energy is not the whole story. To take operating energy into account the rooms in their two different constructions were modeled using ALF, which is a simple energy modeling tool designed specifically for New Zealand conditions (Bassett and Stoeklein, 1998). The rooms were modeled for Wellington, at the bottom end of the North Island, where the climate is temperate and the house heated morning and evening. The results are shown in Table 4.9.

Table 4.9 again shows the importance of operating energy when it comes to lifecycle energy of buildings. From the aspect of total lifecycle energy the straw bale

Table 4.8 Comparison of straw bale and timber frame rooms when first constructed

Element	EE coefficient	EE Year 0 (GJ)	Replacement cycle	EE Year 50 (GJ)	EE Year 100 (GJ)
Foundation	2019 MJ/m^3	6.0	–	6.0	6.0
Slab	2019 MJ/m^3	3.5	–	3.5	3.5
Hardcore	65 MJ/m^3	0.1		0.1	0.1
Insulation polystyrene	1401 MJ/m^3	2.5	–	2.5	2.5
DPM	51 MJ/kg	0.4	–	0.4	0.4
Subtotal ground works		**12.5**		**12.5**	**12.5**
Softwood framing (67 lengths 50 × 100)	2204 MJ/m^3	1.8	–	1.8	1.8
Insulation fiberglass	1026 MJ/m^3	4.3	–	4.3	4.3
Weatherboards 18 mm	2204 MJ/m^3	1.3	–	1.3	1.3
Plasterboard	7080 MJ/m^3	1.9	–	1.9	1.9
Exterior paint water-based	115 MJ/L	0.3	10 years	1.5	3.0
Interior paint water-based	115 MJ/L	0.2	10 years	1.0	2.0
Subtotal walls		**9.8**		**11.8**	**14.3**
Subtotal roof		**8.9**		**14.7**	**22.7**
Subtotal elements		**3.0**		**6.0**	**9.0**
Total timber frame room		**34.2**		**45.0**	**58.5**
Total straw bale room		**36.0**		**46.1**	**58.4**

room seems to be the better choice. However, whatever the materials of the room, 50 years into the life of the room the operating energy becomes the larger component of the total. The other thing to note is that although both structures meet the NZ Building Code, the insulating value of the straw bale wall ($R = 8.0\ m^2\ °C/W$) is much higher than that of the timber-framed structure ($R = 2.3\ m^2\ °C/W$). For the same width of foundations assumed for the 100 mm timber wall (200 mm wide) the studs in the timber frame could be doubled and two thicknesses of insulation installed ($R = 4.6\ m^2\ °C/W$). With everything else the same, this produces a new embodied energy for the walls as shown in Table 4.10.

Table 4.9 Embodied and operating energy of straw bale and timber frame room over 100 years

Straw bale room					
EE Year 0 (GJ)	36.0	EE Year 50 (GJ)	46.1	EE Year 100 (GJ)	58.4
OE Year 0 (GJ)	1.6	OE Year 50 (GJ)	80.0	OE Year 100 (GJ)	160.0
Total Year 0 (GJ)	**37.6**	**Total Year 50 (GJ)**	**126.1**	**Total Year 100 (GJ)**	**218.4**
OE as % total	4.3		63.4		73.3
Timber-framed room					
EE Year 0 (GJ)	34.2	EE Year 50 (GJ)	45.0	EE Year 100 (GJ)	58.5
OE Year 0 (GJ)	2.0	OE Year 50 (GJ)	100.0	OE Year 100 (GJ)	200.0
Total Year 0 (GJ)	**36.2**	**Total Year 50 (GJ)**	**145.0**	**Total Year 100 (GJ)**	**258.5**
OE as % total	5.5		69.0		77.4

Table 4.10 Embodied energy of better-insulated timber frame room as first constructed

Element	EE coefficient	EE Year 0 (GJ)	Replacement cycle	EE Year 50 (GJ)	EE Year 100 (GJ)
Subtotal ground works		**12.5**		**12.5**	**12.5**
Softwood framing (134 lengths 50 × 100)	2204 MJ/m^3	3.6	–	3.6	3.6
Insulation fiberglass	1026 MJ/m^3	8.6	–	8.6	8.6
Weatherboards 18 mm	2204 MJ/m^3	1.3	–	1.3	1.3
Plasterboard	7080 MJ/m^3	1.9	–	1.9	1.9
Exterior paint water-based	115 MJ/L	0.3	10 years	1.5	3.0
Interior paint water-based	115 MJ/L	0.2	10 years	1.0	2.0
Subtotal walls		**15.9**		**17.9**	**20.4**
Subtotal roof		**8.9**		**14.7**	**22.7**
Subtotal elements		**3.0**		**6.0**	**9.0**
Total insulated timber frame room		**40.3**		**51.1**	**64.6**

Table 4.11 Comparison of embodied and operating energy of straw bale and better-insulated timber frame room over 100 years

Straw bale room					
EE Year 0 (GJ)	36.0	EE Year 50 (GJ)	46.1	EE Year 100 (GJ)	58.4
OE Year 0 (GJ)	1.6	OE Year 50 (GJ)	80.0	OE Year 100 (GJ)	160.0
Total Year 0 (GJ)	**37.6**	**Total Year 50 (GJ)**	**126.1**	**Total Year 100 (GJ)**	**218.4**
OE as % total	4.3		63.4		73.3
Better-insulated timber-framed room					
EE Year 0 (GJ)	40.3	EE Year 50 (GJ)	51.1	EE Year 100 (GJ)	64.6
OE Year 0 (GJ)	1.6	OE Year 50 (GJ)	80.0	OE Year 100 (GJ)	160.0
Total Year 0 (GJ)	**41.9**	**Total Year 50 (GJ)**	**131.1**	**Total Year 100 (GJ)**	**224.6**
OE as % total	3.8		61		71

Table 4.11 sets out the embodied and operating energy comparisons for the straw bale and better insulated wall timber-framed rooms.

Redesigning the timber frame wall to double the thickness of fiberglass insulation has made the operating energy for both straw bale and better insulated timber frame wall the same. However, the embodied energy in the fiberglass insulation means the lower embodied energy of the straw bale produces the lower lifecycle energy at all life stages of the room, so straw bale is still slightly the better choice, though the margin of difference has decreased. The only way to reduce the better insulated timber frame would be to choose an insulation with lower embodied energy, such as cellulose fiber, but this would mean redesigning the wall so the material could be blown in damp and then set (Vale and Vale, 2000, p. 168).

The other issue to consider given the impact of concrete is the CO_2 emissions associated with the two types of construction. Both straw and timber have the ability to lock up carbon so this will also have to be taken into account. This examination was limited to CO_2 emissions only, ignoring other gases with global warming potential, and using Alcorn's (2003) coefficients where possible. Negative coefficients indicate the level of CO_2 lock-up in the building component. The comparison was only between the walls and foundation elements, as the slab, roof, and windows and doors are common in both designs for the room. The results are shown in Table 4.12.

It is perhaps surprising that the ordinary timber frame wall and foundation elements lock up more CO_2 than is taken to produce them until year 100 when the effect of the repainting is large enough to be included, even though more CO_2 is locked up in the straw wall than the timber wall. The timber frame room could be further improved by cladding the interior in ply instead of plasterboard. For the straw bale wall the cement, lime, and sand plaster could be replaced with an earth plaster, but this would need a waterproof coating if used externally, and preferably with a good overhang to the roof to protect the walls, which would mean using more materials in the roof structure and for exterior painting (Minke, 2006, p. 93).

Table 4.12 Comparison of CO_2 emissions for the straw bale and timber frame rooms

Straw bale room	CO_2 coefficient	CO_2 Year 0 tonne	Replacement cycle	CO_2 Year 50 tonne	CO_2 Year 100 tonne
Foundations	268 kg/m^3	4.7	–	4.7	4.7
Straw bale walls (1967 kg)	−1.35 kg/kg[a]	−2.3	–	−2.3	−2.3
Cement, lime sand render 2.25 m^3 550 kg/m^3	180 kg/tonne[2b]	0.2	30 years	0.4	0.6
Wood framing					
Softwood	−567 kg/m^3	−0.1	–	−0.1	−0.1
Plywood	−619 kg/m[3c]	neg (−0.02)		neg	neg
Subtotal		2.5		2.7	2.9
Timber frame room					
Foundations	268 kg/m^3	0.8	–	0.8	0.8
Softwood framing (134 lengths 50 × 100)	−567 kg/m^3	−0.9	–	−0.9	−0.9
Insulation fiberglass	24.6 kg/m^3	0.2	–	0.2	0.2
Weatherboards 18 mm	−567 kg/m^3	−0.3	–	−0.3	−0.3
Plasterboard	404 kg/m^3	0.1	–	0.1	0.1
Exterior paint water-based 1.09 kg/L	5.062 kg/kg[c]	neg (0.02)	10 years	0.1	0.2
Interior paint water-based 1.09 kg/L	5.062 kg/kg[c]	neg (0.01)	10 years	neg (0.05)	0.1
Subtotal		**−0.1**		**−0.1**	**0.2**

[a]Sodagar et al. (2011).
[b]Schlegel and Shtiza (2014).
[c]Mithraratne et al. (2007).

Although lime produces CO_2 during its manufacture a lime render or mortar has the property of slowly absorbing CO_2 over time, in a process known as carbonation and mixing the lime with cement can aid this curing process (Arandigoyen and Alvarez, 2006). Schlegel and Shtiza (2014) suggest this can be measured as a negative CO_2 coefficient. However, this is to an extent arguing over the merits of a small part of the total global warming potential of the construction while ignoring the large part, which is the concrete in the foundations. What this series of exercises shows is that choosing a natural material—straw—does not automatically produce a sustainable building, and many other aspects have to be taken into account. One that has not yet been considered is size.

For a home office a floor area of 3.5×5 m (17.5 m^2) is quite large and one possible way of reducing environmental impact would be to make a smaller building. Using a floor area of 3×3.5 m (10.5 m^2) and the standard timber frame construction, the embodied energy is reduced on a pro rata basis and compared with the modeled operating energy in Table 4.13.

The percentage reduction from using straw bale in place of normal timber frame is compared with that of reducing the size of the room in Table 4.14. This is a 40% reduction in the floor area.

These tables suggest that making a building larger than strictly necessary has a much greater effect on the environmental impact than choosing straw bale over timber-framed walls. This again points to the problem of thinking about sustainable building materials without also accounting for the whole building. The smallest possible building will need less energy to run it and this, as Table 4.14 shows, has a significant effect on the overall environmental impact of the building.

Given the discussion of recycling of components it is also useful to see the effect of this in the lifecycle energy accounting. Table 4.15 sets out components that could be from recycled sources in the straw bale room.

Looking first at the components that cannot be from reclaimed sources, the plastic membranes are an obvious problem. Fiberglass insulation can be recycled into acoustic and thermal insulation boards (Hart, 2016) but this is a manufacturing process. In the United States it is possible to buy reclaimed and surplus expanded polystyrene at

Table 4.13 Comparison of timber frame and small timber frame rooms over 100 years

Timber-framed room					
EE Year 0 (GJ)	34.2	EE Year 50 (GJ)	45.0	EE Year 100 (GJ)	58.5
OE Year 0 (GJ)	2.0	OE Year 50 (GJ)	100.0	OE Year 100 (GJ)	200.0
Total Year 0 (GJ)	**36.2**	**Total Year 50 (GJ)**	**145.0**	**Total Year 100 (GJ)**	**258.5**
Small timber-framed room					
EE Year 0 (GJ)	20.5	EE Year 50 (GJ)	27.0	EE Year 100 (GJ)	35.1
OE Year 0 (GJ)	0.8	OE Year 50 (GJ)	40.0	OE Year 100 (GJ)	80.0
Total Year 0 (GJ)	**21.3**	**Total Year 50 (GJ)**	**67.0**	**Total Year 100 (GJ)**	**115.1**

Table 4.14 Comparison of using straw bale or building small

	EE + OE Year 0 (GJ)	EE + OE Year 50 (GJ)	EE + OE Year 100 (GJ)
Timber frame	36.2	145.0	258.5
Straw bale	37.6	126.1	218.4
% reduction	N/A	13.0	15.5
Small timber frame	21.3	67.0	115.1
% reduction	41.2	53.8	55.5

Table 4.15 Possible use of recycled components in straw bale and timber frame rooms

Component	Material	Use of reclaimed product
Walls	Straw bale	N/A
	Wood framing	Recycled timber and plywood
	Cement, lime, sand render to both sides of wall	
Foundations	Concrete	Recycled aggregate
Slab	Concrete	Recycled aggregate
	Hardcore	Recycled aggregate
	Under slab insulation	Recycled exp. polystyrene?
	DPM	N/A
	Hardcore with sand blinding	Recycled
Flat roof	Plasterboard ceiling	Recycled ply
	Timber structure	Recycled timber
	Insulation	N/A
	Decking	Recycled ply
	Butyl membrane	N/A
Window	Aluminum and timber Double-glazed argon-filled	Recycled but probably not the same specification
Door	Aluminum and timber Double-glazed argon-filled	Recycled but probably not the same specification

50%–75% savings on the price of the original (GreenInsulationGroup, 2011) and recycled polystyrene sheets can be bought from Poly Palace in Porirua, Wellington (Poly Palace, 2016). According to an Australian industry study the embodied energy of recycled aggregate is 21 MJ/t (Sustainable Aggregates South Australia, n.d.), compared with 40 MJ/t (Alcorn, 2003). The energy coefficient of the concrete used in the room foundations is 0.9 MJ/kg (2019 MJ/m^3) (Alcorn, 2003) or 900 MJ/t. Using recycled aggregate would reduce this to 881 MJ/t. It is assumed that reclaimed windows and doors of the same specification can be obtained over the life of the building. This gives the following reduced embodied energy values in Table 4.16.

Table 4.16 Effect of using recycled components

Element	EE coefficient	EE Year 0 (GJ)	Replacement cycle	EE Year 50 (GJ)	EE Year 100 (GJ)
Foundations	881 MJ/t	14.0	–	14.0	14.0
Slab[a]	881 MJ/t	3.5	–	3.5	3.5
Hardcore	21 MJ/t	neg (0.05)		neg	neg
Insulation reclaimed polystyrene	–	–	–	–	–
DPM (0.27 kg/m^2)	51 MJ/kg	0.4	–	0.4	0.4
Subtotal ground works		**17.9**		**17.9**	**17.9**
Straw bale walls	30.5 MJ/m^3	0.5	–	0.5	0.5
Cement, lime sand render 2.25 m^3 550 kg/m^3	1.09 MJ/kg	1.3	30 years	2.6	3.9
Wood framing Softwood Plywood		–	–	–	–
Subtotal walls		**1.8**		**3.1**	**4.4**
Roof timber		–	–	–	–
Ply decking		–	–	–	–
Insulation fiberglass	1026 MJ/m^3	1.8		1.8	1.8
Butyl membrane	149 MJ/m^2	4.0	25 years	8.0	16.0
Ceiling (five sheets)		–	–	–	–
Subtotal roof		**5.8**		**9.8**	**17.8**
Window		–	45 years	–	–
Door		–	45 years	–	–
Subtotal elements		**0.0**		**0.0**	**0.0**
Total EE		**25.5**		**30.8**	**40.1**

[a]The slight reduction disappears with rounding.

Table 4.17 puts together the embodied energy of the timber frame building, the same with its reclaimed and recycled components, and the operating energy. The latter has not changed as the assumption is the reclaimed and recycled components give the same performance.

As mentioned before, the impact of using recycled and reclaimed components reduces with the life of the building. This simulation also assumes all timber is

Table 4.17 Summary of lifecycle energy for timber frame room with and without recycled components

Timber-framed room					
EE Year 0 (GJ)	34.2	EE Year 50 (GJ)	45.0	EE Year 100 (GJ)	58.5
OE Year 0 (GJ)	2.0	OE Year 50 (GJ)	100.0	OE Year 100 (GJ)	200.0
Total Year 0 (GJ)	**36.2**	**Total Year 50 (GJ)**	**145.0**	**Total Year 100 (GJ)**	**258.5**
Timber-framed room with recycled components					
EE Year 0 (GJ)	25.5	EE Year 50 (GJ)	30.8	EE Year 100 (GJ)	40.1
OE Year 0 (GJ)	2.0	OE Year 50 (GJ)	100.0	OE Year 100 (GJ)	200.0
Total Year 0 (GJ)	**27.5**	**Total Year 50 (GJ)**	**130.8**	**Total Year 100 (GJ)**	**240.1**
% reduction	24.0		9.8		7.1

recycled, which in current circumstances could be difficult to achieve as it may not be available exactly when needed. Had the reclaimed door and window been of a lesser specification (single-glazed timber) this would have increased the operating energy. All this suggests that simply using a few recycled components is not enough to achieve a sustainable building. The aim must first be to design to reduce the energy in use and then worry about reducing the embodied energy. The longer the building lasts the less important a large part of the embodied energy becomes, so designing and building for long life and low maintenance is also vital in making sustainable buildings.

4.3 Caveat: building users

There is one issue not yet mentioned, which is the much explored field of the role of users of sustainable and low-energy buildings. Consideration of this began in the 1970s with the discussion of Alternative Technology, which was essentially technology over which the user had control (Dickson, 1975). Research has since shifted to the social side of building energy performance (Lutzenhiser, 1993; Rohracher, 2003; Wilhite, 2008; Karvonen, 2013). For instance, Wilhite (2008) argues that "...efficient technologies may well increase the efficiency of energy throughput but that promised reductions in energy demand seldom pan out." Often the energy savings from living in a zero or low-energy building are recouped in heating to higher temperatures and having improved comfort, especially in lower-income households, or the financial savings are used to purchase items like an overseas flight or a move to a larger energy-efficient house (Bourrelle, 2014). In an Austrian study of seven buildings retrofitted to meet the Passiv Haus standard measured heating demand was larger than predicted in six of the buildings "...by a factor of 4–6" (Housez et al., 2014). Obviously this is something over which the designers of sustainable buildings have no control. However, it should be borne in mind when worrying over small differences in the energy LCA performance between, say, the straw bale and timber-framed examples in this section. It also seems that involvement in the process has an influence on the use of sustainable technologies (Ornetzeder and Rohracher, 2006; Bahho et al, 2016), which is coming back to the idea of Alternative Technology and its focus on the "hands-on" approach through empowering people to be more self-sufficient for basic commodities like food and energy. This suggests that building or assisting in procuring and building a low-energy house or workplace will affect how users feel about the building in use. This leads to a whole new field of choosing materials that are appropriate for making self-build sustainable buildings, again suggesting simple timber frame and straw bale might be better than concrete tilt-slab and cavity brickwork, but this is an area for further research.

4.4 Choosing healthy and low-impact materials

The aim of Part II, made up of Chapter 3, and this chapter, was to investigate what materials might be best when it comes to making sustainable buildings from the viewpoint of those engaged in design and specification. As shown in Part I, to be truly sustainable the building needs to acknowledge and work with the eco-system services in which it is sited, as happened in the vernacular traditions of the past. Part II has accepted that what needs to happen now is that all decisions are directed at making the most sustainable buildings possible, both new build and refurbishment. The results are not unsurprising. The goal should be to ask first whether the building is needed, then how small it can be and still function well, both in the present and future. The next goal is to make it as zero energy in operation as possible, given that the users will always be an unknown quantity. Following this comes choosing materials for long life, low maintenance, and low embodied energy. However, as the section on lifecycle energy analysis shows this approach needs to be thought of in a holistic way, so that the impact of choices can be explored. Wherever possible the lowest embodied energy materials should be specified, and as shown in Part III, these plant-based materials are also good for the health of the occupants.

To paraphrase Pollan's (2008) words about food, the simple advice becomes:

Build only what is needed, keep it small, make it zero energy in operation, use mostly plants.

References

Adalberth, K., 1994. The energy demand from cradle to grave for three single family houses in Sweden, Case study buildings paper 4. In: Proceedings of First International Conference on Buildings and the Environment, CIB task Group 8, BRE Watford.

Alcorn, A., 1996. Embodied Energy Coefficients of Building Materials. Victoria University of Wellington, Centre for Building Performance Research, Wellington.

Alcorn, A., 2003. *Embodied Energy and CO_2 Coefficients of NZ Building Materials.* Victoria University of Wellington, Centre for Building Performance Research, Wellington.

Aquino, J.T. (Ed.), 1995. *Waste Age/Recycling Times' Recycling Handbook.* Lewis Publishers, Florida.

Arandigoyen, M., Alvarez, J.I., 2006. Pore structure and carbonation in blended lime-cement pastes. Mater. Constr. 56 (282), 17–30.

Asif, M., Davidson, A., Muneer, T., 2001. Embodied energy analysis of aluminium-clad windows. Build. Serv. Eng. Res. Technol. 22 (3), 195–199.

Bahho, M., Vale, B., Milfont, T., 2016. Buildings that teach: a strategy for sustainable design. In Eco-Architecture 2016. Alicante, 13–15 July 2016, 358–369.

Bassett, M., Stoeklein, A., 1998. A new thermal design guide for New Zealand houses. In: Proceedings IPENZ Conference, Auckland. Available from: http://www.branz.co.nz/cms_show_download.php?id=f0397fd3e8741535f102469142d5d53fbb34f40c (accessed 15.07.16.).

Berge, B., 2000. The Ecology of Building Materials. Architectural Press, Oxford.

Blengini, G.A., 2009. Life cycle of buildings, demolition and recycling potential: a case study in Turin, Italy. Build. Environ. 44, 319–330.

Bourrelle, J.S., 2014. Zero energy buildings and the rebound effect: a solution to the paradox of energy efficiency? Energy Build. 84, 633–640.

Chang, Y., Huang, Z., Ries, R.J., Masanet, E., 2016. The embodied air pollutant emissions and water footprints of buildings in China: a quantification using disaggregated input-output life cycle inventory model. J. Clean. Prod. 113, 274–284.

Concrete Services, 2015. Calculation of the amount of cement, sand and gravel for making concrete. Available from: http://www.zhitov.ru/en/beton/ (accessed 19.07.16.).

Dickson, D., 1975. The Politics of Alternative Technology. Universe Books, New York, NY.

Downton, P., 2013. Your home: straw bale. Available from: http://www.yourhome.gov.au/materials/straw-bale (accessed 12.08.16.).

Eaton, K.J., Amato, A., 1998. A Comparative Environmental Life Cycle Assessment of Modern Office Buildings. The Steel Construction Institute, Ascot.

Franklin Associates, 1991. Comparative energy evaluation of plastic products and their alternatives for the building and construction and transportation industries. Available from: https://greenbuildingsolutions.org/Main-Menu/Resources/Comparative-Energy-Evaluation-of-Plastic-Products-1991.pdf (accessed 6.07.16.).

Fung, J., 2010. Life cycle costs and analysis of roof cladding systems, Study report 248, Wellington, BRANZ.

GreenInsulationGroup, 2011. Supplies reclaimed, recycled, and surplus rigid insulation of all types. Available from: http://www.greeninsulationgroup.com/ (accessed 20.07.16.).

Hammond, G.P., Jones, C.I., 2008. Embodied energy and carbon in construction materials. In: Proceedings of the Institution of Civil Engineers, Energy 161, 87–98. Available from: http://opus.bath.ac.uk/12382/1/Hammond_%26_Jones_Embodied_energy_%26_carbon_Proc_ICE-Energy_2008_161(2)_87-98.pdf (accessed 16.07.16.).

Harris, J., 2007. Moving Rooms: The Trade in Architectural Salvages. Yale University Press, New Haven.

Hart, G.H., 2016. Recycling fiberglass insulation into commercial board products. Available from: http://www.insulation.org/articles/article.cfm?id=IO010701 (accessed 19.07.16.).

Hinkle Charitable Foundation, n.d. How do we contribute individually to global warming? Available from: http://www.thehcf.org/emaila5.html (accessed 2.09.16.).

Hojgrova, K., 2012. Morrow Royal Pavilion, Las Vegas, United States. Available from: http://www.worldarchitecturenews.com/project/2012/21305/realm-of-design/morrow-royal-pavilion-in-las-vegas.html (accessed 15.07.16.).
Housez, P.P., Pont, U., Mahdavi, A., 2014. A comparison of projected and actual energy performance of buildings after thermal retrofit measures. J. Build. Phys. 38 (2), 138–155.
IPCC, 2013. Working group 1: the scientific basis. Available from: <https://www.ipcc.ch/ipccreports/tar/wg1/137.htm> (accessed 9.01.17.).
Janez, A., Llera, E., Aranda, A., Scarpellini, S., Zabalza, I., 2013. Use of LCA as a tool for building ecodesign. A case study of a low energy building in Spain. Energies. 6 (8), 3901–3921.
Kajaste, R., Hurme, M., 2016. Cement industry greenhouse emissions—management options and abatement cost. J. Clean. Prod. 112 (5), 4041–4052.
Karvonen, A., 2013. Towards systematic domestic retrofit: a social practices approach. Build. Res. Inf. 41 (5), 563–574.
Khajehzadeh, I., Vale, B., 2015. Life cycle energy and large and small housing in New Zealand. In: Proceedings 49th International Conference of the Architectural Science Association, December 2015, Melbourne.
Kluender, E.J., 2013. Quantification of water footprint: calculating the amount of water needed to produce steel. J. Purdue Undergrad. Res. 3, 50–57. Available from: http://docs.lib.purdue.edu/cgi/viewcontent.cgi?article=1074&context=jpur (accessed 15.07.16.).
Landmark Trust, 2016. Shelwick court. Available from: http://www.landmarktrust.org.uk/search-and-book/properties/shelwick-court-11906 (accessed 20.07.16.).
Leah, A., 2015. Lifestyle and Embodied Energy: A Proposed Hybrid Analysis Method for Housing, PhD thesis. Victoria University of Wellington, Wellington.
Lutzenhiser, L., 1993. Social and behavioural aspects of energy use. Ann. Rev. Energy Environ. 181 (1), 247–289.
McCormack, M., Treloar, G.J., Palmowski, L., Crawford, R., 2007. Modelling direct and indirect water requirements of construction. Build. Res. Inf. 35 (2), 156–162.
Minke, G., 2006. Building with Earth: Design and Technology of a Sustainable Architecture. Birkhäuser, Basel.
Mithraratne, N., Vale, B., Vale, R., 2007. Sustainable Living: The Role of Whole Life Costs and Values. Butterworth-Heinemann, Oxford.
National Slag Association, 2013. Common uses for slag. Available from: http://www.nationalslag.org/common-uses-slag (accessed 19.07.16.).
Ochsendorf, J., Norford, L.K., Brown, D., Durschlag, H., Hsu, S.L., Love, A., Santero, N., Swei, O., Webb, A., Wildnauer, M., 2011. Methods, impacts, and opportunities in the concrete building life cycle, research report R11-01. MIT. Available from: http://www.cement.ca/images/stories/MIT_LifeCycleAssessment_on_Buildings.pdf (accessed 19.07.16.).
Ornetzeder, M., Rohracher, H., 2006. User-led innovations and participation processes: lessons from sustainable energy technologies. Energy Policy. 34 (2), 138–150.
Pawley, M., 1975. Garbage Housing. Architectural Press, London.
Pollan, M., 2008. *In Defense of Food: an Eater's Manifesto*. Penguin Press, New York, NY.
Poly Palace, 2016. Polystyrene for a Green World. Available from: http://www.polypalace.com/ (accessed 14.09.16.).
Rohracher, H., 2003. The role of users in the social shaping of environmental technologies. Innovat. European J. Soc. Sci. Res. 16 (2), 177–192.
Schlegel, T., Shtiza, A., 2014. Environmental footprint study of mortars, renders and plasters formulations with no, low or high hydrated lime content. In: 9th International Masonry Conference at Guimaraes, Portugal 7-9 July. www.eula.eu/file/516/download?token = 0GhdTtCy (accessed 20.07.16.).

Shen, S., 2012. The Environmental Impact of Expositions: A Study of Some Contributing Factors, PhD Thesis. Victoria University of Wellington.

Sodagar, B., Deepak, R., Jones, B., Wihan, J., Fieldson, R., 2011. The carbon-reduction potential of straw-bale housing. Build. Res. Inf. 39 (1), 51–65.

Song, J., 2016. The phenomena of China's disposable buildings. Epoch Times. Available from: http://www.theepochtimes.com/n3/1932144-the-phenomena-of-chinas-disposable-buildings/ (accessed 18.09.16.).

Storey, J., Gjerde, M., Charleson, C., Pedersen, M., 2003. The State of Deconstruction in New Zealand. Centre for Building Performance Research, Victoria University, Wellington.

Sustainable Aggregates South Australia, n.d. Recycled aggregates bring carbon reduction benefits. Available from: http://www.sustainableaggregates.com.au/docs/SASA_Env_Info_Sheet.pdf (accessed 20.07.16.).

Vale, B., Vale, R., 2000. *The New Autonomous House: Design and Planning for Sustainability*. Thames and Hudson, London.

Wackernagel, M., Rees, W., 1996. Our Ecological Footprint. New Society Publishers, Gabriola Island.

West, J., Atkinson, C., Howard, N., 1994. Embodied energy and carbon dioxide emissions for building materials. In Proceedings of First International Conference on Buildings and the Environment, CIB task Group 8, BRE Watford.

Wilhite, H., 2008. New thinking on the agentive relationship between end-use technologies and energy-using practices. Energy Efficiency. 1 (2), 121–130.

Part III

Indoor Toxicity from Building Materials

A lack of recognition of potential health risks from building materials

5

Emina Kristina Petrović
Victoria University of Wellington, Wellington, New Zealand

5.1 Introduction

This part of the book focuses on indoor toxicity from building materials. Although use of harmful substances in buildings, and in everyday life, is not new, in the last several decades a number of changes has intensified the need to consider indoor toxicity more carefully. In this chapter these changes are discussed under the subheadings of six distinct challenges, jointly contributing to a complex *problem*. This *problem* includes low recognition of the indoor air toxicity, increases in use of synthetic chemicals, issues of increased proportions of time spent indoors, problems with obtaining sufficient evidence to trigger change, delays in implementation, and the reliance of the construction literature on regulations. This chapter proposes that because of these challenges, issues with indoor air toxicity tend to be poorly recognized and the changes can be slow. Yet, it is essential to push against such trends if there is a desire to prevent buildings from adversely contributing to human health.

Part III proposes that as part of the sustainable consideration of materials it is essential to assess the totality of health impacts that these materials can pose on human health. What is the point in saving energy, and protecting the natural ecosystems, if the same choice happens to be damaging the health of the building users? This is especially relevant as many of the choices which are good for human health are also good for the natural ecosystems, by the simple extension that humans belong to the natural ecosystems themselves. Therefore, by carefully developing approaches which prioritize what is simultaneously good for the humans and the ecosystems, it should be possible to develop thinking able to elevate the real sustainability of material choices, and possibly also achieve energy savings.

Later sections of this chapter propose a new approach to this problem, by evaluating the patterns in existing recognition of issues associated with the toxicity of building materials. A framework of three stages of recognition of associated health risks is introduced, setting the scene for the subsequent three chapters.

Materials for a Healthy, Ecological and Sustainable Built Environment: Principles for Evaluation.
DOI: http://dx.doi.org/10.1016/B978-0-08-100707-5.00005-8

5.2 The *problem*

5.2.1 Challenge 1: low recognition of indoor air toxicity

Discussion of the *problem* opens with the issues with low recognition of indoor air toxicity, as the most fundamental obstacle for more intense discussion in this area. In 1993 and 2000, two analyses of the actual causes of death in the US placed toxic agents as high as the fifth place (after smoking, poor diet and physical inactivity, alcohol consumption, and microbial agents; McGinnis and Foege, 1993; Mokdad et al., 2004). Both assessments considered all toxic exposures and estimated that 2%–3.5% of total mortality was associated with toxic agents, excluding tobacco smoke exposure (McGinnis and Foege, 1993; Mokdad et al., 2004). Other assessments suggest that the health impact related to indoor air pollution costs an estimated US$400 billion per year in the United States (Birkeland, 2008, p. 105). Much of these adverse impacts relate to complex reactions of human bodies to indoor toxicants. Globally, societies are witnessing an increase in the prevalence of modern diseases such as cancer, multiple allergy syndrome, allergies, asthma, autism, and attention deficit disorder (Thompson, 2004; Armstrong et al., 2007). Unfortunately, the totality of the implications of indoor toxicants are still unclear, which makes this a less obvious problem.

There is a relatively modest recognition of the *problem* because although clearly tangible, the health impacts seem modest against other health priorities. For example, the World Health Organization (WHO) assessment of the total worldwide impact of indoor air quality on human health revealed that indoor air pollution can be seen as responsible for 2.7% of the global burden of disease (WHO, 2002). Yet, indoor air pollution from the combustion of solid fuels (coal, firewood) was noted as the main component of this impact and it accounted for 1.6 million deaths and 39 million DALYs (disability-adjusted life years, which considers years of life lost due to both disability and death) (WHO, 2002). Pollution from solid fuels is recognized as one of the 10 most important worldwide threats to public health (WHO, 2007). This is more the problem in less developed countries, however, even in Europe as many as 2 million DALYs per year, or 2 million years of healthy life are lost annually due to the total calculated burden of diseases attributed to indoor air quality (Jantunen et al., 2011). Similar reports place risks from building materials and the associated indoor volatile organic compounds (VOCs), only after a number of other important factors, such as combustion particles, either from indoor solid fuel combustion or tobacco smoke, building dampness, and outdoor air pollution (Jantunen et al., 2011). Yet, respiratory issues are listed as three out of the top six causes of death in high-income countries, and many cancers are in the top 20 causes of death (WHO, 2016). It is this balance between the issues with toxicity indoors being a clearly tangible and real problem, yet not as acute as some other health concerns, which makes the problem challenging to fully grasp for many organizations and individuals. Recent research in environmental psychology suggests that such confusing and conflicting information can result in general passivity on the issue (Milfont, 2010; Gifford, 2011).

Concurrently, increases in related issues of asthma and allergic diseases have been clearly evident worldwide. The WHO estimates that 15 million DALYs are lost annually due to asthma, representing 1% of the total global disease burden and leading to 250,000 annual deaths worldwide (Balachandran et al., 2010). Similarly, it was estimated that by 1990, 20% of the population suffered from an allergic disease (Spengler and Chen, 2000). Interestingly, the first documented case of hay fever was recorded in the mid-19th century by a British physician who had to collect data for another 10 years before he found seven additional cases (Spengler and Chen, 2000). Thus, hay fever was first recognized as a health condition around the time of industrialization in the country where this process was most accelerated. In 2004, Sundell reported on multidisciplinary reviews of all scientific literature conducted in the Nordic countries and Europe, which established that the increasing incidence of asthma and allergy throughout the developed world in the past 30 years was probably due to environmental changes, as the period was not long enough for change through genetic evolution (Sundell, 2004). Sundell (2004) also pointed out that the then available scientific means could still not provide sufficient explanation for this change, which they interpreted as a general decrease in health.

Some of the worst prognoses expect that in 50 years everybody will have some form of cancer (Saunders, 2002), despite more than 90% of cancers probably being avoidable (Curwell and March, 2002). People are repeatedly informed of the wisdom of changing different facets of modern lifestyles to improve the chances of a healthy life, and some consider these educational efforts understate the role of the indoor environment (Armstrong et al., 2007). In 2002, Saunders illustrated these discouraging facts with the image of contemporary humanity acting like a frog in this story:

> *A frog jumps into a pot of water which is gradually being heated. As the water gets warmer, the frog adjusts its body temperature and continues to adjust to the increasing water temperature until, ultimately, the frog gets boiled alive.*
>
> *(Saunders, 2002, p. 5).*

Thus, in terms of impact on health, it can be said that indoor air quality presents a constant background "noise" contributing to total burden, although with relatively moderate currently quantifiable direct impact on mortality. From the perspective of the construction industry however, this is significant as it presents one of the rare areas where our choices can have a direct positive or negative impact on the health of buildings users. While it could take medical professionals longer before their focus can turn to health initiatives specifically designed to alleviate indoor toxicity problems, it is in our professional mandate to start making these changes without waiting for such interventions and especially before being forced by law (see Chapter 17). With the current changes towards a more sustainable construction, it is very important to integrate better consideration of the health of inhabitants as part of these changes. While we are developing processes for assessing the impact construction has on the natural environment, it is also paramount to consider the implications on human health as much if not more. Only health-neutral or even health-improving construction activities will be truly sustainable both for human life and the natural environment.

5.2.2 Challenge 2: rise of synthetic chemicals

The issue of indoor air toxicity is reasonably new compared to a long history of human involvement with building, thus there could be inherited preconceptions within the profession based on past situations which underestimate the relevance of these issues. However, because of accelerated changes in the production of building materials since WWII, it is essential to reconsider such views (Saunders, 2002). The total production of synthetic chemicals, made artificially by chemical reaction, has increased more than 10-fold: from less than 10 million metric tons in 1945 to over 110 million tons by the early 2000s (Baker-Laporte et al., 2008, p. 272). This trend has impacted many aspects of human life and the natural environment, one strand of which is chemicals used in building and furnishing materials, where synthetic chemicals can be used on their own or in various mixtures with other components that could be more natural in origin. Natural substances, such as lead and asbestos, can also present health risks. However, the main increases have been in the development of petroleum-based synthetic chemicals. The accelerated increase in total exposure has resulted in previously unprecedented problems.

The majority of products that present risk to human health have been developed relatively recently, particularly since the second half of the 19th century, and with great diversification since WWII. Around 2000, it was estimated there were more than four million registered human-made chemicals in the world with 60,000–80,000 in common use, and 1000 being added every year (Pearson, 1998, p. 61; Saunders, 2002, p. 9; Thompson, 2004, pp. 14–15). It is estimated that fewer than 2% of these synthetic chemicals have been tested for their effects on human health and more than 70% have not been tested at all (Snyder in Saunders, 2002, p. 9). It is also suggested that insufficient information exists for health assessments of 95% of chemicals used in construction products (Pacheco-Torgal, 2012). However, development of synthetic chemicals is still accelerating, with the suggestion in 2011 that "more new materials have been developed in the last 20 years than in the rest of history combined," and these newly developed materials present about half of the materials in current use (Schörpfer, 2011, p. 19; Brownell, 2006, p 6). In addition, there is a lack of information and research on the additive and synergistic effects of combinations of chemicals (Armstrong et al., 2007, p. 61).

As a result, many new chemicals appeared in buildings after WWII and together with the tightening of building envelope in response to energy conservation needs, poorer health outcomes and indoor air quality problems were soon observed (Andersen and Gyntelberg, 2011; IBN, 2011). In the early 1980s, this culminated with the recognition of sick building syndrome (Andersen and

Gyntelberg, 2011; Rostron, 1997). Although building and furnishing materials are not unique in their negative contribution to indoor air quality and health, they are recognized as important contributors in this area (Andersen and Gyntelberg, 2011; Clausen et al., 2011). Since the 1970s, the increasing air-tightness of buildings, leading to insufficient air exchange, has led to the build-up of indoor toxicants (Clausen et al., 2011, p. 221; Sundell et al., 2011). Consequently, research on indoor air quality and the impact of building and furnishing materials on human health has developed steadily since WWII, being especially proactive in the last 20–30 years (Clausen et al., 2011). Unfortunately, research is not considered to be keeping up with the need for new knowledge in this area (Clausen et al., 2011, pp. 225–226), with concerns that control of conditions in domestic spaces is even harder to achieve (Howden-Chapman and Carroll, 2004).

5.2.3 Challenge 3: indoors society

Indoor air quality is additionally important because of the increased recognition that most people spend between 80% and 93% of their time indoors (Liu and Little, 2012; Guieysse et al., 2008). In 2011, Delgado-Saborit et al. (2011) studied the time 100 British subjects spent in different environments and confirmed that 87%–91% was indoors. Another study claimed nearly 25% of US residents are affected by poor indoor air quality either at work or home (Guieysse et al., 2008). This differentiation between workplace and home is significant, because Delgado-Saborit et al. (2011) found on average for their participants that 62% of their time indoors was at home, 16% at work, 12% in other indoor environments, and 5%–7% commuting, with the average time outdoors ranging between 2% and 5%. These figures translate to an average of 14–15 hours/day at home and only 30–75 minutes/day outdoors. These proportions signal that indoor air quality at home could be an important factor in people's health.

Some research has focused on indoor air in relation to children's health (Phipps, 2008; Howden-Chapman et al., 2008; Thompson, 2004). Toxicology studies recognize that generally the body's ability to biotransform and excrete toxins is less efficient at the extremes of life, as for young children such processes are not fully developed until 6–12 months of age, and sometimes as late as 6 years of age (Philp, 2001). Children also have higher metabolic rates, which increases their exposure (Philp, 2001), making them more vulnerable than adults. Furthermore, a number of toxic substances indoors are either found in household dust (Chapter 6 and Chapter 7: How substances get regulated against in the building industry: formaldehyde, phthalate plasticizers in polyvinyl chloride/vinyl) or tend to linger close to the floor (see Chapter 7), which can increase exposure for children. In addition, children, especially the very young, spend larger proportions of their time at home indoors (Phipps, 2008; Thompson, 2004), increasing the relevance of indoor air quality in homes.

Another significant development since the recognition of sick building syndrome has been investigations of ventilation rates, which have established a need for an increase first to 10 l/s per person and subsequently to a recommended 20 l/s per person (Wargocki et al., 2002). In 2002, a multidisciplinary overview of the literature established that outdoor air supply rates lower than 25 l/s per person in commercial and institutional buildings correlated with an increased risk of sick building syndrome and more short-term sick leave and reduced productivity (Wargocki et al., 2002; Sundell, 2004). It has also been suggested that improving the indoor environmental quality in all US office buildings could prevent sick building syndrome and reduce absences for millions of US office workers, while also improving work performance, with a potential annual economic benefit of $17–26 billion (Fisk et al., 2011). As a result, there has been much development in ventilation systems and better understanding of indoor air quality needs.

The workplace has been the primary focus of developments in ventilation and occupational health (ASHRAE Handbook, 2009). However, more recent studies are indicating indoor air quality in homes is generally poorer than in offices and other public buildings (Nielsen et al., 2013; Wolkoff and Nielsen, 2010). The focus on improving ventilation and air filtration systems still relies on removing the indoor air pollution once it happens, rather than removing the pollution sources. Spaces with poorer ventilation protocols, like dwellings, have not been addressed to the same level, despite the fact as stated above that people spend most of their time at home. Potentially, high reliance on mechanical ventilation is not a good approach for domestic environments, as it is impossible to regulate and insist on regular use of such systems. For many dwellings installation of such systems could be either technically difficult or financially impossible. Therefore, elimination or encapsulation of the source could prove a more effective strategy in domestic spaces.

5.2.4 Challenge 4: insufficient evidence to trigger change

One of the most significant challenges in this area is the reliance on regulations which tend to require solid evidence of adverse effects before any changes are made, and even then the changes can be rather slow. There is much debate on how to define what is unhealthy for humans. Richard Philp in his book *Ecosystems and Human Health: toxicology and environmental hazards* (2001) explains that toxicological studies have developed research methods to establish thresholds for acute toxicity of many substances. However, it is much harder to evaluate the impact from low-level exposure, especially if these are chronic exposures or ones that have a cumulative effect. Additionally, there is much development in understanding reproductive and geno-toxicity, when a substance is more damaging to reproductive organs or alters the genetic makeup of an unborn child. One of the factors that complicates many studies in this area is the fact that the human body has the ability to self-heal and contemporary medicine is still unable to explain why in some people or situations these self-healing mechanisms seem to be active or inactive (Philp, 2001). The discussion in the whole of Part III should be considered against this backdrop.

While such challenges have been known for a long time now, in recent years better understanding of more sophisticated underlying mechanisms in health issues related with inhalation have emerged providing some fairly new approaches to the problem. Studies on the relationship between genes and adverse reactions of inhaled pollution suggest that there is more than one genetic marker which makes an adverse reaction more likely, although the cataloguing of this is still in the early stages (Gaffney and Christiani, 2015). Similarly, research into vaccines against well-known cancers associated with inhaled pollution (such as asbestos) is still to arrive at any tangible conclusions (Tan et al., 2014). The researchers in such areas warn of the possibilities of genetic discrimination when employment in high-risk industries in concerned, which could lead to social injustice (Gaffney and Christiani, 2015). Thus, even the researchers involved with these efforts recommend the focus should still remain on exposure reduction, as increases in exposure still hold a reliably lineal relationship with increases in adverse reactions.

This inability of current knowledge to explain fully the health mechanisms underlying human health reactions to exposures to many toxicants, points towards an alternative position: a proportion of authors suggest "prudent avoidance" or the precautionary principle (Pearson, 1998; Saunders, 2002; Thompson, 2004). The precautionary principle was established as Principle 15 of the 1992 Rio Declaration on Environment and Development and states "where there are threats of serious or irreversible damage, lack of full scientific certainty shall not be used as a reason for postponing cost-effective measures to prevent environmental degradation" (Philp, 2001). Applied to domestic indoor spaces, the precautionary principle implies avoidance of building and furnishing materials suspected to be harmful to human health without waiting for proof of their harmfulness (Pearson, 1998; Saunders, 2002; Thompson, 2004).

In a sense, the healthiest approach would be to use only substances that are proven to be harmless. This approach would echo well with the definition of health by the WHO as the enjoyment of the highest attainable "state of complete physical, mental and social well-being and not merely the absence of disease or infirmity" (WHO, 2014). A similar approach can be observed in some newer legislation which defines harmlessness as "proven to be safe" rather than "not proven to be harmful." The New Zealand Psychoactive Substances Bill, which came into action in August 2013, stipulates that no substances are to be sold under this Bill until proven by the manufacturers to be safe, for example.

Advancements in new, human-made chemicals and materials are one outcome of the same processes that have produced better testing abilities, better understanding of possible risks, better medical ability to help people cope with and recover from different conditions, and generally much better regulatory controls, which at least theoretically aspire to protect society at large. The problem with contemporary regulations and legislation is that most are based on proof or solid scientific evidence, and until this exists it is often hard to make changes, therefore an alternative way of thinking about these issues is needed, and Part III of this book proposes one such approach.

5.2.5 Challenge 5: delays in implementation

An additional problem is that even when the need for change is clear, implementation could be quite slow. For example, although the late 1970s or early 1980s are generally seen as the time when lead was banned, its use continued past this point, and this played out differently in different industries. Because lead was historically used in water supply pipes, the regulations of the threshold for lead in drinking water were only gradually reduced, from 300 μg/L in the 1970s, to 50 μg/L in the 1980s, 25 μg/L in the 1990s, and just 10 μg/L in 2013 (Pacheco-Torgal, 2012). This gradual and consecutive set of reductions reflected the high cost of total replacement of lead pipework, which was estimated to cost in Europe around 200,000 million euros (Pacheco-Torgal, 2012). Similarly, only in 2015 was lead completely phased out as an additive to polyvinyl chloride (PVC) (Zhang et al., 2015). Similar delays are occurring in many similarly needed changes.

Part of the issue influencing delays is the great financial implications such changes would have for certain industries. A good example of this is the two articles on vinyl published in the same issue of the *Indoor Air* journal in 2013. Carlstedt et al. (2013) asserted that the use of soft PVC as flooring material may increase the human uptake of phthalates in infants. Published in the same issue was an article by Blakey et al. (2013) (all authors are employed in the vinyl industry) which challenged the result based on a small number of procedural imperfections this research had. Because any early work is likely to suffer from procedural imperfections, this type of swift rebuttal financially supported by the industry in question clearly presents a tangible challenge to knowledge development. Similar discussions between researchers financed by the asbestos industry and those asserting the need to recognize the health issues with the use of the material are still facing the negotiations of reparation costs for past asbestos workers (Landrigan, 2016). The costs and financial loss appear to figure strongly in resistance to and delays in implementation of change.

Finally, many existing changes rely on regulations to influence removal from the market of products containing harmful substances. The assumption appears to be that through such intervention the change will filter down to all situations. The experiences with lead and asbestos indicate that such an approach deals poorly with existing or built-in materials.

5.2.6 Challenge 6: issues in construction literature

The lack of clarity and conclusions, discussed above, puts the construction literature into an extremely difficult position. The construction industry is regulated by a number of region-specific professional standards and in many ways most construction-related activities could be seen as either related to implementation of such standards or in some cases their development. This makes it especially difficult to integrate suspicions which have not been regulated as yet. Thus, not surprisingly, construction industry literature tends to consider indoor toxicity by close following of the local regulations at the time. The problems with this approach are obvious, as the delays are magnified.

In 1986 and 2002 two different editions of *Hazardous Building Materials* were published (Curwell and March, 1986; Curwell et al., 2002). Although the two editions differed significantly, reflecting the changes in both regulations and materials production over the period, they jointly asserted that "[e]pidemiological research must of necessity lag behind material developments which inevitably means that the population's health may be at risk in the intervening period" (Burgess et al., 1986, p. 3; Curwell and March, 2002, p. 1). This lag is a perpetual feature of developments in the construction industry, and the two editions of *Hazardous Building Materials* clearly captured the accelerated recognition of the issues associated with lead and asbestos. The 1986 edition operated within the regulatory framework of limited use of asbestos in construction and lead in paint, focusing significant effort on education about these and the need to seek for alternatives, while in the 2002 edition these concerns were only dealt with in terms of past alterations to older houses. Nevertheless, the new edition remarked "we have moved only a short distance since 1985" (Curwell and March, 2002). This perception of the slow accumulation of knowledge in this area seems reflective of the time lag between early indications of health risks with some products and the regulatory elimination of a product.

The same general features can be observed in some of the most comprehensive works on the subject. Published in the last few years are Fernando Pacheco-Torgal, Said Jalali, and Aleksandra Fucic (eds.), *Toxicity of building materials* (2012), Osman Attmann, *Green architecture: advanced technologies and materials* (2010), and John Fernandez, *Material architecture: emergent materials for innovative buildings and ecological construction* (2006). Such works are generally based on the recent research, but are also reflective of the regulative recommendations at the time of writing. While this can be seen as unavoidable, it tends to lead to a relatively conservative representation of the problem: only issues that are almost impossible to disprove are discussed.

The key problem here is that the subject matter is evolving fairly rapidly, making it hard for sources of this kind to keep up. However, it should be acknowledged that this area of study has been experiencing growth for many years now. In the 1950s and 1960s, such works operated as a new field in many countries and often faced not only limitations in existing knowledge, but also technical limitations, such as multiple regression analysis not being available for large data sets due to computational limitations, and no clear publishing opportunities (Andersen and Gyntelberg, 2011). With the establishment of the *Indoor Air* journal in 1991, a focal point for these efforts was established and in the first 20 years of its existence the journal published 843 original research or review articles (Nazaroff, 2011). These works are generally multidisciplinary and scientific in nature, and often face the challenge of providing a sufficiently scientific exploration of relatively uncharted findings. Nevertheless, some specific subdisciplines have successfully developed. For example, studies of indoor air chemistry have experienced growth at an exponential rate between 1991 and 2007, as can be seen by the number of published papers in this area, leading to the conclusion that it can be seen as a subdiscipline in its own right (Weschler, 2011). Nevertheless, there is still much need for more research on indoor air quality.

Thus, it can be concluded that although great knowledge has been developed, and some significant changes are in the pipeline, at this point the architectural or building professions generally rely on sources which reflect the regulations at the time when making decisions about material specifications. Consequently, it is difficult to anticipate likely future changes and to work to standards higher than those that are deemed acceptable by current regulations. This problem becomes even more significant when paired with subsequent regulative changes that are never retrospective as well as the frequent changes manufacturers make to individual products.

Because of these characteristics, Part III proposes a novel approach, focused on mapping out common patterns and trends, which makes it easier to anticipate the likely future progression in recognition of risks associated with certain materials.

5.3 Beyond the challenges: three stages of recognition of health risks

The ability to anticipate likely future changes is especially important in contemporary construction which is trying to respond to calls for more sustainable practices. This process itself is a change which could facilitate uptake of any number of other desirable changes, but also because much of the work in the area of sustainability tends to focus on proposing platforms of principles, such an approach could play an important role in improving awareness and practices when it comes to health issues associated with building and furnishing materials. Thus, it should not be of surprise that many sources on sustainability also briefly explain indoor air quality issues and at least provide some examples of problem substances and available alternatives. However, this is not the rule. This section develops upon the existing literature and proposes a new approach to anticipating the upcoming concerns with indoor air toxicity.

The core research question of this section of the book is whether any clear patterns can be observed in the way knowledge about risk substances is integrated into social practices. Of special interest is whether there are any significant improvements over time, and if so, what form these take.

Rather than providing a detailed overview of all existing knowledge, this part uses the three key stages of progression in the recognition of risks posed by certain substances, devised to organize available information into a clear system for its interpretation, and to illustrate key patterns in the development and assimilation of new scientific knowledge. Table 5.1 summarizes the key features of each stage. Stage 1 records the early recognition of risks associated with a certain substance, Stage 2 reflects the efforts to regulate its use, often leading to a progressive decrease in allowed levels, and finally Stage 3 records the complete ban of the substance.

Table 5.1 The three stages of recognition of health risks and summary of their key features

Stage 1: Recognition of risks	Stage 2: Regulation development	Stage 3: Elimination
• Indications of adverse health effects—absence of complete assessments; • The most obvious health impacts studied—1st generation risks; • Shorter-term observations, e.g., studies that can be done in 1–2 years; • Animal studies more common than human subjects—problems of comparability with human bodies make conclusions difficult (especially for cognitive processes/behaviors); • No proper longitudinal and prospective studies; and • Often assessments of high level exposures, less work on long-term low-level exposures	• Formalized acknowledgement that certain levels of a particular substance are harmful; • Introduction of regulations to limit exposure; • Regulated levels set higher than the subsequent ones; • Subsequent decrease in regulated levels; • In addition to initially recognized risks, new 2nd generation risks are identified and research on these conducted; • More research results available, including a larger numbers of longitudinal and prospective studies; • Precise understanding of the harmful mechanisms, • The available research results are more conclusive; and • For some substances 3rd generation risks are identified, often dealing with issues of much lower levels of exposure which can only be studied after the initial regulations have been set	• Many conclusive results available; • Enough factual information to ban the use of the particular substance; • Results on lower levels of exposure becoming available; • Limitations to existing knowledge are still observed, and more time is needed for more complete understanding; • Future of existing, built-in materials under previous regulations is not dealt with by the ban; and • Knowledge about the problem substance continues developing

Although for specific substances some of these stages could overlap and progress at different speeds, these stages help organize knowledge about the harmful effects of substances commonly found in building and furnishing materials, and illustrate the issues around integrating knowledge about these substances into everyday life.

This part of the book is organized by the three stages, but in a reverse order, starting with the most recognized health hazards in building and furnishing materials. Stage 3 substances, being clearly recognized as health hazards, are especially relevant for this discussion, because only they can help in evaluating the effectiveness of their elimination and how long the whole process took. Therefore, this section starts with the recognized health hazards of lead and asbestos as examples of substances in Stage 3, and these are discussed in Chapter 6.

Discussed in Chapter 7, Stage 2 substances provide good examples of the contradictions and difficulties associated with the process of starting to eliminate substances recognized as harmful. This stage could take long periods of time and include many substages. Although there is a significant qualitative difference between these substages in terms of the allowable thresholds for exposure, one reliable feature of Stage 2 is that the differentiation between these substages tends to be blurred if developments are more accelerated. This blurring is the reason they are discussed here as one stage. This blurring is especially clear for two main examples for Stage 2: formaldehyde and phthalate plasticizers. In order to provide background for Stage 2 these examples are discussed within the context of other VOCs and general issues associated with indoor air quality. The key question posed while conducting the review of Stage 2 substances is: can an acceleration of the progression from early regulations to complete ban be observed in recent years?

Finally, Chapter 8, provides a brief overview of Stage 1 substances, and evaluates the processes which surround the decisions around introducing or diversifying the use of substances that are only suspected of causing harmful effects. This is a short chapter, because for many substances this is a very quiet phase. However, it raises an important issue: is there any observable avoidance of the introduction of new chemicals into production when some concerns have been noted on the potential health impact of these? In contrast to removal, which tends to have financially adverse implications, at the point of introduction finances are invested into the application of the new substance and selecting a substance which is less likely to need subsequent elimination could be seen as financially desirable. Thus, Chapter 8, evaluates whether it is possible to observe indications of such activities becoming more common. Nanotechnology and recently developed chemicals, or those with only somewhat recognized ongoing risks are the key examples for this chapter.

Discussion of three stages of recognition from Chapters 6, 7 and 8, is followed by Chapter 9, which provides quick and indicative reviews on where many chemicals and materials sit in terms of a progression of recognition of their potential harmfulness. This highlights the importance of including systematic examination of the impacts of building materials on human health as a fundamental requirement for the development of a more sustainable built environment.

5.3.1 How hard is it to find out where the problems are?

Rather than providing a comprehensive coverage of all current issues, Part II discusses three stages of recognition of risks and uses those to illustrate the issues. The intention is to educate the reader on relevant principles and empower them to make their own similar assessments. This reflects an aspiration to provide a framework which makes it easy to move beyond practice compliant with the current regulations, and instead towards more aspirational and enduring built environments.

In order to review the impact on human health from building and furnishing materials, it was important to consider medical and other scientific primary sources. Because in many cases the product names were irrelevant for these analyses, in many cases it was important to look into specific substances or chemicals which are commonly found in well-known materials. The PubMed database was recognized as an important source when searching for an effective method of acquiring information on the established medical understanding of the health risks associated with substances used in building and furnishing materials. It is hosted by the US National Library of Medicine and comprises more than 23 million citations for biomedical literature from MEDLINE, life science journals, and online books (PubMed, 2016). Other journal articles have used PubMed as their main search tool for systematic review of literature on medical subjects. For example, Checkoway et al. (2012) searched the PubMed database of publications between 1966 and 2012 to review the association between formaldehyde exposure and lymphohematopoietic malignancies (Checkoway et al., 2012). In addition to the PubMed data base, complementary information was sourced from websites of official organizations or regulating bodies, such as WHO and the (US) Environmental Protection Agency (EPA), and some published detailed overviews, such as *Toxicity of Building Materials*, edited by Pacheco-Torgal, Jalali, and Fucic (2012). Thus, the chapters that follow mainly summarize the knowledge available in medical sources on subjects relevant for architectural construction materials.

5.4 Conclusion

This chapter introduced the problems associated with indoor air toxicity and proposed the use of three stages of recognition of health risks as useful in assessing the likely future regulative treatment of any specified substances. It is hard to imagine that anybody in the construction industry would intentionally set up to harm the health of either the builders or building users, and compliance with regulatory frameworks is strong in our professions. Yet, it is possible that harmful materials approved and commonly specified today could be installed in a building which would last well beyond the subsequent elimination of the same materials from everyday use. Cases of lead paint and asbestos sheets installed while still allowed are painful reminders that simple compliance with the existing regulations can lead to subsequent issues. By using the proposed stages of recognition, it is possible to anticipate likely future regulations against chemicals and make smart material choices now. Integrating such evaluations with the considerations based on Parts I and II will greatly strengthen sustainability of any built project.

References

American Society of Heating, Refrigeration and Air-Conditioning Engineers (US) (ASHRAE) (2009), 2009. ASHRAE Handbook: Fundamentals. ASHRAE, Atlanta.

Andersen, I., Gyntelberg, F., 2011. Modern indoor climate research in Denmark from 1962 to the early 1990s: an eyewitness report. Indoor Air. 21, 182–190.

Armstrong, L., Wordsworth, A., Dauncey, G., 2007. Cancer: 101 Solutions to a Preventable Epidemic. New Society Publishers, Gabriola Island, British Columbia.

Attmann, O., 2010. Green Architecture: Advanced Technologies and Materials. McGraw-Hill, New York, NY.

Baker-Laporte, P., Elliott, E., Banta, J., 2008. *Prescriptions for a Healthy House: A Practical Guide for Architects, Builders and Homeowners*. third ed. New Society Publishers, Gabriola Island, Canada.

Balachandran, A., Shivbalan, S.O., Gowrishankar, N.C., 2010. Acute asthma: advances in management. In: Kabra, S.K. (Ed.), Allergy and Asthma: ECAB Clinical Update Paediatrics. Reed Elsevier India, New Delhi, pp. 29–44.

Birkeland, J., 2008. Positive Development: From Vicious Circle to Virtuous Cycles Through Built Environment Design. Earthscan, London.

Blakey, A., Ott, K., Peters, T., Risotto, S., 2013. Letter to the editor: anti-PVC advocacy disguised as science. Indoor Air. 23 (1), 85–86.

Brownell, B., 2006. Transmaterial: A Catalogue of Materials that Redefine Our Physical Environments. Princeton Architectural Press, New York, NY.

Burgess, R.A., Curwell, S.S., March, C.G., 1986. Introduction. In: Curwell, S.S., March, C.G. (Eds.), Hazardous Building Materials: A Guide to the Selection of Alternatives. E. & F.N. Spon, London, pp. 3–8.

Carlstedt, F., Jönsson, B.A.G., Bornehag, C.-G., 2013. PVC flooring is related to human uptake of phthalates in infants. Indoor Air. 23 (1), 32–39.

Checkoway, H., Boffetta, P., Mundt, D.J., Mundt, K.A., 2012. Critical review and synthesis of the epidemiologic evidence on formaldehyde exposure and risk of leukaemia and other lymphohematopoietic malignancies. Cancer Causes Control. 23, 1747–1766.

Clausen, G., Bekö, G., Corsi, R.L., Gunnarsen, L., Nazaroff, W.W., Olesen, B.W., et al., 2011. Reflections on the State of Research: indoor environmental quality. Indoor Air. 21, 219–230.

Curwell, S.S., March, C.G., 2002. Introduction. In: Curwell, S.S., et al., (Eds.), Hazardous Building Materials: A Guide to the Selection of Environmentally Responsible Alternatives. Spon Press, London; New York.

Curwell, S.S., March, C.G., 1986. Hazardous Building Materials: A Guide to the Selection of Alternatives. E. & F.N. Spon, London.

Delgado-Saborit, J.M., Aquilina, N.J., Meddings, C., Baker, S., Harrison, R.M., 2011. Relationship of personal exposure to volatile organic compounds to home, work and fixed site outdoor concentrations. Sc. Total Environ. 409, 478–488.

Fernandez, J., 2006. Material Architecture: Emergent Materials for Innovative Buildings and Ecological Construction. Architectural Press, Boston, MA.

Fisk, W.J., Black, D., Brunner, G., 2011. Benefits and costs of improved IEQ in U.S. offices. Indoor Air. 21, 357–367.

Gaffney, A., Christiani, D.C., 2015. Gene environment interaction from international cohorts. Semin. Respirat. Crit. Care Med. 36 (3), 347–357.

Gifford, R., 2011. The dragons of inaction: psychological barriers that limit climate change mitigation and adaptation. Am. Psychol. Assoc. 66 (4), 290–302.
Guieysse, B., Hort, C., Platel, V., Munoz, R., Ondarts, M., Revah, S., 2008. Biological treatment of indoor air for VOC removal: potential and challenges. Biotechnol. Adv. 26 (5), 398–410.
Howden-Chapman, P., Carroll, P., 2004. Housing and Health: Research, Policy and Innovation. Steele Roberts, Wellington.
Howden-Chapman, P., Pierse, N., Nicholls, S., Gillespie-Bennett, J., Viggers, H., Cunningham, M., et al., 2008. Effects of improved home heating on asthma in community dwelling children: randomised controlled trial. Br. Med. J. 337, a1411.
Jantunen, M., Oliveira Fernandes, E., Carrer, P., Kephalopoulos, S., 2011. Promoting actions for healthy indoor air (IAIAQ). European Commission Directorate General for Health and Consumers: Luxembourg.
Landrigan, P.J., 2016. Comments on the 2014 Helsinki consensus report on asbestos. Ann. Global Health. 82 (1), 217–220.
Liu, Z., Little, J.C., 2012. Materials responsible for formaldehyde and volatile organic compound (VOC) emissions. In: Pacheco-Torgal, F., Jalali, S., Fucic, A. (Eds.), Toxicity of Building Materials. Woodhead Publishing, Cambridge, pp. 76–121.
McGinnis, M.J., Foege, W.H., 1993. Actual causes of death in the United States. J. Am. Med. Assoc. JAMA. 270 (18), 2207–2212.
Milfont, T.L., 2010. Global warming, climate change and human psychology. In: Corral-Verdugo, V., Gracia-Cadena, C.H., Frias-Armenta, M. (Eds.), Psychological Approaches to Sustainability: Current Trends in Theory, Research and Applications. Nova Science Publishers, New York, NY, pp. 19–42.
Mokdad, A.H., Marks, J.S., Stroup, D.F., Gerberding, J.L., 2004. Actual causes of death in the United States, 2000. J. Am. Med. Assoc. JAMA. 291 (10), 1238–1245.
Nazaroff, W.W., 2011. Editorial: commemorating 20 years of *Indoor Air*. Indoor Air. 21, 177–178.
Nielsen, G.D., Larsen, S.T., Wolkoff, P., 2013. Recent trend in risk assessment of formaldehyde exposures from indoor air. Arch. Toxicol. 87, 73–98.
Pacheco-Torgal, F., Jalali, S., Fucic, A. (Eds.), 2012. Toxicity of Building Materials. Woodhead Publishing, Cambridge.
Pacheco-Torgal, F., 2012. Introduction: types of potentially toxic building materials. In: Pacheco-Torgal, F., Jalali, S., Fucic, A. (Eds.), Toxicity of Building Materials. Woodhead Publishing, Cambridge, pp. xv–xvii.
Pearson, D., 1998. The New Natural House Book: Creating a Healthy, Harmonious, and Ecologically Sound Home. Fireside, New York, NY.
Philp, R.B., 2001. Ecosystems and Human Health: Toxicology and Environmental Hazards. Lewis Publishers, Boca Raton, FL, E-book.
Phipps, R., 2008. Healthy homes for children. Build57–58, October/November.
PubMed, 2016. Available from: www.ncbi.nlm.nih.gov (accessed September 2016).
Rostron, Jack (Ed.), 1997. Sick Building Syndrome: Concepts, Issues and Practices. E&FN Spon, London.
Saunders, T., 2002. The Boiled Frog Syndrome: Your Health and the Built Environment. Wiley, Chichester.
Schörpfer, T., 2011. The alternative approach: observation, speculation, experimentation. In: Schörpfer, T. (Ed.), Material Design: Informing Architecture by Materiality. Birkhäuser GmbH, Basel, pp. 10–33.

Spengler, J.D., Chen, Q.(Y., 2000. Indoor air quality factors in designing a healthy building. Ann. Rev. Energy Environ. 25, 567–600.

Sundell, J., 2004. On the history of indoor air quality and health. Indoor Air. 14 (Suppl. 7), 51–58.

Sundell, J., Levin, H., Nazaroff, W.W., Cain, W.S., Fisk, W.J., Grimsrud, D.T., et al., 2011. Ventilation rates and health multidisciplinary review of the scientific literature. Indoor Air. 21, 191–204.

Tan, Z., Zhou, J., Cheung, A.K.L., Yu, Z., Cheung, K.-W., Liang, J., et al., 2014. Vaccine-elicited DC8$^+$T cells cure mesothelioma by overcoming tumour-induced immunosuppressive environment. Cancer Research. 74 (21), 6010–6021.

The Institute for Building Biology + Ecology/Institut für Baubiologie + Ökologie/ Neubeuern (IBN), 2011. Available from: www.baubiologie.de (accessed October 2011.).

Thompson, A., 2004. Homes that Heal: And Those That Don't. New Society Publishers, Gabriola Island.

Wargocki, P., Sundell, J., Bischof, W., Brundrett, G., Fanger, P.O., Gyntelberg, F., et al., 2002. Ventilation and health in non-industrial indoor environments: report from a European multidisciplinary scientific consensus meeting (EUROVEN). Indoor Air. 12 (2), 113–128.

Weschler, C.J., 2011. Chemistry in indoor environments: 20 years of research. Indoor Air. 21 (3), 205–218.

Wolkoff, P., Nielsen, G., 2010. Non-cancer effects of formaldehyde and relevance for setting an indoor air guideline. Environ. Inter. 36, 788–799.

World Health Organization (WHO), 2002. Indoor air pollution, health and the burden of disease: Indoor air thematic briefing 2. Available from: http://www.who.int/indoorair/info/briefing2.pdf (accessed August 2016.).

World Health Organization (WHO), 2007. Indoor Air Pollution: National Burden of Disease Estimates. World Health Organization, Geneva.

World Health Organization (WHO), 2016. Media centre: fact sheets: the top 10 causes of death. Available from: http://www.who.int/ (accessed August 2016.).

Zhang, M., Buekens, A., Jiang, X., Li, X., 2015. Dioxins and polyvinylchloride in combustion and fires. Waste Manage. Res. 33 (7), 630–645.

Persisting issues with the most recognized building material health risks: lead and asbestos

6

Emina Kristina Petrović
Victoria University of Wellington, Wellington, New Zealand

6.1 Introduction

This chapter takes a detailed look at lead and asbestos, two examples of substances in the third stage of recognition of health risks: they are well known, have been known about for a long time, and have even been legislated against for a reasonably long time. In fact, when the health risks associated with building materials are mentioned many people immediately think of these two, well-known hazardous materials. Both lead and asbestos have been used since prehistoric times. The adverse health effect of lead was clearly observed during the time of the Ancient Greeks, while for asbestos the effects were only noted in the late 19th century. Thus, both lead and asbestos were already in the first stage of the early recognition of risks before the start of the 20th century. Yet, despite this, both materials have seen prolific increase in use during the 20th century. This historical reality, or paradox, of a high increase in the use of materials which were already in part recognized as health hazards is very important. It clearly establishes the idea that during much of the 20th century human health was not valued as much as the physical properties both materials offered. As this chapter shows, the evolution of the increased recognition of the risks they posed and gradual regulation against them also tells an important story of repeatedly slow reactions, tendencies to dismiss strong indications of a health problem as being based on insufficient evidence, and general continuation of favoring the physical properties of material and economic efficiency over human health. These are good examples of a disregard for the precautionary principle.

6.2 Issues with lead

6.2.1 Lead: general overview

Lead (Pb) has been used by human society for at least 4000–5000 years (Brown and Margolis, 2012). Lead is a naturally occurring metal which, historically, has been used in architecture for its pliability in forming the glazing beads (cames) in lead windows, lead roofs, flashings, and plumbing. During the 20th century it was mainly used in architecture as a stabilizer for lead-based paints.

Materials for a Healthy, Ecological and Sustainable Built Environment: Principles for Evaluation.
DOI: http://dx.doi.org/10.1016/B978-0-08-100707-5.00006-X

As early as 370 BC, Hippocrates made some of the earliest recorded observations of health issues related to lead and other heavy metals (Philp, 2001, Chapter 6). Some sources have proposed that lead poisoning influenced the fall of the Roman Empire (Hayes, 2012). Better understanding of the risks eventually led to the passing of the first Food and Drugs Act by the British Parliament in 1875, but despite this cast lead soldiers and other lead toys were fairly common until the late 1940s and early 1950s (Philp, 2001, Chapter 6). Therefore, it is possible to observe that lead was already in the first stage of recognition of health risks before the start of the 20th century, but that the real action against it, indicative of transition to the second stage, did not start until mid-20th century.

Lead can negatively affect almost every organ, system, and process in the human body, including the cardiovascular, gastrointestinal, hemolymphatic, urinary, immune, nervous, and reproductive systems, and can cause tumors in laboratory animals (Carlisle et al., 2009). Of all the systems listed, the main target for lead toxicity is the central nervous system (Sanders et al., 2009). For painters with high lead exposure peripheral neuropathy and cognitive impairment have been recorded, starting with problems with the upper limbs (Krishnan et al., 2012). However, lower levels of on-going exposure can lead to inhibition of several enzymes involved in heme synthesis, influencing functions of the peripheral and central nervous system and increasing blood pressure (Jakubowski, 2011).

Young and unborn children are more at risk from lead than adults because their brains and nervous systems are developing, and because their bodies absorb much higher proportions of ingested lead than adult bodies (Sanders et al., 2009). The blood–brain barrier recognizes one common form of lead in the body as calcium thus allowing its entry into the brain, where lead can take over the functional role of calcium, altering neurochemistry and behavior, especially in the immature brain (Sanders et al., 2009). Acute lead poisoning in children starts with vertigo and irritability, progressing to delirium, vomiting, and convulsions (Philp, 2001, Chapter 6). Studies have shown that lead exposure in children persists into adulthood, and has been associated with brain damage, mental retardation, behavioral problems, developmental delays, violence, and death at high levels of exposure (Sanders et al., 2009).

The half-life for lead in blood is generally 30–35 days, while in bones it is 25 years; bones, hair, and teeth are where most of the free blood lead deposits (Philp, 2001, Chapter 6; Carlisle et al., 2009). Lead in the bones may be remobilized back into circulation at times of stress and tension, such as pregnancy, illness, traumatic life events, and aging (Zahran et al., 2009; Machida et al., 2009). Although the human body can eliminate lead, most of it stays in the body for a long time, contributing to the total body load.

Lead can come in different forms, as elemental lead, in organic lead compounds, and inorganic lead compounds. It can also change forms, e.g., from organic to inorganic (American Cancer Society, 2014). In buildings the most likely form of lead is in inorganic lead compounds. Although, carcinogenicity of substances differs from their toxicity, lead is recognized as a risk in both areas, and all forms of lead are toxic. However, the International Agency for Research on Cancer (IARC) recognizes different forms of lead as varying in their carcinogenicity: they classify inorganic lead compounds in Group 2A (probably carcinogenic to humans), elemental lead in Group 2B (possibly carcinogenic to humans), while organic lead compounds are classified as Group 3 (not classified for carcinogenicity to humans) (IARC, 2016).

6.2.2 Regulations against lead

During the 1970s, systematic actions against lead exposure started in many parts of the world. In this period the main focus was leaded petrol and lead-based paints. Recognitions of the health risks associated with lead in petrol increased during the 1970s leading to removal of leaded petrol in many countries during the 1980s and 1990s (O'Grady and Perron, 2011). In the United States, the Lead Paint Poison Prevention Program was introduced in 1970 (Philp, 2001, Chapter 6). This was followed with the 1977/78 US ban of lead-based paint (Roberts et al., 2012). The level of restriction varied in different countries at different times, e.g., the 1976 Canadian limit for lead in interior paints was set eight times higher than the US restrictions 2 years later (O'Grady and Perron, 2011). In France, the sale of lead carbonate, which was used in paint, was officially only banned in 1993 (Lucas et al., 2012). Therefore, for many developed countries during the 1970s lead has gone through the second stage of recognition of risk, leading to bans from the late 1970s.

Unfortunately, these trends are not shared by all countries. In the last 10 years, high levels of lead in new paints have been found in 20 countries from five continents (Ewers et al., 2011). Most are developing countries where legislation limiting lead content either does not exist or is ineffectively enforced, but some are more developed countries such as India and Taiwan (Khan et al., 2010; Ewers et al., 2011). For example, the study of Ewers et al. (2011) tested newly purchased paints in Taiwan and found that the median lead concentration was more than 30 times the current US standard and less than half of the paint samples had lead content within the current US standard (Ewers et al., 2011). They also observed that when samples from China were tested these had lower levels of lead.

6.2.3 Impact of low levels of lead on human health

Since the introduction of more strict regulations against lead, new studies have measured the impact of low levels of lead on human health. Evaluation of such low levels was generally not possible before regulative action, because most people already had higher levels in their system. Therefore, this section shows the issues associated with attempts at comprehensive assessments of a problem substance when the exposure level in the general population is already elevated: the subtler impacts are impossible to study. This further indicates the importance of action to limit exposure even before a more sophisticated understanding of the issues is possible.

Current research indicates that even very low levels of blood lead in children can cause a measurable negative impact on their development. Levels of safe exposure changed in the later decades of the 20th century. In 1991, the US Centers for Disease

Control and Prevention decreased earlier action blood lead levels to 10 μg/dL for children under the age of six (Roberts et al., 2012), and the same measure was established in many other countries soon after (Oulhote et al., 2011). (10 μg/dL equates to 100 μg/L, both measures are found in the literature.) Good indicative examples of earlier action blood lead levels of concern include the 1960 US level of 60 μg/dL (Jakubowski, 2011) and the 1987 Canadian level of 25 μg/dL (O'Grady and Perron, 2011), clearly indicating a general trend towards lowering acceptable levels.

The action blood lead level of 10 μg/dL for children reflected the knowledge available at that time, which demonstrated a decline in IQ points with every incremental elevation of blood lead levels greater than 10 μg/dL (Roberts et al., 2012, p. 1). In 1978, the US national blood lead level in children aged 1–5 years was over 10 μg/dL for 13.5 million children or almost 90% of American children (Brown and Margolis, 2012). By the early 1990s that figure had dropped to about 10% and continued dropping to approximately 1.5% of all American children by 2007–08 (Fig. 6.1; Brown and Margolis, 2012). Similarly, in the late 1970s, the geometric mean blood lead level for all American children was 15 μg/dL, and in 2012 it was under 2 μg/dL (Brown and Margolis, 2012). This significant decrease in blood lead has enabled more subtle studies, and also indicates that in some countries, even after

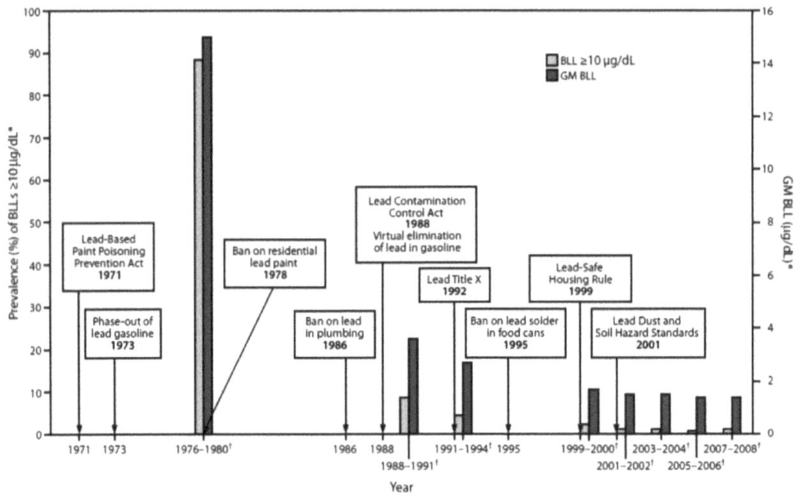

Figure 6.1 Timeline of lead poisoning prevention policies and blood lead levels in children aged 1–5, by year, from National Health and Nutrition Examination Survey (NHANES), United States, 1971–2008. *BLL*, blood lead level; *GM*, geometric mean. *National estimtes for GM BLLs and prevalence of BLLs ≥10 μg/dL, by NHANES survey period and sample size of children aged 1–5 years: 1976–80: N=2372; 1988–91: N=2232; 1991–94: N=2392; 1999–2000: N=723; 2001–02: N=898; 2003–04: N=911; 2005–06: N=968; 2007–08: N=817. †NHANES survey period.

Source: Brown, M.J., Margolis, S., 2012. Lead in drinking water and human blood lead levels in the United States. In: Centers for Disease Control and Prevention. Morbidity and Mortality Weekly Report 61 (Suppl. Aug):1–9, p. 3.

the official bans on use of lead as a sign of the third stage of recognition, some of the features of the second stage continued, creating a need to decrease regulated levels to help children specifically at risk. Nevertheless, more recent estimates still assert that in the United States 25% of the population are at risk from high lead exposure (Jordan et al., 2003), and lead poisoning in children costs the nation $43.4 billion a year, compared to $2.0 billion for childhood asthma (Ferguson et al., 2012).

Research since the 1970s has demonstrated that the greatest rate of decline in IQ occurs in blood lead levels of between 3 and 8 μg/dL (Roberts et al., 2012, p. 1). Therefore, in 2000 the World Health Organization Regional Office for Europe recommended efforts be made to ensure that at least 98% of an exposed population of all ages has blood lead levels under 10 μg/dL, which is supported by the current median blood lead level of under 5.4 μg/dL (Jakubowski, 2011). A very similar level and rationale were set in 2012 by the US Centers for Disease Control (Roberts et al., 2012, p. 1). A French survey from 2007 involved 3800 children and estimated a geometric mean for blood lead level of 1.5 μg/dL, with 0.11% of children having a higher blood lead level of above 10 μg/dL (Oulhote et al., 2011). A study from 2007 to 2008 compared blood lead levels for children in six European cities (mainly central European) and three non-European cities. It showed that blood lead levels in children varied very little between European countries but were noticeably higher in the non-European cities they studied, these being in China, Ecuador, and Morocco (Hrubá et al., 2012). This implies that many parts of the developed world have lead exposure under reasonable control, but that this is not a trend shared worldwide.

Earlier studies, such as one from 1995, suggested a mean decrease in full-scale IQ of the order of 2 IQ points for a change in mean blood lead level from 10 to 20 μg/dL (Jakubowski, 2011). Subsequent studies have established that in fact much of the decrease of IQ occurs for exposures under 10 μg/dL: for a change of exposure from 2.4 to 10 μg/dL, IQ tends to change by 3.9 points, while for a change of exposure from 20 to 30 μg/dL, IQ changes by 1.1 points (Jakubowski, 2011). Similarly, in the 1990s, research established strong links between IQ and blood lead levels in children of 2 years of age, while subsequent research, even of the same cohort, suggests that exposures of school-age children to lead may be more strongly related to performance in cognitive testing (Mazumdar et al., 2011). Several studies have found that poorer performance in these types of test correlate to higher levels of lead, even for levels as low as 2 μg/dL (Jakubowski, 2011; Zahran et al., 2009; Amato et al., 2012).

The current low and moderate blood lead levels in many parts of the Western population have also been shown to have a significant impact on other aspects of children's development. Moderate blood lead levels are correlated to lower growth rate, lower height, and lower weight for age (Yang et al., 2012). Levels of lead from 5 μg/dL have been found to alter reproductive hormones in young girls (Gollenberg et al., 2010). Lead has contributed to attention-deficit hyperactivity disorder (ADHD), and this was even a trend with children who are asymptomatic for ADHD (Winneke, 2011). It has been suggested that lead and other heavy metal prenatal exposure impairs the development of visual processing (Ethier et al. 2012).

Further, blood lead level has been found to be correlated to increased stress responses at currently "normal" levels under 10 μg/dL (Gump et al., 2009). Concerns have been raised that lead poisoning can cause nerve damage to the sense organs and nerves controlling the body, increasing the chance of neurodegenerative diseases like Alzheimer and Parkinson diseases later in life (Sanders et al., 2009; Li et al., 2010).

While the negative impact of lead on memory, learning, and IQ has often been studied, there is also evidence that lead influences other behaviors such as mood (depression), anxiety, schizophrenia, and violence/aggression (Sanders et al., 2009). It has been noted that lead-exposed 4–5-year-old children exhibit an increase in aggression, and from there lead exposure has been associated with juvenile delinquency and criminal behavior (Sanders et al., 2009, Narag et al., 2009). Between 1979 and 1984, a Cincinnati-based study recruited pregnant women living in impoverished neighborhoods with a high concentration of older, lead-contaminated housing (Wright et al., 2009). It measured prenatal and early childhood exposure to lead, and subsequently compared that data with the local criminal justice records on arrests of the same children by 2005, then 19–24 years of age. Increased blood lead levels before birth and during early childhood were associated with higher rates of arrest than the control group. Also, average childhood blood lead was significantly associated with higher risk of arrest involving violent crime (Wright et al., 2009). In a similar New Zealand study, 1265 Christchurch children were studied from birth (in 1977) to age 21 (Fergusson et al., 2008). Their dental lead levels at ages 6–9 were significantly associated with both officially recorded violence/property convictions and self-reported violence/property offending.

More recent research on lead has attempted to identify more subtle ways in which lead is harmful for human bodies. For example, lead has also been researched in relation to body processes such as oxidative stress, cell membrane biophysics, and signaling and neurotransmission (Verstraeten et al., 2008). Another study has established that the presence of lead in cells can trigger the sequence of preprogrammed bodily processes called apoptosis, which leads to the elimination of cells without releasing harmful substances into the surrounding area (Yedjou et al., 2010). While this is a healthy process that protects the body from unhealthy cells or substances, Yedjou et al. noted that the mechanisms by which lead induces this elimination remain unclear. However, their study has demonstrated how the presence of lead nitrate can trigger cellular processes that lead to the formation of human leukemia (HL-60) cells (Yedjou et al., 2010). A 2011 study established that there are genetic biomarkers of increased neurotoxic risk, indicating that boys with a particular genetic expression are most likely to be affected (Sobin et al., 2011). Research of this nature could lead to development of targeted prevention programs, particularly for those who are more at risk.

More recent studies have also made much progress with identifying both possible sources of high exposure, such as certain food plants that tend to absorb and carry higher concentrations of lead (for mushroom varieties particularly prone to absorbing lead see Petkovšek and Pokorny, 2013) or foods, such as brown rice, that might be able to help reduce mammal lead-induced toxicity (Zhang et al., 2010). Furthermore, it has been observed that blood lead levels seem to fluctuate in relation to the seasons, yet without a proper understanding of what influences these fluctuations (Havlena et al., 2009).

The results of these studies of more recent, postregulation issues associated with lead clearly indicate that when the general level of exposure is very high, it is very hard for the research community to evaluate the important thresholds for adverse effects on human health accurately. This conclusion is very disturbing, because it indicates that researchers could be facing similar inability to fully evaluate adverse effects for any other substances which are highly prevalent within the population. Thus, the same could be now taking place with chemicals which are in the second or even in the first stage of recognition of risk.

6.2.4 Issues facing recovery from lead contamination

Some significant sources of lead have continued even after its official bans. The US Center for Disease Control and Prevention lists these as: lead-contaminated soil, food, tap water, lead paint, folk remedies, pottery, and dust (Brown and Margolis, 2012). During the late 1970s and early 1980s, when recognition of the risks associated with lead exposure culminated in various bans, all of these sources were likely to contain lead. Since then, there has been much improvement: lead in tinned food and lead in new paint have been eliminated in developed countries, and much work has been done to reduce the prevalence of lead pipes in public water supplies (Brown and Margolis, 2012). However, the historical use of lead has left a legacy of contamination in the natural environment, since elemental lead does not degrade (Carlisle et al., 2009). Furthermore, lead continues to be newly harvested, and is often used in electronic items, which creates new potential lead pollution at recycling points for electrical waste (Yang et al., 2012). In addition, due to poorer regulations in less developed countries, imported items manufactured in such countries can introduce new lead contamination (Guney and Zagury, 2012). Unfortunately, even in very recent years hidden lead has been found in internationally available toys and low-cost jewelry (Guney and Zagury, 2012).

Nevertheless, the most significant concern associated with lead in developed countries is lead in paint in older housing. In the United States this is increasingly recognized as a problem that comes together with poverty, because US children from low-income families are about three times more likely to have high blood lead levels than children from middle- and high-income families; and similarly, children from the inner city, with concentrations of older housing, are four times more likely to have elevated blood lead levels than children in other areas (Sanders et al., 2009). One comprehensive US intervention specifically targeted windows because they have the highest likelihood of containing lead paint and the highest amounts of lead dust (Dixon et al., 2012). Surveying 189 homes from four cities, they had a group where all windows were replaced, a group where some were replaced, and a group with no replacement windows. All windows that were not replaced were repaired and maintained. Their results show that 12 years after the intervention homes with all windows replaced had interior floor dust lead and window sill dust lead levels almost half those in nonreplacement homes, even when controlling for other factors (Dixon et al., 2012). The homes where some windows were replaced also exhibited a decrease, which was about half the decrease from complete replacement (Dixon et al., 2012). Furthermore, one study found a child with a blood lead level of 23 μg/dL who lived in a house with lead-containing paint which was observed to have excessive dust on the windowsills, dirt on the carpet, and a dirt-covered yard but no evidence of peeling paint (Roberts et al., 2012), indicating that the overall level of cleaning in older houses can contribute to the problem.

While window replacement is a fairly expensive strategy and can often be impracticable, improvements in management of household cleaning can come at no or very low cost, and some studies have evaluated the effectiveness of different educational initiatives on lead risks and strategies to reduce exposure, including effective and safe cleaning in housing with lead paint (Ferguson et al., 2012; Jordan et al., 2003, 2004, 2007; Brown et al., 2006Hilts et al., 1998).

Early studies on the effectiveness of educational interventions related to lead were in highly polluted areas. For example, during the 1990s, the town Trail in Canada has been actively monitored because of the active lead/zinc smelter in the area, and because many children had blood lead levels at the individual intervention level, set then at 15 μg/dL (Hilts et al., 1998). The actions on site included indoor dust control trials, ongoing soil treatment experiments, close monitoring of community blood lead levels, and included community education (Hilts et al., 1998). Even when controlled for other factors, the researchers concluded that community education was making a measurable difference in children's blood lead levels, especially when comparing outcomes of years with and without such programs (Hilts et al., 1998).

When comprehensive peer studies of educational interventions on children's blood lead levels were conducted, findings were mixed, with many pointing to measurable reductions (Hilts et al., 1998; Jordan et al., 2003; Ferguson et al., 2012), and others not observing such a clear difference (Brown et al., 2006). For example, when children's blood and household lead dust levels were comprehensively monitored and compared in paired groups either receiving the standard level of education or a more intense 20-visit intervention, the latter reduced the risk of blood lead levels over 10 μg/dL by approximately 34% (Jordan et al., 2003). Others have observed no such difference between the intervention and nonintervention groups and concluded that once the blood lead level is elevated, it is hard for educational interventions to help to reduce it (Brown et al., 2006).

Research on the outcomes of educational interventions has identified a wide range of factors that influence possible outcomes, decreasing causal clarity. For example, the participants of the nonintervention group generally also receive additional information about their lead exposures, either through measured blood lead levels, or through measured lead content in household dust, or both (Jordan et al., 2003; Brown et al., 2006), and some participants have reported this information as inspiring positive change (Jordan et al., 2003, 2004). Additionally, it is hard to evaluate all possible influential factors, which are as diverse as the level of rainfall in the months leading to the study, which influences the lead in outdoor dust (Hilts et al., 1998), or the type of floor cleaner used for floors painted with lead paint, which can influence greater or lesser release of lead (Jordan et al., 2004).

Some educational interventions monitored neither blood lead level nor household dust, but self-reported change. For example, a broad-scale educational intervention on lead exposure in Arkansas with surveys immediately after the educational intervention and 36 months later showed 45% of participants reported some changes to prevent exposing a child to lead, while 53% reported planning to change something (Ferguson et al., 2012). Such programs have been recognized as increasing the level of lead awareness through education by empowering individuals with knowledge that can aid in reducing blood-lead levels in at-risk children (Ferguson et al., 2012). Such programs are also very cost-effective (Ferguson et al., 2012).

Based on this review it is possible to conclude that the issues facing recovery from lead contamination still seem complex. Lead contained in existing buildings is at the core of this problem and more education and more financial assistance are needed to support a decrease in health risks from lead exposure. Absence of regulations and monitoring of activities in homes, however, makes systematic changes difficult, because many activities from basic DIY projects to cleaning practices can influence the release of lead.

6.2.5 Lead: summary

This overview shows that early regulations against lead exposure were delayed after the articulated recognition of the risks. Although it is hard to give a clear date when the first stage of recognition started for lead, because it probably dates to prehistoric times, it is easy to assert that lead was established in the first stage of recognition before the start of the 20th century. Yet, no significant efforts towards regulation can be observed in the first half of that century. During the 1970s many Western countries experienced the greatest change in treatment of lead through the proposed elimination of leaded petrol and leaded paints, and this decade can be seen as an accelerated second stage of recognition of risk. The third stage then emerged by the end of the 1970s with the ban on leaded paint, which was followed during the 1980s–90s with the ban on leaded petrol. Despite these bans in Western countries, many less developed countries are still in the very early stages of risk recognition.

Furthermore, for lead, research and subsequent decreasing "action levels," generally characteristic of the second stage, are still on-going, long after the bans. It is now known that the average blood lead levels common in US children of the 1970s have probably negatively impacted many aspects of their development, which was impossible to observe while the overall level of lead pollution was much higher. Similarly, the main current source of contamination is old housing with lead paint, which in the past was a relatively minor risk, compared to new application of such paints, lead water pipes, and the pollution from leaded petrol. Furthermore, the total impact is still unknown, because the release of blood lead stores in the later stages of life has generally not yet happened for the children of the 1970s. More sophisticated understanding of the adverse effect lead has on the human body will continue to be produced in years to come. For instance, a recent cost–benefit analysis suggests that for every dollar spent to reduce lead hazards \$17–\$220 is saved, which compares well to immunization costs (Brown and Margolis, 2012). Therefore, lead still does not appear to be a cold case.

6.3 Issues with asbestos

6.3.1 Asbestos: general overview

Asbestos use began 4500 years ago, but significantly increased during the 20th century, and peaked during the mid-1970s (Park et al., 2012). Asbestos is the general commercial name for a group of naturally occurring mineral silicate fibers of the serpentine and amphibole series (Fig. 6.2; Park et al., 2012). Asbestos minerals are crystalline with weaknesses in the crystal structure which cause long thin fibers to be released along fracture planes and become airborne when the rock is stressed (Donaldson and Poland, 2012). Normally, asbestos is defined as a mineral fiber with lengths of 5–10 μm, proportionally small fiber diameters of usually under 1 μm, and with an aspect ratio >3:1 (Park et al., 2012). While its mineral chemical composition gives asbestos its recognized properties against fire, in the building

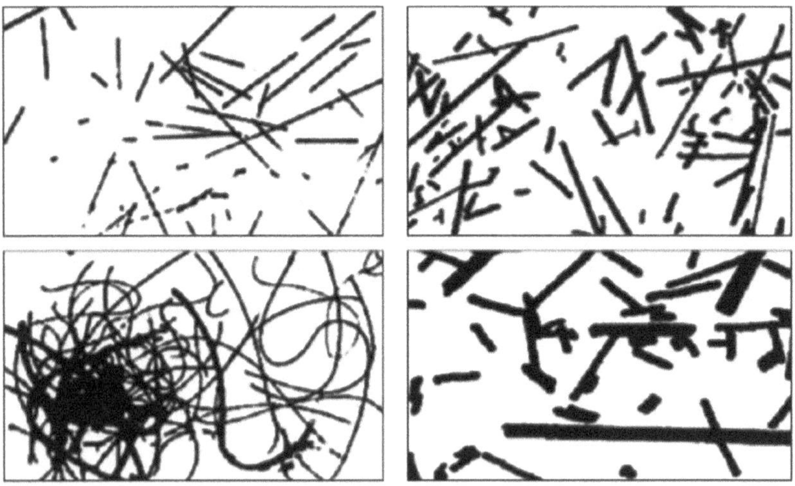

Figure 6.2 Asbestos fiber types. While most types have straight, rod-like fibers, chrysotile has curvilinear fibers. Top left: crocidolite; top right: amosite; bottom left: chrysotile; bottom right: anthrocite.
Source: Liu, G., Cheresh, P., Kamp, D.W., 2013. Molecular basis of asbestos-induced lung disease annual reviews. Pathol. Mech. Dis. 8:161–187.

industry it was also often used to reinforce a surrounding material, such as cement, due to the properties of these mineral fibers. In many Western countries, asbestos-containing products were widely used in construction between the 1920s and mid-1980s, in products such as roof tiles, wall claddings, vinyl flooring, sprayed fire protection, decorative ceilings, roofing membranes, adhesives, and paints (Level, 2016).

Unfortunately, the same feature that gave asbestos fiber good applicability in the construction industry is the foundation of the health problems it causes, and in contrast to lead, this somewhat narrows down the range of impacts asbestos has on the human body. Although much research has been undertaken over the last few decades, the precise molecular mechanisms involved with asbestos are not yet fully understood (Liu et al., 2013). However, it is well known that asbestos exposure can lead to a series of different lung diseases, such as pulmonary fibrosis (asbestosis), pleural abnormalities (effusion and plaques), and malignancies (bronchogenic carcinoma and mesothelioma) (Liu et al., 2013). Further, due to their small diameter, proportionally long length, and biopersistence, these fibers are easy to inhale, and once in the body they can cause a number of processes in the lung that could lead to carcinogenesis (Donaldson and Poland, 2012). Long biopersistent fibers can generate free radicals chronically that directly damage DNA, leading to long accumulation of dose and interaction with cells of the immune system. In addition, the long fibers create a series of processes that inhibit positive cell functioning and trigger the defensive mechanism, which becomes chronic due to cellular inability to expel pollution of the size and proportion of asbestos fibers (Donaldson and Poland, 2012).

One of the challenges with research on asbestos is the long latency period between exposure to asbestos and presentation of health concerns, ranging from 15 to 40 years (Liu et al., 2013). This is related to the damage which asbestos fibers create due to the body's inability to eliminate them effectively, which keeps certain protective and eliminative processes in a state of chronic overuse, thus gradually creating health problems (Sanchez et al., 2009; Donaldson and Poland, 2012). Delays like this have historically made conclusive research more difficult and contributed to blurring of the recognition of the problem.

The evaluation of the health impact of asbestos is further complicated by its natural variability which, at least on a theoretical level, justifies the assumption that its impact could vary to reflect this. Generally, chrysotile asbestos has been seen as potentially less harmful due to various estimates that once in the body it disintegrates, which makes it less pathogenic than the amphiboles (Donaldson and Poland, 2012). This could potentially be very significant, given that chrysotile asbestos represented as much as 95% of the asbestos used worldwide since 1900 (Park et al., 2012). Unfortunately, recent reviews strongly assert this is a misconception (Kanarek, 2011), and concur with the IARC, that all forms of asbestos are recognized human carcinogens and should be banned (Paglietti et al., 2012). Furthermore, in 2009, having asserted the need for a complete ban on all asbestos use, the IARC concluded that several nonpulmonary forms of cancer can be caused by exposure to asbestos, and that asbestos has been associated with risk of systemic autoimmune disease (Park et al., 2012).

There is also recognition that occupational exposure to asbestos means the substance has often made its way into workers' homes through the dust brought home on clothing, which then exposes families to risk (Peretz et al., 2008). More recently, cases of long-latent asbestos-related health problems are becoming apparent in adults who were exposed to asbestos as children through their parents' work (Reid et al., 2013; Peretz et al., 2008). Similarly, longitudinal studies of people who were exposed to asbestos outdoors as children in the 1970s because of local mining are now giving conclusive results that childhood exposure to asbestos increases cancer incidence rates, mainly but not exclusively due to an increase in rates of mesothelioma (Reid et al., 2013). The number of individual cases of this nature is likely to continue increasing because of the long latency period.

In contrast to lead where evaluations of bodily exposures are possible (as seen through measured blood lead level and bone lead level), none of the articles found during this review reported such measures for asbestos. Rather, many research articles used questionnaires and interviews to obtain descriptive accounts of possible asbestos exposure (Peretz et al., 2008; Olsen et al., 2011), or, when possible, measured prevalence of asbestos in air (Reid et al., 2013). This, together with the long latency period, makes accurate studies very difficult, and this is especially the case for lower levels of exposure. Recent research has, through studies of the children of asbestos workers or children who lived in affected neighborhoods, established that their exposure, which was probably much lower than that of the asbestos workers themselves, has also led to adverse health effects (Reid et al., 2013; Peretz et al., 2008). Thus, it seems that for asbestos indirect indications will be needed to evaluate the extent of adverse effects from lower levels of exposure. Studies of removal work carried out now could become especially useful in showing the impact even lower levels of exposure have on human health, but this is likely to mean waiting for 40–50 years before enough evidence accrues given the long latency period.

6.3.2 Regulations against asbestos

Around 1977, when the demand for asbestos peaked internationally, some 25 countries produced a total of 4.8 million metric tons per year, to be used in 85 countries (Park et al., 2012). Although the use of asbestos has since been banned in many industrialized countries, early in the 21st century, global use was still 2.1 million metric tons per year, primarily in Asian and less developed countries, where most of it is used to manufacture asbestos cement building materials (Park et al., 2012). This prolific industrial use of asbestos during the 20th century should be seen in contrast with early recognitions of its negative health impact. In 1898, early in the period of the industrialized use of asbestos, Adelaide Anderson, Principal Lady Inspector of Factories in the British Home Office, accurately noted that "[t]he sharp jagged edge of the insoluble mineral dust had undoubtedly occasioned much illness and death from respiratory disease" (Department of Labour, 2006, p. 38). Early in the 20th century, the relationship between asbestos dust exposure and fibrosis of the lungs and asbestosis was already recognized, and by the 1950s and early 1960s the relationship between lung cancer and mesothelioma and asbestos exposure was confirmed (Department of Labour, 2006, p. 38). Yet, the production was at that time still internationally increasing. This makes asbestos a good example of a material that has been used for a long time since first being recognized as a hazard, thus since the first stage of recognition of risk.

Although early regulations that deal with asbestos date back to 1931 with the formation of the British Asbestos Industry Regulations (Department of Labour, 2006, p. 38), it was in the 1980s that much change took place and many countries started to ban the use of blue and brown asbestos (Department of Labour, 2006, pp. 35–36). However, the complete ban of asbestos generally took place later: Italy in 1992 (Paglietti et al., 2012); New Zealand in 2002 (Waikato District Health Board, 2012); Australia in 2003 (Olsen et al., 2011); and Japan in 2005 (Park et al., 2012). By 2012, 52 countries had completely banned use of any form of asbestos (Paglietti et al., 2012). Therefore, for asbestos the start of the third stage came in two stages. For many countries there were about 20 years between the start of banning and complete bans, which often occurred at the start of the 21st century, more than 100 years since articulation of the first stage of recognition of risk.

6.3.3 Issues facing recovery from asbestos contamination

Although still in use in many less developed countries, it is now possible to start assessing the legacy of asbestos since its official ban in many developed countries. Unfortunately, this reveals new types of concern. Olsen et al. (2011) discuss the impact of asbestos in three waves: first the workers mining and milling raw asbestos and manufacturing asbestos products; second the workers who used asbestos products in industry, such as the building industry; and finally the third wave consisting of people diagnosed with asbestos-related diseases after a short-term and/or low-level exposure in the home or workplace, often due to activities related to home maintenance or renovation involving asbestos-containing building products.

The amounts of asbestos used varied greatly between products. For example, sprayed asbestos varied between 5% and 95% asbestos content (Dumortier and De Vuyst, 2012). Similarly, the prevalence of asbestos-containing products in different countries more or less reflected the ease of its availability. For example, in Australia, where naturally occurring asbestos was mined until the 1960s, 25% of all new homes were clad in asbestos cement (Olsen et al., 2011). Generally, asbestos-containing materials that are in good condition and are not the subject of everyday wear and tear, pose little risk. In other cases, the risk can be very high. For example, asbestos-vinyl flooring tiles have been shown to suffer the on-going release of asbestos fibers in normal use (Sebastien et al., 1982). Thus, they are a health hazard both while installed and if removed. These great differences make it hard to make any general estimates on possible exposure levels.

The problem is that even after the new use of asbestos was banned, much of it remained in existing structures, often as asbestos cement materials used in cladding, roofing, or piping (Olsen et al., 2011), but also in or underneath vinyl and linoleum floors and as a sprayed compound (Level, 2016). Thus, not surprisingly, the clearance of installed asbestos from buildings is now recognized as a major construction business in developed countries (Dumortier and De Vuyst, 2012), and one of the problems is that much of that removal work is carried out by low-skilled casual labor (Department of Labour, 2006, pp. 25–26). In 1992, an audit of floor sanders in Christchurch, New Zealand, revealed work practices which included a failure to prevent spread of dust to other rooms, poor respiratory protection for sanders, sanders transferring asbestos dust to their vehicle and home on their work clothes and tools, lack of a thorough "clean up" after sanding, and even the careless disposal of sanding dust (Department of Labour, 2006, p. 25). Similar problems with poor compliance with established and regulated procedures for the safe removal of asbestos have been noted in other countries (Dumortier and De Vuyst, 2012). Unskilled DIY removal of asbestos can be a particular problem because it is very hard to regulate and monitor.

6.3.4 Other asbestos-related risks

Current knowledge focuses on risks associated with inhalation of asbestos. Ingestion and dermal exposure have so far received very limited research attention. One area of concern is asbestos in drinking water (Kjærheim et al., 2005; Koumantakis et al., 2009; Wei et al., 2013). Consideration of the presence of asbestos in water is especially important because between the 1920s and the 1980s, asbestos cement was commonly used in mains water supply systems in many countries (Browne et al., 2005; Department of Labour, 2006, p. 36). To remediate problems with taste, asbestos-cement pipes were sometimes lined with vinyl-based coatings (Spence et al., 2008). As these networks age they become prone to frequent pipe bursts. Additionally, asbestos can enter water systems from mines or natural occlusion making its way into the rivers, lakes, and water table reserves in measurable quantities (Koumantakis et al., 2009; Wei et al., 2013). However, studies to date generally do not provide strong evidence of an association between exposure to asbestos in drinking water and the presence of gastrointestinal or respiratory cancers (Kjærheim et al., 2005; Browne et al., 2005; Koumantakis et al., 2009), but the cases that have been found indicate a 20-year or longer delay between exposure and development of cancer (Kjærheim et al., 2005). Additional problems have also emerged: asbestos in river water has been associated with a negative impact on the behavior of fish and other water life, and results indicate that homes supplied with asbestos-contaminated water are associated with an increase of airborne asbestos (Koumantakis et al., 2009). Therefore, it is possible to anticipate that studies of total risks from asbestos could enter a new phase with a heightened recognition of these other risks.

Replacements for asbestos are generally called manmade mineral fibers and often take a similar form of being long, thin fibers. Although generally these are considered to break transversely, rather than longitudinally like asbestos, concerns have been raised that a similar pathology could be observed with these materials. Carbonari et al. (2011) studied the in vitro impact of glass fibers, ceramic fibers, and Wollastonite fibers (a form of asbestos), and found that naturally occurring

Wollastonite fibers induced blood vessel formation in a similar fashion to that observed for other asbestos, while the glass and ceramic fibers did not exhibit such similarities. They concluded that the size and shape of the fibers and their chemical composition and biopersistence are important factors influencing the development of adverse health effects and that very thin, persistent fibers are the most harmful (Carbonari et al., 2011). When evaluating the impact of dermal exposure to ceramic fibers introduced as an asbestos replacement, a high level of dermal irritation was observed (Keić-Świerczyńska and Wojtczak, 2000).

6.3.5 Asbestos: summary

The limitations in the available knowledge about risks associated with asbestos indicate that it is still a less recognized and understood health risk than lead. An understanding of the important mechanisms of response to asbestos within the human body is still missing. Once these are more understood, this new knowledge could change the way asbestos is considered. This situation reinforces the reality that many discoveries associated with the second stage of risk recognition continue into the third stage, but also that it is essential to take actions to reduce exposure to hazards based on early knowledge, rather than waiting for the full picture to emerge.

Furthermore, just as with lead, asbestos poses concern in Western countries through the "third wave" of exposures to asbestos already built into structures. Because exposure to asbestos is harder to measure, and personal reports are often used to evaluate exposure, it is even harder than it is for lead to quantify the prevalence and extent of exposures from unskilled DIY, or even simple maintenance actions (such as abrasive washing of sheet materials containing asbestos). Yet, the examples of lead and asbestos clearly suggest that once hazardous materials are contained in the existing housing stock they remain a permanent issue for the life span of these structures.

6.4 Conclusion

The examples of lead and asbestos show that many issues remain unresolved more than three decades since serious regulative efforts have attempted to control their use in Western countries. As discussed, scientific research is still revealing the exact mechanisms of the adverse health impacts these substances have. Some of the longer-term and across-generation observations are only now becoming available for exposure to the substances in the 1970s. As seen for asbestos the latency period could be 40 years, which makes it difficult to obtain clear causal evidence, and thus, conclusive evidence. In both cases newer research shows that much lower levels and shorter exposures than assumed in the 1970s or 1980s also have strong adverse health effects. The levels that were used and built-in during the 20th century now seem both high and hazardous.

Therefore, it is clear that the most significant pattern observed when evaluating lead and asbestos, the two substances most recognized for their adverse effect on human health, is delays in responding to recognition of significant indications of high health risks. As Olsen et al. (2011) explained, many Western countries are now facing the "third wave" of adverse health impacts from both lead and asbestos, through the removal, maintenance, or poor use of already built-in materials. Although these do not impact on everybody in contemporary society, the potential of hidden sources of exposure is present throughout society. Poor regulative control of DIY activities and a high presence of built-in hazards regularly contribute to the creation of new, although completely preventable, diseases through exposure to these well-known hazards. Current regulations are often powerless here.

Further successful removal of health hazards from already built-in materials that contain lead and asbestos is unlikely to be achieved through the use of regulation. Regulative changes have been excellent in removing hazards from new building, but existing built-in hazards also have to be addressed. It is irresponsible to allow continued unregulated, unpredictable exposures from well-known hazards through unskilled activities. To make improvements in this area on a broad-scale education is more likely to be effective in changing the attitudes people have towards the materials in their homes.

Furthermore, the way 20th century society has approached the use and regulation of lead and asbestos shows serious problems with the paradigm of waiting for a scientific proof, and only then regulating. Solid indications existed long before complete evidence, and in the case of lead and asbestos these indications were clear even before the start of the 20th century. The time lapse, of about a century, between solid indications and action, enabled prolific increased use over a very long period, thus unjustifiably increasing the levels of exposure of many people. For the health of the general population a more precautionary principle has to be adopted. It is essential to develop a way of recognizing the risks as being potentially serious much sooner. This chapter proposes the examples of lead and asbestos are evidence for advocating for a paradigm shift, towards adopting a more precautionary principle attitude.

Finally, the examples of lead and asbestos are in many ways easy to discuss. They are well known as harmful for the human body in almost any form of exposure. They are also very persistent, both having a very long half-life, and are very hard to eliminate from the environment properly. While these characteristics make them highly undesirable, they also make that undesirability very obvious. Therefore, the only partial elimination of the health hazards associated with lead and asbestos is very disappointing, and sets a grim agenda for some of the other substances considered here. The next chapter deals with the less obvious hazards of formaldehyde, plasticizers, and other volatile organic compounds. These chemicals are more reactive and much faster to change their form, making it hard to collect permanent data on them. These features make obtaining accepted scientific knowledge even more difficult.

References

Amato, M.S., Moore, C.F., Magzamen, S., Imm, P., Havlena, J.A., Anderson, H.A., et al., 2012. Lead exposure and educational proficiency: moderate lead exposure and educational proficiency on end-of-grade examinations. Ann. Epidemiol. 22 (10), 738–743.

American Cancer Society, 2014. Available from: www.cancer.org (accessed October 2014).

Brown, M.J., Margolis, S., 2012. Lead in drinking water and human blood lead levels in the United States. Centers for Disease Control and Prevention. Morbidity and Mortality Weekly Report. 61 (Suppl. Aug), 1–9.

Brown, M.J., McLaine, P., Dixon, S., Simon, P., 2006. A randomized, community-based trial of home visiting to reduce blood lead levels in children. Pediatrics. 117 (1), 147–153.

Browne, M.L., Varadarajulu, D., Lewis-Michl, E.L., Fitzgerald, E.F., 2005. Cancer incidence and asbestos in drinking water, Town of Woodstock, New York, 1980–1998. Environ. Res. 98, 224–232.

Carbonari, D., Campopiano, A., Ramires, D., Strafella, E., Staffolan, S., Tomasetti, M., et al., 2011. Angiogenic effect induced by mineral fibres. Toxicology. 288, 34–42.

Carlisle, J.C., Dowling, K.C., Siegel, D., Alexeeff, G.V., 2009. A blood lead benchmark for assessing risks from childhood lead exposure. J. Environ. Sci. Health. 44 (12), 1200–1208.

Department of Labour (NZ), 2006. Asbestos exposure in New Zealand 1992 to 2005. Department of Labour, Wellington, New Zealand.

Dixon, S.L., Jacobs, D.E., Wilson, J.W., Akoto, J.Y., Nevin, R., Clark, C.S., 2012. Window replacement and residential lead paint hazard control 12 years later. Environ. Res. 113, 14–20.

Donaldson, K., Poland, C.A., 2012. Inhaled nanoparticles and lung cancer – what we can learn from conventional particle toxicology. Swiss Med. Weekly. 142, w13547.

Dumortier, P., De Vuyst, P., 2012. Asbestos exposure during uncontrolled removal of sprayed-on asbestos. Ann. Occupat. Hygiene. 56 (1), 49–54.

Ethier, A.-A., Muckle, G., Bastien, C., Dewailly, E., Ayotte, P., Arfken, C., et al., 2012. Effects of environmental contaminant exposure on visual brain development: a prospective electrophysiological study in school-age children. NeuroToxicology. 33, 1075–1085.

Ewers, L., Clark, C.S., Peng, H., Roda, S.M., Menrath, B., Lind, C., et al., 2011. Lead levels in new residential enamel paints in Taipei, Taiwan and comparison with those in mainland China. Environ. Res. 111, 757–760.

Ferguson, A., Gilkey, B., Kern, D., Jasmine, D., 2012. An education initiative concerning lead in an Arkansas community: results from pre- and post-surveys. J. Commun. Healthy. 37, 599–609.

Fergusson, D.M., Boden, J.M., Horwood, L.J., 2008. Dentine lead levels in childhood and criminal behaviour in late adolescence and early adulthood. J. Epidemiol. Commun. Health. 62 (12), 1045–1050.

Gollenberg, A.L., Hediger, M.L., Lee, P.A., Himes, J.H., Buck Louis, G.M., 2010. Association between lead and cadmium and reproductive hormones in peripubertal U.S. Girls. Environ. Health Perspect. 118, 1782–1787.

Gump, B.B., Reihman, J., Stewart, P., Lonky, E., Granger, D.A., Matthews, K.A., 2009. Blood lead (Pb) levels: further evidence for an environmental mechanism explaining the association between socioeconomic status and psychophysiological dysregulation in children. Health Psychol. 28 (5), 614–620.

Guney, M., Zagury, G.J., 2012. Heavy metals in toys and low-cost jewelry: critical review of US and Canadian legislations and recommendations for testing. Environ. Sci. Technol. 46, 4265–4274.

Havlena, J., Kanarek, M.S., Coons, M., 2009. Factors associated with the seasonality of blood lead levels among preschool Wisconsin children. Wiscon. Med. J. 108 (3), 151–155.

Hayes, C.R., 2012. Heavy metals: lead. In: Pacheco-Porgal, F., Jalali, S., Fucic, A. (Eds.), Toxicity of Building Materials. Woodhead, Cambridge, pp. 283–296.

Hilts, S.R., Bock, S.E., Oke, T.L., Yates, C.L., Copes, R.A., 1998. Effect of interventions on children's blood lead levels. Environ. Health Perspect. 106, 79–83.

Hrubá, F., Strömberg, U., Černá, M., Chen, C., Harari, F., Harari, R., et al., 2012. Blood cadmium, mercury, and lead in children: an international comparison of cities in six European countries, and China, Ecuador, Ibelings and Morocco. Environ. Inter. 41, 29–34.

International Agency for Research on Cancer (IARC), 2016. IARC monographs on the evaluation of carcinogenic risks to humans. Available from: http://monographs.iarc.fr/ENG/Classification/index.php (accessed August 2016).

Jakubowski, M., 2011. Low-level environmental lead exposure and intellectual impairment in children – the current concepts of risk assessment. Inter. J. Occupat. Med. Environ. Health. 24 (1), 1–7.

Jordan, C.M., Yust, B.L., Robison, L.L., Hannah, P., Deinard, A.S., 2003. A randomized trial of education to prevent lead burden in children at high risk for lead exposure: efficacy as measured by blood lead monitoring. Environ. Health Perspect. 111, 1947–1951.

Jordan, C.M., Lee, P.A., Hampton, R., Pirie, P.L., 2004. Recommendations from lead poisoning prevention program participants: best practices. Health Promotion Practice. 5, 429–437.

Jordan, C.M., Lee, P.A., Olkon, R., Pirie, P.L., 2007. Messages from moms: barriers to and facilitators of behavior change in a lead poisoning preventive education project. J. Health Commun. Inter. Perspect. 12 (8), 771–786.

Kanarek, M.S., 2011. Mesothelioma from chrysotile asbestos: update. Ann. Epidemiol. 21, 688–697.

Keić-Świerczyńska, M., Wojtczak, J., 2000. Occupational ceramic fibres dermatitis in Poland. Occupat. Med. 50 (5), 337–342.

Khan, M.I., Ahmad, I., Mahdi, A.A., Akhtar, M.J., Islam, N., Ashquin, M., et al., 2010. Elevated blood lead levels and cytogenetic markers in buccal epithelial cells of painters in India: Genotoxicity in painters exposed to lead containing paints. Environ. Sci. Pollut. Res. 17 (7), 1347–1354.

Kjærheim, K., Ulvestad, B., Martinsen, J.I., Andersen, A., 2005. Cancer of the gastrointestinal tract and exposure to asbestos in drinking water among lighthouse keepers (Norway). Cancer Causes Control. 16, 593–598.

Koumantakis, E., Kalliopi, A., Dimitrios, K., Evangelos, G., 2009. Asbestos pollution in an inactive mine: determination of asbestos fibres in the deposit tailings and water. J. Hazardous Mater. 167, 1080–1088.

Krishnan, A.V., Park, S.B., Huynh, W., Lin, C. S.-Y., Henderson, R.D., Kiernan, M.C., 2012. Impaired energy-dependent processes underlie acute lead neuropathy. Muscle Nerve. 46 (6), 954–956.

Level: The Authority on Sustainable Building (NZ), 2016. Available from: www.level.org.nz (accessed August 2016).

Li, Y.-Y., Chen, T., Wan, Y., Xu, S., 2010. Lead exposure in pheochromocytoma cells induces persistent changes in amyloid precursor protein gene methylation patterns. Environ. Toxicol. 27 (8), 495–502.

Liu, G., Cheresh, P., Kamp, D.W., 2013. Molecular basis of asbestos-induced lung disease, Ann. Rev. Pathol. Mech. Dis., 8. pp. 161–187.

Lucas, J.-P., Le Bot, B., Glorennec, P., Etchevers, A., Bretin, P., Douay, F., et al., 2012. Lead contamination in French children's homes and environment. Environ. Res. 116, 58–65.

Machida, M., Sun, S.-J., Oguma, E., Kayama, F., 2009. High bone matrix turnover predicts blood levels of lead among perimenopausal women. Environ. Res. 109, 880–886.

Mazumdar, M., Bellinger, D.C., Gregas, M., Abanilla, K., Bacic, J., Needleman, H.L., 2011. Low-level environmental lead exposure in childhood and adult intellectual function: a follow-up study. Environ. Health. 10, 24.

Narag, R.E., Pizarro, J., Gibbs, C., 2009. Lead exposure and its implications for criminological theory. Crimin. Justice Behav. 36 (9), 954–973.

O'Grady, K., Perron, A., 2011. Reformulating lead-based paint as a problem in Canada. Am. J. Public Health. 101 (Suppl. 1), S: 176-187.

Olsen, N.J., Franklin, P.J., Reid, A., de Klerk, N.H., Threlfall, T.J., Shilkin, K., et al., 2011. Increasing incidence of malignant mesothelioma after exposure to asbestos during home maintenance and renovation. Med. J. Austr. 195 (5), 271–274.

Oulhote, Y., Le Bot, B., Poupon, J., Lucas, J.-P., Mandin, C., Etchevers, A., et al., 2011. Identification of sources of lead exposure in French children by lead isotope analysis: a cross-sectional study. Environ. Health. 10, 75.

Paglietti, F., Malinconico, S., Di Molfetta, V., Giangrasso, M., 2012. Guidelines for asbestos remediation at Italian superfund sites. J. Environ. Sci. Health Part C. 30, 253–286.

Park, E.-K., Takahashi, K., Jiang, Y., Movahed, M., Kameda, T., 2012. Elimination of asbestos use and asbestos-related diseases: an unfinished story. Cancer Sci. 103 (10), 1751–1755.

Peretz, A., Van Hee, V.C., Kramer, M.R., Pitlik, S., Keifer, M.C., 2008. Pleural plaques related to "take-home" exposure to asbestos: An international case series. Inter. J. General Med. 1, 15–20.

Petkovšek, S.A.S., Pokorny, B., 2013. Lead and cadmium in mushrooms from the vicinity of two large emission sources in Slovenia. Sci. Total Environ. 443, 944–954.

Philp, R.B., 2001. Ecosystems and Human Health: Toxicology and Environmental Hazards. Lewis Publishers, Boca Raton, FL.

Reid, A., Franklin, P., Olsen, N., Sleith, J., Samuel, L., Aboagye-Sarfo, P., et al., 2013. All-cause mortality and cancer incidence among adults exposed to blue asbestos during childhood. Am. J. Ind. Med. 56, 133–145.

Roberts, J.R., Allen, C.L., Ligon, C., Reigart, J.R., 2012. Are children still at risk for lead poisoning? Clin. Pediatr. XX (X), 1–6.

Sanchez, V.C., Pietruska, J.R., Miselis, N.R., Hurt, R.H., Kane, A.B., 2009. Biopersistence and potential adverse health impacts of fibrous nanomaterials: what have we learned from asbestos? WIREs Nanomed. Nanobiotech. 1, 511–529.

Sanders, T., Liu, Y., Buchner, V., Tchounwou, P.B., 2009. Neurotoxic effects and biomarkers of lead exposure: a review. Rev. Environ. Health. 24 (1), 15–45.

Sebastien, P., Bignon, J., Martin, M., 1982. Indoor airborne asbestos pollution: from the ceiling and the floor. Science. 216, 1410–1413.

Sobin, C., Parisi, N., Schaub, T., Gutierrez, M., Ortega, A.X., 2011. δ-Aminolevulinic acid dehydratase single nucleotide polymorphism 2 and peptide transporter 2*2 haplotype may differentially mediate lead exposure in male children. Arch. Environ. Contamin. Toxicol. 61, 521–529.

Spence, L.A., Aschengrau, A., Gallagher, L.E., Webster, T.F., Heeren, T.C., Ozonoff, D.M., 2008. Evaluation of the Webler-Brown model for estimating tetrachloroethylene exposures from vinyl-lined asbestos-cement pipes. Environ. Health. 7, 24.

Verstraeten, S.V., Aimo, L., Oteiza, P.I., 2008. Aluminium and lead: molecular mechanisms of brain toxicity. Arch. Toxicol. 82, 789–802.

Waikato District Health Board, 2012. Available from: http://www.waikatodhb.govt.nz/page/pageid/2145875708 (accessed August 2013).

Wei, B., Ye, B., Yu, J., Jia, X., Zhang, B., Zhang, X., et al., 2013. Concentrations of asbestos fibers and metals in drinking water caused by natural crocidolite asbestos in the soil from a rural area. Environ. Monitor. Assessment. 185, 3013–3022.

Winneke, G., 2011. Developmental aspects of environmental neurotoxicology: lessons from lead and polychlorinated biphenyls. J. Neurol. Sci. 308, 9–15.

Wright, J.P., Dietrich, K.N., Ris, M.D., Hornung, R.W., Wessel, S.D., Lanphear, B.P., et al., 2009. Lead concentrations with criminal arrests in early adulthood. PLoS Med. 2008. 5 (5), e101:0732-0740.

Yang, H., Huo, X., Yekeen, T.A., Zheng, Q., Zheng, M., Xu, X., 2013. Effects of lead and cadmium exposure from electronic waste on child physical growth. Environ. Sci. Pollut. Res. 20 (7), 4441–4447.

Yedjou, C.G., Milner, J.N., Howard, C.B., Tchounwou, P.B., 2010. Basic apoptotic mechanisms of lead toxicity in human leukemia (Hl-60) cells. Inter. J. Environ. Res. Public Health. 7 (5), 2008–2017.

Zahran, S., Mielke, H.W., Weiler, S., Berry, K.J., Gonzales, C., 2009. Children's blood lead and standardized test performance response as indicators of neurotoxicity in metropolitan New Orleans elementary schools. Neurotoxicology. 30 (6), 888–897.

Zhang, R., Lu, H., Tian, S., Yin, J., Chen, Q., Ma, L., et al., 2010. Protective effects of pre-germinated brown rice diet on low levels of Pb-induced learning and memory deficits in developing rat. Chemico-Biol. Interact. 184, 484–491.

How substances get regulated against in the building industry: formaldehyde, phthalate plasticizers in polyvinyl chloride/vinyl

7

Emina Kristina Petrović
Victoria University of Wellington, Wellington, New Zealand

7.1 Introduction

Compared to Stage 3 substances, discussed in Chapter 6, there is much less clarity about substances in all other stages. As explained in Chapter 5, there are a number of substages in Stage 2. In this chapter the broad group of volatile organic compounds (VOCs) is used to show the complexities inherent with the presence of organic chemicals in indoor air. VOCs, formaldehyde, and some phthalate plasticizers, are also considered a good example of the complexities that surround substances in Stage 2 of recognition. The focal question of this review is whether the process of recognizing health risks has accelerated since the very slow responses to the hazards of lead and asbestos. The main objective is to illustrate some of the challenges in progressing from a suspicion that chemicals are harmful to a more complete understanding of their effects and their elimination.

7.2 Overview of volatile organic compounds

There is no agreed definition of VOCs, with definitions and lists of substances somewhat varying between key international organizations monitoring their possible impact on human health (Guieysse et al., 2008). Generally, VOCs are based on carbon chains or rings (Guieysse et al., 2008). These are organic chemical compounds which have reasonably low boiling point, meaning that they readily evaporate in normal indoor air conditions (EPA, 2016a,b). Depending on boiling point and volatility, the World Health Organization (WHO) categorizes indoor organic pollutants as very volatile (VVOCs), volatile, and semivolatile (SVOCs) (Table 7.1).

Materials for a Healthy, Ecological and Sustainable Built Environment: Principles for Evaluation.
DOI: http://dx.doi.org/10.1016/B978-0-08-100707-5.00007-1

Table 7.1 **Classification of volatile organic pollutants**

Description	Abbreviation	Boiling point range (°C)	Example compounds
Very volatile (gaseous) organic compounds	VVOC	<0 to 50–100	Propane, butane, methyl chloride
Volatile organic compounds	VOC	50–100 to 240–260	Formaldehyde, D-Limonene, toluene, acetone, ethanol (ethyl alcohol) 2-propanol (isopropyl alcohol), hexanal
Semivolatile organic compounds	SVOC	240–260 to 380–400	Pesticides (DDT, chlordane), plasticizers (phthalates), fire retardants (PCBs, PBB)

Source: Based on Environmental Protection Agency (US) (EPA), 2016a. Available from: www.epa.gov (Accessed 2013-2016.), Environmental Protection Agency (US) (EPA), 2016b. Technical Overview of Volatile Organic Compounds. United States Environmental Protection Agency. Available from: https://www.epa.gov/indoor-air-quality-iaq/technical-overview-volatile-organic-compounds#2 (accessed August 2016.).

According to this definition, in normal indoor conditions, VOCs readily evaporate, VVOCs are gaseous, while SVOCs somewhat resist evaporation. The EPA acknowledges this classification as slightly arbitrary because all of these volatile compounds fall within the broad definition of the VOCs (EPA, 2016a,b). In this chapter they are, therefore, considered as one complex group. Already the official definitions of VOCs indicate some of the inconsistencies in the study of these chemicals, closely followed by inconsistent use of names of the same chemical (ATSDR, 2013).

Taking the broadest definition, VOCs provide a good basis for an evaluation of the impact of the numerous chemicals developed since World War Two. The US Agency for Toxic Substances and Disease Registry (ATSDR) lists 43 chemicals as VOCs with reasonably well-established adverse health effects (ATSDR, 2013). Almost half are specifically acknowledged as synthetic in origin; of the remainder, some come from crude oil or petrol processing, others are synthetic copies of chemicals that naturally appear in very small amounts in the oceans or volcanic eruptions, with only one-quarter being natural in origin (ATSDR, 2013).

VOCs can cause various health effects like eye, nose, and throat irritation, headaches, loss of coordination, nausea, and damage to liver, kidneys, and the central nervous system (Katsoyiannis et al., 2008; EPA, 2016a,b). Some of these chemicals "can cause cancer in animals; some are suspected or known to cause cancer in humans" (EPA, 2016a,b), with others having toxic effects (ATSDR, 2013). VOCs can enter homes through the use of paints, varnishes, solvents, and building and furnishing materials, as well as via many hobby, home office, and art and craft materials, combustion, and cleaning products (EPA, 2016a,b). Thus, one part of VOC exposure in homes is related to building and furnishing materials, while the other is related to everyday activities in the space. The focus of this discussion is the former.

The most recognized hazardous VOCs for human health include formaldehyde and the aromatic compounds of benzene, toluene, ethylbenzene, xylenes, and styrene (on VOCs in carpets see Table 8.2). Benzene, and more recently formaldehyde, are recognized by official organizations as high risk both for toxicity and carcinogenesis and are given the highest priority on both scales (ECHA, 2013; IARC, 2013; IRIS, 2013; PubChem Compound, 2013; ATSDR, 2013). Toluene, styrene and xylenes still engender much discussion on their exact level of toxicity and carcinogenesis, which has influenced their current lower level of recognition as risks (ECHA, 2013; IARC, 2013; IRIS, 2013; PubChem Compound, 2013; ATSDR, 2013). From this it is possible to consider benzene and formaldehyde as currently in the later substages of Stage 2 recognition of risks, while toluene, styrene, and xylenes are probably in earlier substages, given that official organizations are already acknowledging a growing body of knowledge on their harmful effects (CalEPA, 2013; IRIS, 2013; PubChem Compound, 2013; ATSDR, 2013), without as yet much action being taken against these.

Currently there is no method capable of measuring all VOCs present in a space at once, rather different measuring approaches are used for different VOCs. For example, benzene and toluene are measured using a different method than for formaldehyde and other similar compounds (EPA, 2016a,b). This means different researchers tend to focus on a selection of VOCs when assessing indoor air, making the results less comparable. For example, two studies evaluating VOCs from carpets used different sampling techniques and only one VOC was tested in both studies, making comparisons difficult (Wilke et al., 2004; Katsoyiannis et al., 2008). Measurements of SVOCs are even more complex and take longer because of the sorption of the measuring instruments (Xu et al., 2012; Wilke et al., 2004).

There are other difficulties with assessing levels of VOCs in indoor air, as their concentrations depend on the indoor air conditions. For example, Pegas et al. (2011) have demonstrated that in schools VOC emissions increase for most groups of substances during winter, which they propose could be attributed to a decrease in ventilation during colder periods (Pegas et al., 2011). However, there are important exceptions to this general rule, as formaldehyde emissions are highest in warmer periods (Pegas et al., 2011). Room temperature, relative humidity, and ventilation rate all directly influence the emission of VOCs.

Efforts have been made to establish which VOC chemicals are particularly problematic, or if the total mass of VOCs (TVOCs) is the main problem (Sundell, 2004). Unfortunately, to date neither approach has provided desired simplifications to this area (see discussion on TVOCs in Chapter 8). Rather, more recent research has focused on recognition of indoor chemical reactions and more sophisticated understanding of reactions within human bodies to air pollutants, indicating that the level of complexity will increase before any certain conclusions can be reached.

With the absence of firm definitions of VOCs, various measuring methods, and many factors influencing fluctuations in VOCs levels indoors, it is not surprising that research papers often provide results that are only partially comparable. This lack of consistency indicates the relative infancy of research into VOCs, and is a significant feature of any discussion in this area.

7.2.1 Reduction options

In contrast to lead and asbestos which once built-in get released only if mechanically disturbed, there are many chemicals in domestic environments that readily get released under normal air conditions. Therefore, any discussion of VOCs is inseparably related to indoor air quality.

When considering reducing indoor contamination from VOCs, the American Society of Heating, Refrigeration and Air-Conditioning Engineers (ASHRAE) proposed three options: source elimination, dilution with ventilation air, and air filtration (ASHRAE, 2009, 10.10). As an organization they have focused on the two latter. In addition to increasing ventilation rates, there has been experimentation in new ways of removing VOCs from indoor air using filtering and purification systems. Unfortunately, literature in this area indicates available products can successfully remove particulate matter but still struggle to remove gaseous chemicals from air effectively (Schmid et al., 2011; Hart et al., 2011; Guieysse et al., 2008). Furthermore, some could be producing even more harmful chemicals (Schmid et al., 2011; Hubbard et al., 2005). It is claimed some textiles can remove VOCs from indoor space (Drapalux brand, for example) but no independent confirmation of such results is available. The most noticeable measured VOC removal has so far been observed with specific species of pot plants (Pegas et al., 2012; Daniels, 1997), but studies generally record high variability in level of decrease and fall short of making explicit recommendations. Some of the more recent experimental work includes use of genetically modified plants capable of absorbing high levels of formaldehyde from surrounding air (Nian et al., 2013), which unfortunately poses its own set of health concerns associated with any genetic modifications. However, perhaps a more traditional approach that places greater emphasis on elimination of hazards would be a more reliable course of action than any of these experiments.

7.3 Issues with formaldehyde

One VOC that has received much research attention in recent years is formaldehyde. Formaldehyde (HCHO) is a naturally occurring organic chemical; it is a product of one-carbon metabolism and is a toxic compound because it can combine with proteins, nucleic acids, and lipids (Nian et al., 2013). In contrast to lead and asbestos, which are naturally occurring substances unrelated to the natural processes of living organisms, formaldehyde is part of the natural chemistry and metabolism of many living organisms and therefore, as discussed below, it is much harder to talk about complete elimination of formaldehyde. Its reactivity and quick absorption in humans make it additionally difficult to evaluate. Formaldehyde is found in most human and other living cells in varying concentrations as a normal product of the metabolism of serine, glycine, methionine, and chlorine (Checkoway et al., 2012). Formaldehyde is a gas at room temperature (WHO, 2010; Salthammer et al., 2010).

In ambient air, the half-life of formaldehyde is about one hour, because it is either transformed into carbon dioxide through the process of photo-oxidation or reacts with hydroxyl radicals to give formic acid (WHO, 2010).

Scientifically, formaldehyde was first described in 1855, and by 1867 a German chemist achieved its synthesis through dehydration of methanol (Salthammer et al., 2010). The early 20th century saw much development in applications of formaldehyde. In 1907, the combination of phenol and formaldehyde became available to industry under the name "bakelite" (Salthammer, 2013). In 1931, the first glue based on the combination of urea and formaldehyde entered the market, and the first commercial particle board was produced during World War II in Bremen, Germany, and since 1950 has become a common alternative to solid wood (Salthammer et al., 2010). Such products tend to release formaldehyde due to chemical reactions between indoor air and the resins, specifically, due the hydrolysis of urea–formaldehyde resins (Salthammer, 2013).

The recognition of health concerns related to formaldehyde swiftly followed its more popular use. In 1962, indoor air pollution from release of formaldehyde from installed particle board was first observed (Salthammer, 2013; Salthammer, Mentese and Marutzky, 2010). During the 1960s and 1970s, some very high indoor concentrations of formaldehyde were observed, and since then great efforts have been made to reduce its release into indoor air (Salthammer, 2013). In 1977 in Germany an early guideline on formaldehyde specified a value of 0.1 ppm for human exposure in dwellings (Salthammer et al., 2010). By 1981, early regulations of formaldehyde emissions from wood-based materials were established in Germany and Denmark, followed by the United States in the mid-1980s (Salthammer et al., 2010). Since 1981 the IARC has evaluated formaldehyde every few years, and in 2004, reclassified it from group 2B "possibly carcinogenic to humans" to group 1 "carcinogenic to humans" (IARC Monograph 88, 2006), with additional adjustments to this classification in 2012 (IARC Monograph 100F, 2012). Nevertheless, to date there is no universally agreed standard regarding acceptable exposure to formaldehyde.

Formaldehyde can enter the human body through inhalation, ingestion, or dermal absorption (Kim et al., 2011). In addition to formaldehyde emissions in air, it has been used in many cleaning and cosmetic products and even as a food preservative for some Italian cheeses, dried foods, and fish (Kim et al., 2011). Once absorbed, the high solubility of formaldehyde in water causes possible rapid absorption in almost every tissue in the body (Salthammer et al., 2010; Kim et al., 2011). The biological half-life of formaldehyde is extremely short at about 1 minute (Salthammer et al., 2010; Wolkoff and Nielsen, 2010).

The WHO (2010) notes that large amounts of formaldehyde can be released into the natural environment through biomass combustion, such as forest or bush fires, or natural decomposition, or through anthropogenic activities, such as industrial emissions and fuel combustion from traffic, and other combustion processes (WHO, 2010).

When used as glue, as in composite wood, formaldehyde is generally in combination with another active chemical in the form of urea–formaldehyde, phenol–formaldehyde, and melamine–formaldehyde (UF, PF, and MF resins), and these accounted for 70% of global demand for formaldehyde in 2015 (IHS

Chemical, 2015a). More recently, China has become the largest single market for formaldehyde, and in 2015 consumed almost half of world demand (IHS Chemical, 2015a). The world consumption of formaldehyde is forecast to continue growing at an average rate of about 5%, with continuing rapid demand in China and other parts of Asia balancing out moderate growth in North America and Western Europe (Kim et al., 2011). Therefore, even after the significant efforts to reduce exposure to formaldehyde in Western countries, its use is forecast to increase in non-Western countries, following similar trends to the examples of lead and asbestos.

Indoor air sources of formaldehyde include combustion processes such as smoking, heating, cooking, and candle or incense burning (WHO, 2010), although formaldehyde levels from these will vary with lifestyle. Building materials and consumer products that emit formaldehyde are a major source of noncombustion indoor formaldehyde, especially in the first few months after installation (WHO, 2010). The main sources are composite wood products such as medium-density fiberboard, particle board, and plywood, due to use of urea–formaldehyde or phenol–formaldehyde glues, fiberglass insulation containing similar glues, and other products such as carpets and drapes where formaldehyde has been used as a biocide (Wolkoff and Nielsen, 2010). Two additional sources of indoor formaldehyde, which have received more research attention recently, are indoor chemical reactivity and outdoor formaldehyde. The WHO recognizes that chemical reactions in air through the oxidation of VOCs and reactions between ozone and alkenes (especially terpenes) within indoor spaces contribute to secondary formation of formaldehyde (WHO, 2010). They also note that these chemical processes and resulting indoor concentrations are still not fully quantified (WHO, 2010). This is consistent with other sources which have observed that publication of research related to indoor air chemistry has greatly increased in the last 10–15 years, from a position of virtually nothing before the 1990s (Weschler, 2011).

Although formaldehyde can now be seen as being in an early stage of complete elimination, more recent work often focuses on demonstrating that very low-level exposures to it are harmless, arguing against further decreases in regulated levels. In 2010, Nielsen and Wolkoff, at the time on the WHO's team developing *WHO guidelines for indoor air quality: selected pollutants* (2010), concluded that there are levels of formaldehyde where no excess nasopharyngeal cancer is observed (Nielsen and Wolkoff, 2010). Further, they argued for the "no observed adverse effect level" (NOAEL) approach when regulating for exposure to formaldehyde (Nielsen and Wolkoff, 2010). The same approach was adopted by the WHO (2010). Nielsen and Wolkoff argued that based on estimates the risk to the general population from 0.1 to 0.2 ppm of formaldehyde is negligible, and further that the earlier (2000) guideline set by the WHO of 0.08 ppm (0.1 mg/m^3) introduces an additional margin-of-exposure, and that there was no need to change these (Nielsen and Wolkoff, 2010). Nielsen et al. (2013) recommend the conversion factor: 1 ppm = 1.228 mh/m^3 at 25°C and 1 atm.) This level was considered "defendable for prevention of all types of cancer, including lymphohematopoietic malignancies" (Nielsen and Wolkoff, 2010). This should be considered in relation to the formaldehyde levels in outdoor air.

For much of the history of formaldehyde investigations, indoor levels were higher than outdoor levels. With more recent reductions in the former, Salthammer (2013) suggests indoor concentrations of 0.008 ppm (the lower end of WHO-recommended levels) are often unachievable because outdoor concentrations are higher. Experts are now asking whether further reductions of indoor formaldehyde levels are possible due to outdoor formaldehyde levels (Salthammer, 2013). Because formaldehyde can be released from many sources, including most combustion processes, it is likely that zero formaldehyde indoors is impossible without great improvements of outdoor air quality. Thus, it might be useful to focus on complete avoidance of added formaldehyde in building components.

More recent studies suggest that with a number of existing regulations of formaldehyde in building components a general drop in indoor air formaldehyde levels is being achieved in many places (Salthammer, 2013). In Europe, Canada, the United States, and Japan, mean formaldehyde levels in homes and dwellings were generally within 20–40 $\mu g/m^3$ (Nielsen et al., 2013). However, the same review observes much higher mean concentrations in China of 240 $\mu g/m^3$ (Nielsen et al., 2013). Furthermore, in public buildings in Europe, mean concentrations of formaldehyde have generally fallen below 25 $\mu g/m^3$ (Wolkoff and Nielsen, 2010), meaning the general levels in public buildings and offices in Europe and the United States are now lower than in dwellings (Nielsen et al., 2013; Wolkoff and Nielsen, 2010). Nevertheless, these same studies reiterate the important relationship between indoor concentrations of formaldehyde and air exchange rate, and estimate that air exchange rates need to increase by 30% in homes with new formaldehyde sources (Wolkoff and Nielsen, 2010). They also emphasize that in new Californian homes with low formaldehyde-emitting materials, half the conventionally needed exchange rates were sufficient to keep the formaldehyde levels low (Wolkoff and Nielsen, 2010). This reliance on removal of formaldehyde through ventilation is problematic, and suggests that complete avoidance of introduction of formaldehyde indoors is likely to achieve better health and energy outcomes.

Although legislatively there is now agreement that very significant reductions in formaldehyde levels are essential, with regulation in many countries, there is discussion in the research community on the exact safe levels and the adverse health effect of low-level formaldehyde exposure. Although not conclusive, these discussions indicate that the point of significant reduction regulation is not necessarily the point when all conclusions are available. As with lead and asbestos, further adjustment of the limits set might be needed as better understanding of any adverse health effects from low levels of exposure become available. Disagreements in the available literature also reveal the problem of having only partial understanding of the mechanisms formaldehyde exposure triggers inside the human body, suggesting the need for new discoveries in future.

One on-going problem well illustrated by formaldehyde is that reductions in levels (attributed to ventilation) in commercial and public buildings took place much earlier and more successfully than in domestic spaces. Therefore, in addition to strategies proven effective in nondomestic contexts, there is a need to develop effective strategies for indoor domestic spaces. Such a lack is illustrated by the example of New Zealand. Although New Zealand is a leading producer of zero formaldehyde composite wood, due to exports in 2011 the product was only available locally at special request. The 2006 New Zealand standard limits formaldehyde in particle boards to less than 0.5 mg/L for E0 (so-called “zero” boards), less than 1.5 mg/L for E1, and less than 4.5 mg/L for E2 (AS/NZS 1859.1:2004, February 2006 amendments). Often E labels are not displayed in building material retail outlets and the information is difficult to ascertain.

7.4 Issues with vinyl, polyvinyl chloride, and plasticizers

Another significant group of VOCs or SVOCs is related to indoor use of the family of vinyl products, with concerns being raised about the use of polyvinyl chloride (PVC) as a base material and phthalate plasticizers as additives to it. World production of PVC, which is a significant consumer of phthalates, is around 40 million tons, and forms the second largest plastic by volume after polyethylene (PE) (Akovali, 2012). About 70% of PVC consumption worldwide is in the construction industry in applications such as pipes, wiring, cladding, flooring, and wallpaper (IHS Chemical, 2014; Akovali, 2012). Some of these rely on the use of plasticized PVC (pPVC in flooring, wiring, and wallpaper), while others use unplasticized PVC (uPVC, in pipes and cladding). PVC was first invented in 1872, but its importance was not fully recognized until 1926 (Akovali, 2012). In 1933 vinyl flooring was displayed at the Century of Progress Exposition in Chicago, but it did not become commercially available until the end of World War Two. Subsequently, vinyl flooring largely overtook the use of linoleum. The latter was invented and patented in 1845, and first manufactured in Scotland in the 1860s, and is made from renewable materials such as linseed oil, pine rosin, cork dust, wood flour, and mineral fillers (Akovali, 2012).

PVC is produced through polymerization of vinyl chloride monomer (VCM) (Akovali, 2012), and there are serious concerns about the health effect of VCM, which is genotoxic (Kumar et al., 2013), a known human carcinogen, and is toxic for immune and cardiovascular systems, liver, and organ development (ATSDR, 2013). The group Greenpeace has advocated the complete phasing-out of PVC globally because of its manufacturing byproducts and the complex environmental issues which surround its disposal (Akovali, 2012). The IHS 2011 review of the world consumption of PVC reported noticeable reductions (IHS Chemical, 2014). While they interpreted these as reflecting the 2007/8 recession, some of the changes were hard to explain in this way. Unfortunately, at the world scale this makes little change, due to the increased use of PVC in Asian countries, which use 80% of world consumption (IHS Chemical, 2015).

However, for indoor air quality the most significant concern in relation to PVC is plasticizers, which generally come in the form of phthalates. Phthalates are a group of aromatic chemicals containing a phenyl ring with two attached and extended acetate groups (ASTDR, 2013). They are added to PVC or other plastics to increase their flexibility and transparency, and are used in proportions as varied as 10%–60% of final PVC products (Liu and Little, 2012). Flexible, plasticized, polyvinyl chloride (pPVC) accounts for 80%–90% of world plasticizer consumption (IHS Chemical, 2015). Because they are not part of the chain of polymers that make plastics, they can be slowly released from these products (ASTDR, 2013; Liu and Little, 2012). In architecture the main use of plasticizers is for vinyl flooring, while they can be found in a wide variety of everyday products including vinyl upholstery, shower curtains, food containers, cling wraps, toothbrushes, toys, tools, automobile parts, adhesives, and sealers (Liu and Little, 2012; ASTDR, 2013). They are also used in cosmetics, insecticides, and aspirin (ASTDR, 2013). Although in terms of chemistry they form their own subgroup, at a more general level they can be seen as part of the semivolatile organic compounds (SVOCs) group (Liu and Little, 2012; Xu et al., 2012). In addition to phthalates, other plasticizers currently in use include aliphatics, epoxy, terephthalates, trimellitates, polymerics, and phosphates (IHS Chemical, 2015b). In recent years the consumption of phthalates has decreased from 88% of all plasticizers in 2005, to 78% in 2012, and is forecast to further decrease to 75.5% by 2018, with China accounting for nearly 38% of world consumption in 2012 (IHS Chemical, 2015b). Between 2011 and 2018 world consumption of phthalate plasticizers is forecast to have reduced growth of 2.4%, due to lower-molecular-weight phthalates, such as DEHP, being replaced by nonphthalates (IHS Chemical, 2015b).

As with other VOCs, there is much variation in the way the harmfulness of phthalates is acknowledged by different organizations. For example, ECHA includes eight phthalates in their list of substances of very high concern (Table 7.2), while its US counterpart, ATSDR, only lists four (di(2-ethylhexyl)phthalate (DEHP), di-*n*-butyl phthalate, di-*n*-octylphthalate (DNOP), and diethyl phthalate; ATSDR, 2013). Generally, phthalates are recognized as toxic for reproduction, and systematic inclusion of phthalates in the list of substances of very high concern has only started since 2008 (Table 7.2; ECHA, 2013).

Historically, different phthalates have received more prolific use in different periods. Di(2-ethylhexyl)phthalate (DEHP) is a short-chain phthalate frequently found in older PVC products. In 1999, the European Union regulated against its use in toys, and this was followed by regulations for other products (Holmgren et al., 2012). Since then, longer-chain phthalates such as di-iso-nonyl phthalate (DINP) have been used more and there is hope that the 2005 European Union ban of use of all phthalates in toys that can be put in a baby's mouth will have the same positive effect, and that development of nonphthalate plasticizers could follow (Holmgren et al., 2012). Because of this historical background, DEHP is more researched than other phthalates.

There is still limited epidemiological evidence related to phthalates, and decisions to classify these chemicals as being of very high concern are mainly based on animal studies. Studies of the health impact of DEHP on rats have established that it is an endocrine disruptor with antiandrogenic activity, and it suppresses testosterone-related processes (Martinez-Arguelles et al., 2013). Once DEHP is absorbed by the lining of the gut it metabolizes into mono-2-ethylhexyl phthalate (MEHP), which has antiandrogenic activity 10 times greater than DEHP (Martinez-Arguelles et al., 2013). When biomonitoring of concentrations of phthalate

Table 7.2 Summary of key information on phthalates from ECHA's list of substances of very high concern (SVHC)

Phthalate	CAS no.	Year identified as SVHC	Reason for inclusion	More specific reasons
Bis(2-ethylhexyl)phthalate (DEHP)	117-81-7	2008	Toxic for reproduction (2008) Equivalent level of concern having probable serious effects to environment (2014)	May impair fertility; May cause harm to the unborn child. Endocrine disruptor for the environment.
Dibutyl phthalate (DBP)	84-74-2	2008	Toxic for reproduction	May cause harm to the unborn child. Possible risk of impaired fertility. Dangerous for the environment: very toxic to aquatic organisms.
Benzyl butyl phthalate (BBP)	85-68-7	2008	Toxic for reproduction	May cause harm to the unborn child.
Diisobutyl phthalate (DIBP)	84-69-5	2009	Toxic for reproduction	May cause harm to the unborn child. Possible risk of impaired fertility.
Bis(2-methoxyethyl) phthalate	117-82-8	2011	Toxic for reproduction	May damage the unborn child. Suspected of damaging fertility.
n-pentyl-isopentylphthalate	776297-69-9	2012	Toxic for reproduction	May damage fertility. May damage the unborn child.
Diisopentylphthalate (DIPP)	605-50-5	2012	Toxic for reproduction	May damage fertility. May damage the unborn child.
Dipenthyl phthalate (DPP)	131-18-0	2013	Toxic for reproduction	May damage fertility. May damage the unborn child.

Source: Based on European Chemicals Agency (ECHA), 2017. Avaialble from: https://echa.europa.eu/ (accessed March 2017.).

metabolites was undertaken in the general population using blood and urine, the blood results suggested that over 75% of the US population is exposed to phthalates, while for urine this was as high as 95% (Xu et al., 2012). DEHP and its metabolites have been found in semen, saliva, amniotic fluid, umbilical cord blood, human milk, and baby formula, and it is estimated that today most humans are exposed to it at levels of 3–30 μg/kg/day from various products (Martinez-Arguelles et al., 2013). Animal studies have shown that acute exposure to DEHP in utero disrupts the organogenesis of androgen-dependent tissues by inhibiting testosterone production in a dose-dependent manner (Martinez-Arguelles et al., 2013). Due to this mechanism, DEHP has the potential to influence many other aspects of development, and studies have found that in utero exposure to DEHP induces long-term cardiovascular changes in the male offspring, and affects the behavior of young adult and elderly male rats (Martinez-Arguelles et al., 2013). When similar animal studies tested the impact of a combination of bisphenol-A (BPA) and two very common phthalates (DEHP and DBP), they found that the impact increased in subsequent generations, leading to third-generation offspring having pubertal abnormalities, and developing obesity and testis and ovarian disease (Manikkam et al., 2013). This was to such a degree that the researchers concluded that ancestral environmental exposures could be generating transgenerational inheritance of disease, often with adult onset (Manikkam et al., 2013).

As with lead and asbestos, early animal studies have so far tended not to quantify the impact of lower doses and more complex combinations of factors on human health. Consequently, research findings such as those now available for older phthalates should be considered as moving from earlier substages of Stage 2 of recognition, to more developed later substages, where the dose effect is evaluated. From that perspective, it is good that changes have started to be implemented even without conclusive epidemiological evidence for human exposures. However, the fact that some international organizations have classified phthalates as substances of very high risk is only part of the process, as it takes much longer for risk to be integrated into regulations worldwide. For example, an article from 2009 commented on challenges with integrating such regulative changes into the NZ context (Anon, 2009). Echoing the 2005 European ban of phthalates in children's toys that can be put into a baby's mouth, a similar regulation took effect in the United States in 2009, while in NZ at that date phthalates were only regulated in cosmetics (Anon, 2009). The article explained that because different possible sources of phthalates are regulated by different bodies (the Ministry of Consumer Affairs is responsible for toy safety, while food packaging comes under the Food Standards Australia New Zealand, and the New Zealand Food Safety Authority), a series of individual evaluations and decisions must be made before phthalates can be consistently removed from the NZ market (Anon, 2009). Even then, the chances of imported products containing phthalates remain. These are all features of substances that are making the transition from earlier to later substages of Stage 2 of recognition.

Nevertheless, with the classification of phthalates as substances of high concern, some labeling schemes have responded quickly. For example, the Blue Angel label specifies that "no plasticising substances from the class of phthalates may be used in the manufacture of floor coverings," but also that floor coverings "shall not contain more than 0.1% by mass of phthalates as impurities" (Der Blaue Engel, 2011a,b).

When releases of phthalates from pPVC in interiors were evaluated, it was established that because of saturation and sorption processes, and because these establish an equilibrium over a long period of time, a very small amount of pPVC emits almost as many phthalates as a large area, and that this did not change with increased ventilation rates (Afshari et al., 2004). Therefore, Afshari et al. (2004) concluded that "if there is any surface material in an interior that contains plasticisers, it is impossible to avoid the phthalates in indoor air." Similar difficulties with saturation, sorption, and long equilibrium states were observed by others (Xu et al., 2012). Once emitted into the indoor air, phthalates find their way into household dust (Gevao et al., 2012; Bornehag et al., 2005). Studies of dust samples for phthalates have established that vinyl flooring should be considered a source of nondietary exposure to phthalates and that, just as with lead paint dust, children and especially toddlers are more at risk than adults (Gevao et al., 2012). DEHP was generally reported as the most predominant phthalate found in these dust studies (Gevao et al., 2012; Bornehag et al., 2005).

Because the epistemological evidence around the impact of phthalates on human health is still limited, it seems reasonable to expect future discoveries in this area. Current studies confirm that during the second half of the 20th century vinyl flooring contained DEHP in combination with other phthalates (Gevao et al., 2012; Bornehag et al., 2005). Therefore, it seems likely that as the second and third generation of offspring from those exposed to such flooring are born, more adverse health effect on humans could be expected. This is particularly noteworthy, as increases in human obesity and reproductive difficulties are both potentially triggered by phthalates.

7.5 Conclusion

This overview of key issues related to indoor air quality and VOCs, with more detailed consideration of formaldehyde and phthalates, indicates that in the last 10–15 years this area has experienced great changes. Regulations against formaldehyde and phthalates have started to appear in developed countries as part of a shift that has led to recognition of a broader range of health risks. Current regulations are similar in character to those seen in the late 1970s and early 1980s for lead and asbestos, with continuing discussion of safe levels and applications. For lead and asbestos, this was the point when more sophisticated research and better understanding of lower dose impacts started to appear, leading subsequently to more stringent regulations against these substances. Therefore, significant new discoveries are still possible (especially when dealing with lower dose exposure and epistemological evidence of mechanisms of impact on human health) and renegotiation of regulation levels is likely to follow. Consequently, current regulations should be expected to undergo further adjustments.

One of the main conclusions is that the stories of formaldehyde and phthalates are still unfinished, making it harder to assess whether responses to these substances were more accelerated than for lead and asbestos. Since the end of World War Two formaldehyde and phthalates have appeared in everyday objects, and the introduction of regulations against their use in the early 2000s indicates the process took at least half a century, with probably another decade or longer before it is in the final stages. Considering the history of the human use of lead and asbestos, this seems to be an improvement. However, if the comparison is made between the more accelerated industrial use of lead and asbestos (from the late 19th or early 20th century), this acceleration is mild at best. Half a century seems a very long period for exposing people to harmful substances.

Furthermore, formaldehyde and phthalates are not the only new substances to have been introduced into the everyday human environment mid-20th century, and successive regulation against a number of other chemicals could happen in years to come. So far human society has worked through eliminating the most obvious hazards, and if such patterns continue, recognition of substances with less obvious or even less adverse health impacts could continue over a long period. Chapter 8, looks at such substances.

References

Afshari, A., Gunnarsen, L., Clausen, P.A., Hansen, V., 2004. Emission of phthalates from PVC and other materials. Indoor Air. 14, 120–128.

Agency for Toxic Substances and Disease Registry (US) (ATSDR), 2013. Available from: www.atsdr.cdc.gov (accessed 08.13.).

Akovali, G., 2012. Plastic materials: polyvinyl chloride (PVC). In: Pacheco-Torgal, F., Jalali, S., Fucic, A. (Eds.), Toxicity of Building Materials. Woodhead Publishing, Cambridge, pp. 23–53.

American Society of Heating, Refrigeration and Air-Conditioning Engineers (US) (ASHRAE) (2009), 2009. ASHRAE Handbook: Fundamentals. ASHRAE, Atlanta.

Anon, 2009. Plasticizers: Regulations. Consumer NZ. Available from: http://www.consumer.org.nz/reports/plasticisers/regulations (accessed August 2013.).

Blue Angel/Der Blaue Engel, 2011a. Basic Criteria for Award of the Environmental Label: Resilient Floor Coverings RAL-UZ 120. Available from: http://www.blauer-engel.de/en/ (accessed August 2013.).

Blue Angel/Der Blaue Engel, 2011b. RAL-UZ 120 – Resilient Floor Coverings: Basic Criteria for Award of the Environmental Label. Available from: http://www.blauer-engel.de/en/products_brands/survey_basic_award_criteria.php (accessed February 2013.).

Bornehag, C.-G., Lundgren, B., Weschler, C.J., Sigsgaard, T., Hagerhed-Engman, L., Sundell, J., 2005. Phthalates in indoor dust and their association with building characteristics. Environ. Health Perspect. 113, 1399–1404. Available from: http://dx.doi.org/10.1289/ehp.7809.

California Environmental Protection Agency (CalEPA), 2013. Available from: http://www.calepa.ca.gov (accessed August 2013.).

Checkoway, H., Boffetta, P., Mundt, D.J., Mundt, K.A., 2012. Critical review and synthesis of the epidemiologic evidence on formaldehyde exposure and risk of leukaemia and other lymphohematopoietic malignancies. Cancer Causes Control. 23, 1747–1766.

Daniels, K., 1997. The Technology of Ecological Building: Basic Principles and Measures, Examples and Ideas. Birkhäuser, Basel; Boston; Berlin.

Environmental Protection Agency (US) (EPA), 2016a. Available from: www.epa.gov (Accessed 2013-2016.).

Environmental Protection Agency (US) (EPA), 2016b. Technical Overview of Volatile Organic Compounds. United States Environmental Protection Agency. Available from: https://www.epa.gov/indoor-air-quality-iaq/technical-overview-volatile-organic-compounds#2 (accessed August 2016.).

European Chemicals Agency (ECHA), 2017. Avaialble from: https://echa.europa.eu/ (accessed March 2017.).

European Chemicals Agency (ECHA), 2013. Available from: https://echa.europa.eu/ (accessed April 2013.).

Gevao, B., Al-Ghadban, A.N., Bahloul, M., Uddin, S., Zafar, J., 2012. Phthalates in indoor dust in Kuwait: implications for non-dietary human exposure. Indoor Air. 23 (2), 126–133.

Guieysse, B., Hort, C., Platel, V., Munoz, R., Ondarts, M., Revah, S., 2008. Biological treatment of indoor air for VOC removal: potential and challenges. Biotechnol. Adv. 26 (5), 398–410.

Hart, J.F., Ward, T.J., Spear, T.M., Rossi, R.J., Holland, N.N., Loushin, B.G., 2011. Evaluating the effectiveness of a commercial portable air purifier in homes with wood burning stoves: a preliminary study. J. Environ. Public Health2011: ID 324809.

Holmgren, T., Persson, L., Andersson, P.L., Haglund, P., 2012. A generic emission model to predict release of organic substances from materials in consumer goods. Sci. Total Environ. 437, 306–314.

Hubbard, H.F., Coleman, B.K., Sarwar, G., Corsi, R.L., 2005. Effects of an ozone-generating air purifier on indoor secondary particles in three residential dwellings. Indoor Air. 15, 432–444.

International Agency for Research on Cancer (IARC), 2013. IARC Monographs on the evaluation of carcinogenic risks to humans. Available from: http://monographs.iarc.fr/ENG/Classification/index.php (accessed 01–08.13.).

IHS Chemical, 2014. Chemical Economic Handbook: IHS Chemical: Polyvinyl Chloride (PVC) Resins. Available from: www.ihs.com (accessed August 2016.).

IHS Chemical, 2015. Available from: https://www.ihs.com/industry/chemical.html (accessed 05.15.).

IHS Chemical, 2015a. Chemical Economic Handbook: IHS Chemical: Formaldehyde. Available from: www.ihs.com (accessed August 2016.).

IHS Chemical, 2015b. Chemical Economic Handbook: IHS Chemical: Plasticizers. Available from: www.ihs.com (accessed September 2016.).

Integrated Risk Information System (IRIS), 2013. Available from: http://www.epa.gov/iris/index.html (accessed July 2013.).

International Agency for Research on Cancer (IARC), 2006. IARC Monographs on the Evaluation of Carcinogenic Risks to Humans: Monograph 88. Available from: http://monographs.iarc.fr/ENG/Classification/index.php (accessed September 2016.).

International Agency for Research on Cancer (IARC), 2012. IARC Monographs on the Evaluation of Carcinogenic Risks to Humans: Monograph 100F. Available from: http://monographs.iarc.fr/ENG/Classification/index.php (Accessed September 2016.).

Katsoyiannis, A., Leva, P., Kotzias, D., 2008. VOC and carbonyl emissions from carpets: A comparative study using four types of environmental chambers. J. Hazard. Mater. 152 (2), 669–676.

Kim, K.-H., Ara Jahan, S., Lee, J.-T., 2011. Exposure to formaldehyde and its potential human health hazards. J. Environ. Sci. Health (Part C). 29, 277–299.

Kumar, A.K., Balachandar, V., Arun, M., Khan, S.A., Ahamed, M., Kumar, S.S., et al., 2013. A comprehensive analysis of plausible genotoxic covariates among workers of a polyvinyl chloride plant exposed to vinyl chloride monomer. Arch. Environ. Contam. Toxicol. 64 (4), 652–658.

Liu, Z., Little, J.C., 2012. Semivolatile organic compounds (SVOCs): phthalates and flame retardants. In: Pacheco-Torgal, F., Jalali, S., Fucic, A. (Eds.), Toxicity of Building Materials. Woodhead Publishing, Cambridge, pp. 122–137.

Manikkam, M., Tracey, R., Guerrero-Bosagna, C., Skinner, M.K., 2013. Plastics derived endocrine disruptors (BPA, DEHP and DBP) induce epigenetic transgenerational inheritance of obesity, reproductive disease and sperm epimutations. PLoS One. 8 (1), e55387.

Martinez-Arguelles, D.B., McIntosh, M., Rohlicek, C.V., Culty, M., Zirkin, B.R., Papadopoulos, V., 2013. Maternal in utero exposure to the endocrine disruptor di-(2-ethylhexyl) phthalate affects the blood pressure of adult male offspring. Toxicol. Appl. Pharmacol. 266 (1), 95–100.

Nian, H., Meng, Q., Zhang, W., Chen, L., 2013. Overexpression of the formaldehyde dehydrogenase gene from brevibacillus brevis to enhance formaldehyde tolerance and detoxification of tobacco. Appl. Biochem. Biotechnol. 169, 170–180.

Nielsen, G.D., Wolkoff, P., 2010. Cancer effects of formaldehyde: a proposal for an indoor air guideline value. Arch. Toxicol. 84, 423–446.

Nielsen, G.D., Larsen, S.T., Wolkoff, P., 2013. Recent trend in risk assessment of formaldehyde exposures from indoor air. Arch. Toxicol. 87, 73–98.

Pegas, P.N., Alves, C.A., Evtyugina, M.G., Nunes, T., Cerqueira, M., Franchi, M., et al., 2011. Seasonal evaluation of outdoor/indoor air quality in primary schools in Lisbon. J. Environ. Monit. 13 (3), 657–667.

Pegas, P.N., Alves, C.A., Nunes, T., Bate-Epey, E.F., Evtyugina, M., Pio, C.A., 2012. Could houseplant improve indoor air quality in schools? J. Toxicol. Environ. Health Part A: Current Issues. 75 (22-3), 1371–1380.

PubChem Compound, 2013. Available from: http://www.ncbi.nlm.nih.gov/pccompound (accessed August 2013.).

Salthammer, T., 2013. Formaldehyde in the ambient atmosphere: from an indoor pollutant to an outdoor pollutant? Angew. Chem. Int. Ed. 52, 3320–3327.

Salthammer, T., Mentese, S., Marutzky, R., 2010. Formaldehyde in the indoor environment. Chem. Rev. 110, 2536–2572.

Schmid, S., Jecklin, M.C., Zenobi, R., 2011. Degradation of volatile organic compounds in a non-thermal plasma air purifier. Chemosphere. 79, 124–130.

Sundell, J., 2004. On the history of indoor air quality and health. Indoor Air. 14 (Suppl. 7), 51–58.

Weschler, C.J., 2011. Chemistry in indoor environments: 20 years of research. Indoor Air. 21 (3), 205–218.

Wilke, O., Jann, O., Brödner, D., 2004. VOC- and SVOC-emissions from adhesives, floor coverings and complete floor structures. Indoor Air. 14 (Suppl. 8), 98–107.

Wolkoff, P., Nielsen, G., 2010. Non-cancer effects of formaldehyde and relevance for setting an indoor air guideline. Environ. Inter. 36, 788–799.

World Health Organisation (WHO), 2010. WHO Guidelines for Indoor Air Quality: Selected Pollutants. Copenhagen: World Health Organisation. Available from: www.who.int (accessed September 2013.).

Xu, Y., Liu, Z., Park, J., Clausen, P.A., Benning, J.L., Little, J.C., 2012. Measuring and predicting the emission rate of phthalate plasticizer from vinyl flooring in a specially-designed chamber. Environ. Sci. Technol. 46 (22), 12534–12541.

New and less recognized risks with building materials: volatile organic compounds, replacement chemicals, and nanoparticles

8

Emina Kristina Petrović
Victoria University of Wellington, Wellington, New Zealand

8.1 Introduction

Following the discussions in Chapter 6 and Chapter 7, this chapter looks into the Stage 1 or with the early recognition of health risks associated with a particular substance, when early indications of plausible or likely adverse health effects are articulated, this being the point of reasonable "suspicion" that a particular substance could be harmful. Based on such "suspicion" more research is then undertaken to prove or disprove it, which leads to more articulation of health effects, and potentially triggers recognition and progression towards Stages 2 and 3. The key question of this section is whether it is possible to observe any indications in current practices that substances "suspected" of being hazardous are held back from introduction into everyday use. In this section, two types of "suspected" hazards are reviewed: substances that have been "suspected" for a while, and newly developed substances.

8.2 On-going suspicion—volatile organic compounds from carpets

While the risks associated with formaldehyde, phthalates, and PVC, discussed in Chapter 7, are reasonably well established, much more confusion surrounds volatile organic compounds (VOCs) from carpets. During the 1980s, when sick building syndrome was most discussed, carpets were often seen as a contributing factor. In fact, in 1999, one study evaluated the direct impact on workers' productivity of exposure to fumes from a 1980s' carpet sample 20 years later, showing the adverse effect still held strong (Wargocki et al., 1999). One relevant observation from that experiment was that the sample did not present elevated total VOCs (TVOCs), but rather elevated levels of certain individual VOCs, in this case acetone, acetic acid and aldehydes, ketones, and organic acids (Wargocki et al., 1999). Some literature has since concluded that because the irritant potential and toxicity of individual VOCs vary widely, evaluation of VOCs through a

Materials for a Healthy, Ecological and Sustainable Built Environment: Principles for Evaluation.
DOI: http://dx.doi.org/10.1016/B978-0-08-100707-5.00008-3

combined measure of TOVCs can no longer be supported (ASHRAE, 2009, section 10.10). However, in contemporary carpet-related regulations TVOCs are still reported in some sources, some are expressed in relation to individual VOCs, and some combine both approaches, jointly suggesting poor regulative protocols.

Table 8.1 compares leading European voluntary labeling schemes for how they define TVOC levels (Katsoyiannis et al., 2008). In addition to varying levels, these schemes also use different testing chambers, test durations, and report results in different units. To evaluate the importance of such differences, one study tested four carpet samples using various testing chambers and periods, concluding that differences of as much as 75% were possible (Katsoyiannis et al., 2008). They concluded that establishing a clearer pan-European regulating protocol for carpets would be simpler for the users and those conducting tests and could even prove more cost-effective for manufacturers (Katsoyiannis et al., 2008).

There are similar issues in New Zealand, where two voluntary schemes currently operate: one through the Carpet Institute of Australia (2013) and one through Environmental Choice New Zealand (2011 and 2012), with numerous differences between them. Table 8.2 compares levels of individual VOCs listed by these schemes with the official classification of the same chemicals for their impact on human health. One observable feature is that some of the VOCs listed are recognized as substances of very high concern (e.g., benzene and formaldehyde), many have different lower levels of recognition, while some do not appear on any lists (e.g., 4-phenylcyclohexene and 2-ethylhexonoic acid). Therefore, there is great variety around the wide range of chemicals used in carpets in their recognized impact on human health, representing different stages of recognition.

For the purpose of discussion of Stage 1 substances, 4-phenylcyclohexene (4PCH) is a good example of the complexities encountered when evaluating the health impact of less recognized chemicals. Already in the early 1990s, research on emissions from carpets established that 4PCH was one of the VOCs emitted from new carpets, "responsible for new carpet odour" (ASHRAE, 2009, section 11.9), and differences were observed between different technologies used for backing and presence or level of 4PCH emissions (Beekman et al., 1996; Singhvi et al., 1990). Generally, 4PCH is a byproduct of the polymerization process between styrene and butadiene, which are the main components of styrene-butadiene rubber (SBR) (SBR is discussed in more detail in Chapter 9). ASHRAE recommends that it should only be tested for in carpets and fabrics with SBR backing (ASHRAE Standard 189.1-2011). However, 4PCH is found in carpets with SBR backing and laminated fabric backing (Katsoyiannis et al., 2008). Recent studies have found only a small proportion, up to 30%, of carpet samples with no 4PCH and no clear relationship could be observed between different carpet and backing materials and the presence of 4PCH (Katsoyiannis et al., 2008; Wilke et al., 2004). This, together with evaluations of 4PCH in some paper printing technologies (Landy et al., 2004), seems to indicate that the use of 4PCH has expanded from just being related to SBR backing.

Table 8.2 shows the results of a search through official organizations on VOCs from carpets. However, these sources provide very little information on 4PCH (Table 8.2). A PubMed database search returned only a small number of articles which related to animal tests of 4PCH for acute toxicity, and which made no conclusive observations (Beekman

Table 8.1 **Emissions requirements (or test chamber concentrations) of leading European voluntary labeling schemes for carpet materials**

	Leading European voluntary labeling schemes for carpet materials								
	AgBB	**CESATF**	**M1**	**LQAI**	**Nature Plus**	**Blue Angel**		**GUT**	**Emicode EC1**
	Germany	**Rance**	**Finland**	**Portugal**	**Europe**	**Germany/ Europe**	**Austrian Ecolabel**	**Germany/ Europe**	**Europe**
TVOC after 3 days	10,000 μg/m^3	5000 μg/m^3		5000 μg/m^2/h		1200 μg/m^3		300 μg/m^3	(10 days) 500 μg/m^3
TVOC after 28 days	1000 μg/m^3	200 μg/m^3	200 μg/m^2/h	200 μg/m^2/h	200–300 μg/m^3	360 μg/m^3	380 μg/m^2/h		
Formaldehyde (HCHO)	28 days: 120 μg/m^3	28 days: 10 μg/m^3	28 days: 50 μg/m^3	28 days: 10 μg/m^3	28 days: 36 μg/m^3	28 days: 60 μg/m^3	–	10 μg/m^3	1 days: 50 μg/m^3

Source: Katsoyiannis et al. (2009).

Table 8.2 Comparison between regulated levels of VOCs in carpets in New Zealand, and the classification of the same chemicals for impact on human health

Substance description	AU ACCS Max emiss. (24 h) μg/h/m^2	NZ Env. Choice for syn. carpet μg/m^3	US Green label plus Max emiss. (24 h) μg/h/m^2	Classification by: • ECHA • IARC • CAL/OSHA	Health effects	General use
Benzene C_6H_6	55	30	55	• 1 ECHA 2008 • 1 IARC 2012 • Cal classified	Benzene is carcinogenic (IARC 1982). Prolonged exposure at levels below 1 ppm (3.2 mg/m^3) can lead to mutagenicity and carcinogenicity. Concerns for fertility also mentioned	Available in petrol, normal traffic pollution and close to petrol stations. Also in some paints, car interior accessories, and perfumes
Formaldehyde CH_2O	10	16	16	• 1 ECHA 2012 • 1 IARC 2012 • Cal classified	Carcinogenic	Numerous uses in building materials, mainly as part of gluing agents, and chemical treatments. In carpets used as biocide/antimicrobial
Toluene $C_6H_5 - CH_3$	280	150	280	• 2 ECHA 2004 • 3 IARC 1999 • Cal classified	Possible acute, system and specific organ toxicity due to inhalation or dermal exposure, neurological and cardiovascular toxicity, concerns for fertility and developmental effects, and spontaneous abortions due to inhalation	In high-octane blending petrol, as solvent for paints and coating, gums, resins, oils, rubber, and adhesives, as an intermediate in the preparation of many chemicals, dyes, pharmaceuticals, detergents, and explosives, for printing and manual cleaning
Styrene C_8H_8	410	220	410	• Not classified • 2B IARC 2002 • Cal classified	Harmful by inhalation, sensitizer; toxic for nervous system, liver, and eyes; conclusively observed to be toxic to fertility and toxic to reproductive development, some strong relations to carcinogenesis	Used in plastics, latex paints and coatings, synthetic rubbers, polyesters, and styrene-alkyd coatings. In construction in pipes, fittings, lighting fixtures, synthetic marble, flooring, carpet backing, molded furnishing

Acetaldehyde C_2H_4O	20	4.5	130	• Not classified • 2B IARC 1999 • Cal classified	Could be respiratory sensitizer, and carcinogenic for oral and bronchial tumors, possible skin sensitization, aspiration hazard and possible reproductive toxin Other: Probable carcinogen (Carpet Institute of Australia) Endocrine-disrupting chemical, carcinogen, teratogen, and causes allergic airway inflammation (Kawano et al., 2012)	Natural product of combustion and photo-oxidation of hydrocarbons, industrial use as intermediate, used in silvering of mirrors, leather tanning, denaturant for alcohol, fuel mixtures, hardener for gelatine fibers, as flavoring agent, in cosmetics, in glue and casein products, preservative for fish and fruit, in paper industry, in carpets used as biocide/antimicrobial
Vinyl Acetate $C_4H_6O_2$	400	100	190	• Not classified • 2B IARC 1995 • Cal classified	Acute toxicity oral and dermal; skin irritation, respiratory sensitization, reproductive toxicity, suspected carcinogenesis for oral exposure	Used to make polyvinyl acetate adhesives for paper, wood, glass, metals, and porcelain, also in latex water paints, coatings for paper, textile, and leather, base for inks and lacquers, in cosmetics, pharmaceuticals, food additives, and pesticides
Naphthalene $C_{10}H_8$	20	4.5	8.2	• Not classified • 2B IARC 2002 • Cal classified	Limited evidence of adverse health effects, mainly lungs for inhalation	Used in manufacture of phthalic anhydride which is used as intermediate in the production of phthalate plasticizers, resins, dyes, pharmaceuticals, insect repellents, also used in baby oils
Caprolactam $C_6H_{11}NO$	120	100	130	• Not classified • 4 IARC 1999 • Cal classified	No carcinogenic effect observed, possible skin sensitization, aspiration hazard and possible reproductive toxin, more concerns regarding toxicity than carcinogenesis (IRIS)	Primarily used in manufacture of synthetic fibers and resins (nylon 6), bristles, film, coating, synthetic leather, plasticizers, paint vehicles, cross-linking agent for polyurethanes
1-Methyl-2-Pyrrolidone C_5H_9NO	300	160	300	• Not classified • NA • Cal classified	Respiratory and skin sensitization, aspiration hazard, concern that it may damage fertility or the unborn child due to oral exposure	Used as a solvent for resins and acetylene, as paint stripper
Nonanal $C_9H_{18}O$	24	13	24	• Not classified • NA • NA	Skin and eye irritation, possible skin sensitization, reproductive toxicity, and germ cell mutagenicity	Used in flavors and perfume production
Octanal $C_8H_{16}O$	24	7.2	13	• Not classified • NA • NA	Skin and eye irritation	Used as flavoring, in perfumery, in preparation of synthetic citrus oils

(*Continued*)

Table 8.2 (Continued)

Substance description	AU ACCS Max emiss. (24 h) μg/h/m^2	NZ Env. Choice for syn. carpet μg/m^3	US Green label plus Max emiss. (24 h) μg/h/m^2	Classification by: • ECHA • IARC • CAL/OSHA	Health effects	General use
2-Ethylhexonoic Acid $C_8H_{16}O$	46	25	46	• NA • NA • NA	Some indications of oral toxicity (PubChem Compound)	Chemical intermediate in manufacture of resins for baking enamels, lubricants, detergents, flotation aids, and corrosion inhibitors; catalyst for polyurethane foaming
4-Phenylcyclohexene $C_{12}H_{14}$	50	2.5	50	• NA • NA • NA		

The Australian Carpet Classification Scheme (ACCS) follows ISO 10580:2010, and all ECS Level carpets have to comply with this; Environmental Choice New Zealand for synthetic carpets (document for wool does not set VOC levels); US Green Label Plus, the Carpet and Rug Institute. The AU and US scales provide a 24-h emission VOC emissions rate immediately after carpet manufacture in μg/h/m^2, while the NZ scale has a loose definition.

IARC evaluates only carcinogenesis to humans (group 1—carcinogenic; group 2A—probably carcinogenic; group 2B—possibly carcinogenic; group 3—not classifiable; group 4—probably not carcinogenic); ECHA and CAL/OSHA classify priority substances due to toxicology and/or carcinogenesis, thus "no classification" means that at this point substance is not recognized as needing classification. Substances that could not be found for either organization were noted as NA (not available).

Source: Carpet Institute of Australia, 2013. Australian carpet classification scheme incorporating the environmental certification scheme: technical guidelines. http://www.carpetinstitute.com.au/environment/index.htm (retrieved August 2013.); Environmental Choice New Zealand (2011 and 2012); Carpet and Rug Institute (US), 2013. Available from: www.carpet-rug.org (accessed August 2013.) (Carpet and Rug Institute (US), 2013); ECHA (2013); International Agency for Research on Cancer (IARC), 2013. IARC Monographs on the Evaluation of Carcinogenic Risks to Humans. Available from: http://monographs.iarc.fr/ENG/Classification/index.php (accessed January-August 2013.) (IARC, 2013); CAL/OCHA 2013; Integrated Risk Information System (IRIS), 2013. Available from: http://www.epa.gov/iris/index.html (accessed August 2013.) (IRIS, 2013); PubChem Compound, 2013. Available from: http://www.ncbi.nlm.nih.gov/pccompound (accessed August 2013.); Agency for Toxic Substances and Disease Registry (US) (ATSDR), 2013. Available from: <www.atsdr.cdc.gov> (accessed August 2013.) (ATSDR, 2013).

et al., 1996). However, search of the PubChem Compound database under the BioActivity summary returned six "inactive" items, which indicate that 4PCH is genotoxic in human embryonic kidney cells (PubChem Compound, 2013). In addition, Toxnet: toxicology data network indicated that reports were produced in the early 2000s evaluating acute toxicity of 4PCH in animals, but these could not be accessed. Therefore, currently it seems unlikely that the 4PCH impact mechanisms on human health are understood. However, the absence of clear adverse effects does not mean that 4PCH should be considered safe, simply because there are only a small number of studies and no clear adverse effects. Furthermore, the available information indicates an adverse health effect is possible, and more research is needed to explain this. Nevertheless, recently 4PCH has been introduced into the food packaging industry (Landy et al., 2004) and therefore exposure of the general population to it is likely to increase.

This treatment of 4PCH is very characteristic of many substances currently in use. While chemicals that produce acute toxicity at relatively low levels are increasingly becoming recognized as unhealthy, with subsequent regulation of these, currently there are no good mechanisms for evaluating the health effects of chemicals that are less aggressively adverse to human health. This means that the lessons from phthalates are still not absorbed, because absence of very clear acute toxicity does not mean the absence of adverse effects. Indications of in utero toxicity for kidney tissue clearly belong to this type of less obvious impact. Therefore, VOCs that are commonly used in carpet production but not found in lists of high-risk chemicals should generally still be approached with caution.

Similar patterns can be observed with a number of other everyday chemicals. For example, water-based paints and varnishes are generally considered safer than their solvent-based equivalents. However, there is an increasing body of research recognizing the toxicity of common solvents used in water-based products, even at very low levels (Lin et al., 2013; Spee et al., 2012). 2-Butoxyethanol, also known as ethylene glycol monobutyl ether ($BuOC_2H_4OH$) is an organic solvent used in some water-based paints and surface-coating products (ASHRAE, 2009, section 11.9). It is listed in California Occupational Safety and Health Regulations as a hazardous substance (CAL/OSHA, 2013), although in 2004 the EPA removed it from its list of hazardous air pollutants (EPA, 2004). Others have observed that 2-butoxyethanol has been related to poor indoor air quality complaints (Rella et al., 2012). Jointly these observations indicate that for the last 10 years, 2-butoxyethanol has been on the point of Stage 1 recognition as risk, while still present in many products. (For a more detailed discussion on paints and varnishes see Chapter 9).

The problem is that most research efforts are focused on more recognized hazards and providing sufficient knowledge to lead to effective removal of these, while borderline hazardous substances receive marginal research attention. Unfortunately, that does not mean that they are safe, but rather that they are not well understood.

8.3 Replacement and new substances

One important area of new development is substances that replace those phased out as hazardous. More recently, replacements for formaldehyde, phthalate plasticizers, and fire-retardants have been developed. For phthalates and fire-retardants early observations show that regulations against one set of such chemicals produced an increase in use of other chemicals from the same family. For phthalates the change was from shorter-chain phthalates (such as DEHP) to longer-chain phthalates (such as DINP) (Holmgren et al., 2012), and to a number of nonphthalate plasticizers (see Chapter 7). For fire-retardants the change was from a polybrominated diphenyl ether (PBDE) fire-retardant mixture PendaBDE to more prolific use of tris(1,3-dichloroisopropyl) phosphate (TDCPP), a suspected human carcinogen (Stapleton et al., 2012). Unfortunately, in both cases the replacement chemicals are simply less researched and less well understood for their health effects (Holmgren et al., 2012; Stapleton et al., 2012). The issue here is the interpretation of the wording: "suspected of adverse health effects." If using the precautionary principle, "suspected" could be seen as meaning "probably" if not "likely," while the manufacturers seem to be interpreting it as "ready to use." This is especially the case with TDCPP which is at least in the early Stage 2 of recognition. This implies that many substances which are introduced as replacements for eliminated risks could also be seen as being in the entry stages of the same process of recognition of health risks. Fig. 8.1 illustrates this almost cyclic process.

Similarly, although many formaldehyde-free products are increasingly available, PubMed currently reports only a very small number of studies on their health impact, indicating replacements are introduced without much evaluation of their health effect. However, there are some exceptions. For example, one study explored the health impact of the change from a phenol-formaldehyde bonding agent in fiberglass insulation to a carbohydrate-carboxylic acid binder, observing that in vitro this binder did not impact the biosolubility of glass wool insulation, but also noted that droplet sizes had an impact (Potter and Olang, 2013). Droplet size of the bonding agent, which

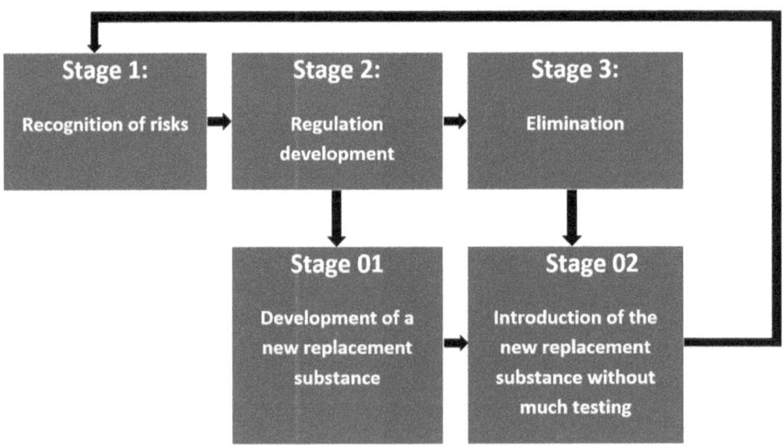

Figure 8.1 Cycle of introduction of replacement substances into testing for adverse impact on human health.

greatly varied in the samples studied, seems controllable by careful manufacturing. This indicates that much more sophisticated research is needed in these early stages of development of new materials, if they are to be reliably safe.

One area of recent development is materials that use nanoparticles. With the technology that enables design at scales unprecedented in the past, a range of completely new material characteristics is becoming available for the first time. Unfortunately, there are already warnings about nanoparticles. Researchers have remarked that nanoparticles are likely to impact the human body similarly to microparticles, the best known of these being the asbestos family (Donaldson and Poland, 2012; Sanchez et al., 2009; Pacheco-Blandino et al., 2012). If learning from the experiences with asbestos, the problem would seem to be the shape and size of particles and their biopersistency within the human body. For asbestos it was the long, thin shape and very long biopersistency that produced inflammation and subsequent onset of disease after a long latency. Because this knowledge is available, it seems reasonable to use it when developing new materials with similar features. However, texts that deal with design and the exciting opportunities nanotechnology offers fail to mention risks associated with such particles (Yeadon, 2011), indicating the same mistakes could be repeated.

8.4 Discussion on Stage 1 and conclusion

Although Chapter 7, recorded some acceleration in the way substances now in Stage 2 of recognition are progressing towards elimination, the overall impression is that this progress is generally still slow. The experience with lead and asbestos showed that it can take a long time before the total impact is fully understood. The experiences with formaldehyde and phthalates could unfortunately be just "the very tip of the iceberg" because they are only the most recognized risks from the great array of chemicals introduced into everyday use since the mid-20th century. The actions observed with Stage 1 substances indicate that potentially very limited learning from past mistakes has occurred. Substances that are "suspected" to be harmful are readily used, and their use often increases while the recognition of risks also increases.

Unfortunately, the scientific understanding of these risks is still partial, and if proof of harmful affect is needed prior to any action, such proofs do not exist for many of the substances. One aspect of concern is the very limited understanding of effects these substances create when people are exposed to their various combinations. One estimate suggests that if only three chemical combinations were to be comprehensively evaluated, it would need 166 million tests, and take 11,000 years, assuming 15,000 tests per year (Armstrong et al., 2007, p. 61). Evaluating 11 chemical combinations would take a million times longer than the universe has existed (Armstrong et al., 2007, p. 61). Therefore, currently assuming there is sufficient scientific knowledge on the health risks associated with building and furnishing materials is unrealistic, and all professional activities have to develop work strategies that acknowledge this limitation.

The solution to this problem is the precautionary principle, and the literature sources that adopt this idea are the popular works related to building biology or those based on their authors direct personal experiences (Baker-Laporte et al., 2008; Thompson, 2004; Hobbs, 2003; Bower, 1989, 2000). In fact, these can act as "whistle blowers" for Stage 1 of recognition, because they do not have to adhere to academic nor regulative protocols, provide the most comprehensive lists of potential risks, and offer useful practical advice on their avoidance. Their only problem is that precisely because of their character, they might not be taken seriously by many. If the predominant logic is that "suspicion" is not enough, but rather a proof is needed prior to any change, these works fail to provide such proofs.

References

Agency for Toxic Substances and Disease Registry (US) (ATSDR), 2013. Available from: www.atsdr.cdc.gov (accessed August 2013.).

American Society of Heating, Refrigeration and Air-Conditioning Engineers (US) (ASHRAE), 2009. 2009 ASHRAE Handbook: Fundamentals. ASHRAE, Atlanta.

American Society of Heating, Refrigeration and Air-Conditioning Engineers (US) (ASHRAE), US Green Building Council (USGBC), 2011. ANSI/ASHRAE/USGBC/IES Standard 189.1-2011: Standard for the Design of High-Performance Green Buildings: Except Low-Rise Residential Buildings. ASHRAE, Atlanta.

Armstrong, L., Wordsworth, A., Dauncey, G., 2007. Cancer: 101 Solutions to a Preventable Epidemic. New Society Publishers, Gabriola Island, British Columbia.

Baker-Laporte, P., Elliott, E., Banta, J., 2008. Prescriptions for A Healthy House: A Practical Guide for Architects, Builders and Homeowners. third ed. New Society Publishers, Gabriola Island, Canada.

Beekman, M.J., Maurissen, J.P., Johnson, K.A., 1996. 4-Phenylcyclohexene: 2-week inhalation toxicity and neurotoxicity studies in Swiss-Webster mice. Food Chem. Toxicol. 34, 873–881.

Bower, J., 1989. The Healthy House: How to Buy One; How to Cure a 'Sick' One; How to Build One. Carol Communications, New York, NY.

Bower, J., 2000. Healthy House Building for the New Millennium: A Design and Construction Guide. The Healthy House Institute, Bloomington.

California Occupational Safety and Health Regulations (CAL/OSHA), 2013. Subchapter 1: Article 5: Hazardous substances information and training. Available from: http://www.dir.ca.gov/title8/339.html (accessed August 2013.).

Carpet and Rug Institute (US), 2013. Available from: www.carpet-rug.org (accessed August 2013.).

Carpet Institute of Australia, 2013. Australian carpet classification scheme incorporating the environmental certification scheme: technical guidelines. http://www.carpetinstitute.com.au/environment/index.htm. Retrieved August 2013.

Donaldson, K., Poland, C.A., 2012. Inhaled nanoparticles and lung cancer – what we can learn from conventional particle toxicology. Swiss Med. Weekly. 142, w13547.

Environmental Protection Agency (US) (EPA), 2004. List of hazardous air pollutants, petition process, lesser quantity designations, source category list; petition to delist of ethylene glycol monobutyl ether. Available from: http://www.epa.gov/EPA-AIR/2004/November/Day-29/a26071.htm (accessed 08.13.).

Environmental Choice New Zealand, 2011. The New Zealand ecolabelling trust licence criteria for wool and wool-rich pile carpets EC-04-11. Available from: www.environmentalchoice.org.nz(accessed August 2013.).

Environmental Choice New Zealand, 2012. The New Zealand ecolabelling trust: licence criteria for synthetic carpets: EC-33-12. Available from: www.environmentalchoice.org.nz (accessed August 2013.).

European Chemicals Agency (ECHA), 2013. Available from: www.eeca.govt.nz (accessed 04.13.).

Hobbs, A., 2003. The Sick House Survival Guide: Simple Steps to Healthier Homes. New Society Publishers, Gabriola Island, British Columbia.

Holmgren, T., Persson, L., Andersson, P.L., Haglund, P., 2012. A generic emission model to predict release of organic substances from materials in consumer goods. Sci. Total Environ. 437, 306–314.

Integrated Risk Information System (IRIS), 2013. Available from: http://www.epa.gov/iris/index.html (accessed August 2013.).

International Agency for Research on Cancer (IARC), 2013. IARC Monographs on the Evaluation of Carcinogenic Risks to Humans. Available from: http://monographs.iarc.fr/ENG/Classification/index.php (accessed January-August 2013.).

Katsoyiannis, A., Leva, P., Kotzias, D., 2008. VOC and carbonyl emissions from carpets: A comparative study using four types of environmental chambers. J. Hazard. Mater. 152 (2), 669–676.

Kawano, T., Matsuse, H., Fakahori, S., Tsuchida, T., Nishino, T., Fukushima, C., et al., 2012. Acetaldehyde at a low concentration synergistically exacerbates allergic airway inflammation as an endocrine-disturbing chemical and as a volatile organic compound. Respiration. 84, 135–141.

Landy, P., Nicklaus, S., Sémon, E., Mielle, P., Guichard, E., 2004. Representativeness of extracting of offset paper packaging and analysis of the main odor-active compounds. J. Agric. Food. Chem. 52, 2326–2334.

Lin, C.-C., Huang, C.-N., Hwang, Y.-H., Wang, J.-D., Weng, S.-P., Shie, R.-H., et al., 2013. Shortened menstrual cycles in LCD manufacturing workers. Occup. Med. (Chic. Ill). 63 (1), 45–52.

Pacheco-Blandino, I., Vanner, R., Buzea, C., 2012. Toxicity of nanoparticles. In: Pacheco-Torgal, F., Jalali, S., Fucic, A. (Eds.), Toxicity of Building Materials. Woodhead Publishing, Cambridge, pp. 427–475.

Potter, R.M., Olang, N., 2013. The effect of a new formaldehyde-free binder on the dissolution rate of glass wool fibre in physiological saline solution. Part. Fibre. Toxicol. 10, 13.

PubChem Compound, 2013. Available from: http://www.ncbi.nlm.nih.gov/pccompound (accessed August 2013.).

Rella, R., Sturaro, A., Vianello, A., 2012. 2-Butoxyethanol from cleaning products responsible for complaints in workplaces: a case study. J. Environ. Monit. 14, 2659–2662.

Sanchez, V.C., Pietruska, J.R., Miselis, N.R., Hurt, R.H., Kane, A.B., 2009. Biopersistence and potential adverse health impacts of fibrous nanomaterials: what have we learned from asbestos? WIREs Nanomed. Nanobiotechnol. 1, 511–529.

Singhvi, R., Burchette, S., Turpin, R., Lin, Y., 1990. 4-Phenylcyclohexene from carpets and indoor air quality. Indoor Air '90. In: The 5th International Conference on Indoor Air Quality and Climate, Toronto, 1990.

Spee, T., van Valen, E., van Duivenbooden, C., van der Laan, G., 2012. A screening programme on chronic solvent-induced encephalopathy among Dutch painters. Neurotoxicology. 33, 727–733.

Stapleton, H.M., Sharma, S., Getzinger, G., Ferguson, P.L., Gabriel, M., Webster, T.F., et al., 2012. Novel and high volume use flame retardants in US couches reflective of the 2005 PentaBDE phase out. Environ. Sci. Technol. 46, 13432–13439.

Thompson, A., 2004. Homes that Heal: and Those that don't. New Society Publishers, Gabriola Island.

Wargocki, P., Wyon, D.P., Baik, Y.K., Clausen, G., Fanger, P.O., 1999. Perceived air quality, sick building syndrome (SBS) symptoms and productivity in an office with two different pollution loads. Indoor. Air. 9 (3), 165–179.

Wilke, O., Jann, O., Brödner, D., 2004. VOC- and SVOC-emissions from adhesives, floor coverings and complete floor structures. Indoor. Air. 14 (Suppl. 8), 98–107.

Yeadon, P., 2011. Materializations of nanotechnology in architecture. In: Schörpfer, T. (Ed.), Material Design: Informing Architecture by Materiality. Birkhäuser GmbH, Basel.

An overview of health hazards from materials: application of principles

9

Emina Kristina Petrović
Victoria University of Wellington, Wellington, New Zealand

9.1 Introduction

This chapter reviews a number of health hazards from materials, ranging from the very harmful to those that are less so. It also applies the principles discussed in Chapters 5, 6, 7 and 8. The objective is to provide a number of brief reviews, which jointly show a general overarching logic for the recognition of issues. The exception is the detailed introduction of petroleum-based polymers, which is provided so readers are aware of the context of these materials, before the discussion of problems with specific polymers.

The core question of this chapter is: Is it possible to identify types of materials which are better for human health? Unfortunately, the answer is that in most cases it is easier to identify broad types of materials which pose concerns, rather than those where no issues can be observed. This is especially true of a number of human-made, mainly polymer-based substances used in a variety of architectural and everyday applications. Many of these substances can be seen as already in the early stages of recognition of health risks, but with only modest acknowledgment that they pose risks for human health. However, even if many polymers are a potential risk for human health, it is impossible to generalize such conclusions in relation to all human-made materials. One subgroup of human-made materials which appears unlikely to have an adverse impact on human health is the highly inert materials, such as glass, steel, and aluminum. These tend to not interact in significant ways with indoor air. However, any use of these materials should include other considerations of their total impacts. As discussed in Parts I and II, glass, steel, aluminum, and other highly inert materials need energy for their manufacture and embody environmental concerns over the extraction of their raw components. Additionally, the total ecosystem and global warming cost of the manufacture of such energy-intensive materials is still in many aspects unknown. Therefore, it is impossible to draw unequivocal conclusions about the suitability of all human-made materials in sustainable construction. Rather, close examination in greater detail, and evaluation of specific design strategies are necessary.

Natural materials can also pose risks to human health, after all, lead and asbestos (see Chapter 6) are two natural materials posing very significant health

Materials for a Healthy, Ecological and Sustainable Built Environment: Principles for Evaluation.
DOI: http://dx.doi.org/10.1016/B978-0-08-100707-5.00009-5

problems. This chapter also reviews natural materials, revealing some potentially less well-known problems. The examples of radon gas and titanium dioxide illustrate this point, whereas linoleum (manufactured from natural materials) and timber are likely to pose few health issues. The example of linoleum shows that part of the problem facing contemporary construction is the neglect during the second half of the 20th century of the development of industries manufacturing more natural materials. Greater investment into the development of these for sustainable construction is needed. The discussion of timber shows that common timber treatments can erode the natural qualities of the wood, and therefore possibly make some timbers less good for human health. This suggests more investigation is needed before thorough human health impact assessments become possible.

For many human-made materials in current contemporary construction the balance of the advantages versus disadvantages is considered to be such that these materials should be used even if there are genuine concerns that these might adversely impact human health. Yet, the core proposition of this chapter is that because of the total increase of health risks in everyday environments, it is essential to start considering these risks as real and work on their reduction and elimination. The first step in this is to increase general awareness of these issues. This chapter aims to provide more information on why change in how materials are regarded is needed. The review of polymers clearly suggests that the issues continue to accumulate and that there are currently no adequate ways to quantify many of the health impacts. What this chapter illustrates more clearly than the previous four in Part III is that many human-made synthetic substances not only pose risks for humans but also for natural ecosystems. Because humans are by implication part of natural ecosystems, there are great opportunities to rethink the way contemporary society engages with the development and introduction of new synthetic substances, to include concurrently more careful consideration of the totality of implications for both humans and natural ecosystems.

9.2 Key commonalities in polymer sources

Plastics are a large group of materials used in many human activities, but there are no indications that these are safe for either people or natural ecosystems. Plastics are discussed here as a broad group of polymers with general similarities. Polymers used in construction are part of the broader range of human-made materials, and in this chapter polymers for construction are discussed as inseparable from the broader trends noted in the whole group. This is important in order to see the full picture, rather than an isolated and unavoidably fragmented part, but also because all of the plastics discussed are commonly found inside buildings and influence indoor air quality. Polymers are introduced here assuming limited prior knowledge in chemistry, which could be useful for readers interested in contemporary construction materials.

As much as 95% of all synthetic substances are presently manufactured from fossil resources (Rass-Hansen et al., 2007). The majority of plastics are produced using seven key hydrocarbons (compound consisting entirely of hydrogen and carbon) as their building blocks (Table 9.1). The key hydrocarbons contain one to eight atoms of carbon: syngas (C1), ethylene (C2), propylene (C3), C4 olefins or butadiene (C4), benzene (C6), toluene (C7), and xylene (C8). The lower hydrocarbons are in gaseous form (C1–C6), while the higher ones are liquids (C7 and C8). All seven key hydrocarbons are petrochemical in origin, being derived from natural gas, crude oil, and coal. Enormous amounts of petrochemical hydrocarbons are produced and consumed every year (Pyl et al., 2012), although these still represent only 10% of total use of fossil fuels (85% of fossil fuel is used for combustion (Christensen et al., 2008)). The majority of these chemicals are generated by breaking up higher carbon chemicals (with greater number of carbon atoms) found in fossil fuels into lower carbon chemicals (with fewer carbon atoms). Steam cracking of crude oil is commonly used for producing these hydrocarbons, and for refining of crude oil into petrol and similar derivatives. Refining of petrol developed early in the 20th century with significant increases in efficiency between the 1920s and 1980s (Vogt and Weckhuysen, 2015). Despite these improvements and consequent production of decreasing proportions of petroleum coke and unconverted slurry oil (Vogt and Weckhuysen, 2015), these processes still produce residues which vary from being harmless to human health and ecosystems (90% carbon (EPA, 2016)) to highly toxic (Alvarez et al., 2004). Many of these residues are subsequently mixed into other fuels (Guercio, 2010), but environmentally safe final disposal is still to be developed for some (USS, 2016; Alvarez et al., 2004).

The most common steam cracking methods include: fluid catalytic cracking (the most prevalent method, used in almost half of the oil refineries worldwide), hydrocracking (prevalence 20%), coking (prevalence 16%), thermal operations (prevalence 12%), and resid hydrotreating (prevalence 5%) (Vogt and Weckhuysen, 2015). In recent years, demand for petrol has been reasonably stable, with greater increases observed in the demand for hydrocarbons for polymerization (Vogt and Weckhuysen, 2015). Steam cracking for hydrocarbons involves bringing feedstocks to a steam cracking furnace where these are combined with steam and "cracked" at temperatures between 790°C and 830°C (White, 2007; Christensen et al., 2008). Hydrocarbon steam cracking plants tend to be designed for heavy (liquid) crackers or light (gas) crackers. Heavy crackers use naphtha, condensates, or gas oils as feedstocks and produce greater quantities of butadiene and heavier coproducts. Light crackers use natural gas to extract ethane and propane, and these become the feedstock for ethylene and propylene production with very low quantities of C4s and heavier coproducts (White, 2007). Unfortunately, in the steam cracking processes, only 20% of the heat input is used in the chemical conversions (Christensen et al., 2008), adding to the carbon footprint of this process. In many cases, additional energy is needed for future processing of these basic ingredients and many are transported globally. This high energy demand and the toxic residues created during their manufacture mean it is hard to see plastics as sustainable materials.

Table 9.1 Seven key petrochemical building blocks used to manufacture polymers discussed in Table 9.2

No.	Group	Name	Carbon content	Description	Background	Based on official wording from PubChem and ECHA	
						Health impact	Disposal
1		Syngas (synthesis gas)[a]	C1, H, CO and CO_2	Hydrogen, carbon monoxide, and carbon dioxide			
2	Olefins	Ethylene (or ethene)	C2, C_2H_4, or $CH_2=CH_2$	Colorless gas with a sweet odor and taste, lighter than air Extremely flammable, requires pressurized storage	Over 130 million metric tons produced in 2013. Mainly used in the production of polyethylene (PE). An important natural plant hormone, used to force the ripening of fruit. Normally produced for industrial use by energy-intensive steam cracking of crude oil. In 2013, about 63% of global demand for ethylene was for producing HDPE, LLDPE, LDPE	May cause drowsiness or dizziness; known for specific target organ toxicity from single exposure. Harmful to aquatic life with long-lasting effects. Inadequate evidence that it is carcinogenic—IARC class 3 carcinogen	Should not be released into the environment
3	Olefins	Propylene (old propene)(Other C3 hydrocarbons of C3, such as propane (C_3H_8) can be included)	C3, C_3H_6, or CH_2CHCH_3	Propylene is a colorless gas with faint petroleum-like odor and is heavier than air Extremely flammable, requires pressurized storage, very volatile organic compound (VVOC)	Found in nature as a byproduct of vegetation and fermentation processes. Normally produced for industrial use by steam cracking of crude oil. Total production of propylene is about half of that of ethylene, or about 85 million metric tons in 2013	May cause drowsiness or dizziness; known for specific target organ toxicity from single exposure. If released indoors during production, it can cause suffocation. May damage fertility or the unborn child. Harmful to aquatic life with long-lasting effects. Inadequate evidence that it is carcinogenic—IARC class 3 carcinogen	Should not be released into the environment, but ECHA comments that release is likely to occur from industrial use. Registration under REACH is required in Europe

4	Olefins	Olefins or hydrocarbons C4, or 1,3-butadiene, butadiene, divinyl, vinylethylene, biethylene, etc.	C4, C_4H_6, or $CH_2=(CH)_2=CH_2$ or $CH_2=CHCH=CH_2$	A set of synthetic colorless gases Extremely flammable; very volatile organic compounds (VVOC)	A family of chemicals based on C_4 which are byproducts of steam cracking used to produce ethylene and other olefins. Mainly unstable and highly reactive. 1,3-butadiene is usually found in urban air from vehicle exhausts. It breaks down quickly in the atmosphere. In 2020, about 13.7 million metric tons of butadiene will be industrially processed globally	1,3-Butadiene is carcinogenic to humans by inhalation—IARC class 1 (eye, nose, throat). It may cause central nervous system disturbances; may cause genetic defects (skeletal abnormalities, decreased fetal weight, and reproductive effects). May cause drowsiness or dizziness; known for specific target organ toxicity from single exposure. Causes damage to organs through prolonged or repeated exposure	Registration under REACH is required in Europe. Environmental regulatory agencies to be contacted before any disposal. Keep materials out of water sources and sewers
5	Aromatics	Benzene (or benzol)	C6, C_6H_6	Volatile, clear, colorless, gas with petrol-like, sweet odor Highly flammable; volatile organic compound (VOC)	Byproduct of coal distillation and crude oil refining processes. Used as a solvent, a chemical intermediate, and in petrol. Formerly used as a parasiticide. Found in air from burning fossil fuels, and in cigarette smoke. Used to make plastics including resins, nylon and synthetic fibers, and also rubbers, lubricants, dyes, detergents, drugs, and pesticides. In 2012, the total market for benzene was approx. US$69 billion	Causes central nervous system and bone marrow damage, and is carcinogenic. May be fatal if swallowed or enters airways. May cause genetic defects (germ cell mutagenicity). Known carcinogenic—IARC class 1. Causes damage to organs through prolonged or repeated exposure. Harmful to aquatic life with long-lasting effects	There is no proven method of disposal for carcinogenic compounds. Destruction by incineration may be the best option
6	Aromatics	Toluene	C7, C_7H_8, or $C_6H_5CH_3$	Clear colorless liquid with a distinctive aromatic odor Highly flammable liquid and vapor; volatile organic compound (VOC) with vapors heavier than air	Toluene mainly occurs as a byproduct in steam cracking or catalytic reforming. Most toluene is processed	May be fatal if swallowed and enters airways. Causes skin and eye irritation. May damage fertility or fetuses.	Toluene may be disposed of by controlled incineration. It should not be

(Continued)

Table 9.1 (Continued)

No.	Group	Name	Carbon content	Description	Background	Based on official wording from PubChem and ECHA	
						Health impact	Disposal
					into benzene and xylene, and from there into fuels. In addition, it is used in lubricants and greases, antifreeze products, biocides, nonmetal surface treatments, inks and toners, leather treatment, polishes and waxes, textile dyes, adhesives, and sealants. In 2013, US$24.5 billion was generated globally from sales of toluene	Causes damage to organs (specific organ toxicity from single exposure), and damage through prolonged or repeated exposure. Toxic to aquatic life, with long-lasting effects. (For more detail, see Table 8.1)	put in sewers because of possibility of exposition
7	Aromatics	Xylenes (a group of chemicals: o- Xylene, m- xylene, p- xylene, etc.)	C8, C_8H_{10}, or $C_6H_4(CH_3)_2$	o-Xylene, m- xylene, and p-xylene are colorless watery liquids with a sweet odor and irritating vapor (less dense than water) Flammable liquid and vapor; volatile organic compound (VOC)	Xylenes are used in the production of plastics (PET), plasticizers (phthalate anhydride), fuels, and solvents. The main application of xylenes is use in the production of paraxylene, which is processed into terephthalic acid (TPA) and used in manufacture of PET. In 2013 this process generated US$51 billion globally. TPA is a substitute for dimethyl terephthalate (DMT)	o-Xylene, m- xylene, and p-xylene have similar health impacts which include acute toxicity if inhaled and damage to organs. Toxic to aquatic life with long-lasting effects	Used in production of fuels, which is their effective disposal

[a]Methanol can also be included as C1, as syngas is used to synthesize it. Methanol is a highly flammable liquid and vapor, which is toxic if swallowed, inhaled, or in contact with skin. It may damage fertility or the unborn child. It damages organs, especially through prolonged or repeated exposure. The main use of methanol is in the production of formaldehyde (see Chapter 7) and acetic acid (used to make PET).

Source: PubChem Compound, 2016. Review of Chemicals by the National Center for Biotechnology Information (NCBI). Available from: www.ncbi.nlm.nih.gov (accessed October 2016.); ECHA, 2016. Chemicals Search from European Chemicals Agency (ECHA). Available from: echa.europa.eu (accessed October 2016.); Ceresana, 2016. Introductory Previews of Ceresana: Market Intelligence, Consulting reports. Available from: www.ceresana.com (accessed October 2016.) (Ceresana, 2016).

Unfortunately, the problem does not end there. When examining the seven key hydrocarbons in Table 9.1, it is clear that all are flammable, some require pressurized storage, and many are volatile organic compounds (VOCs). All pose significant risks, both for human health and the natural environment. In fact, apart from syngas, all of these chemicals are already recognized as toxic and should at least be seen as in the Stage 2 of recognition as health risks. Unfortunately, their use has not been banned; rather, they are readily used in the manufacture of polymers and used as a foundation for the manufacture of polymers. In addition to flammability and toxicity, some of these chemicals have to be contained in specialized pressurized conditions, and others are VOCs or VVOCs (see Chapter 7) that require very careful use. Thus, many strategies are used to keep manufacture with these toxic chemicals safe for the workers and the natural environment. The situation is similar for many intermediate chemicals used in the production of polymers, and reports suggest that such practices succeed in keeping exposure within current regulated limits (Unwin et al., 2013). However, regardless of the robustness of such safety protocols accidents do happen.

The majority of the key hydrocarbons have been implicated in accidents (Levshina et al., 2009; Avis and Hutton, 1993; Luo et al., 1990; Biscaldi et al., 1981). Some of these accidents involved fire, and others the uncontrolled release of toxic chemicals. For example, in 2005 a forklift in a steam cracking factory in Texas accidently removed a small propylene valve (CSB, 2006). The propylene soon caught fire and for the subsequent 5 days explosions and fire dominated the complex (CSB, 2006). Workers, and the wider community, were exposed to hazardous fire and gases, and the complete assessment of the health implications of this accident was undetermined (Anon., 2005). Many of the intermediates in the chemical industry have been implicated in dangerous accidents. The Bhopal disaster in India, e.g., involved a large release of methyl isocyanate gas, which is an intermediate for the production of pesticides and polyurethane rubber (Akovali, 2012b). In this 1984 accident, almost 4000 people died immediately, with 15,000–20,000 premature deaths attributed to it 20 years later, and with an estimated half a million people exposed to the gas (Broughton, 2005). Unfortunately, while these accidents might have led to improvements in workplace safety, the more these dangerous chemicals are used in manufacture, the more likely are future similar accidents. This suggests all workers involved with the manufacture of these chemicals should be considered as being constantly exposed to potential occupational health risks. A similar position is already adopted by the European Chemicals Agency (ECHA) which labels all of the seven key hydrocarbons (Table 9.1) as release into the natural environment is likely to occur from industrial use while used in industry (ECHA, 2016). Thus, all of these seven key hydrocarbons present either potential risks or active hazards for both human health and the natural environment.

As this section shows, at the core of all petroleum-based polymers is the use of the seven toxic key hydrocarbons, which in itself is sufficient to question whether any use of materials based on such chemistry can be considered sustainable.

9.3 Plastics—more solid polymers

Because of the variety of polymers currently in use, the best way to review their characteristics is to compare some of the most common ones. Table 9.2 presents one such analysis. Some of the polymers currently used widely are made as a direct result of the polymerization of the seven key hydrocarbons: polyethylene is a polymer of ethylene, polypropylene is a polymer of propylene, while others use more complex monomers. For example, polystyrene is a polymer of styrene which in turn is derived from benzene. Polymerization can occur through a variety of methods from simple, free radical polymerization, to the use of catalysts, such as the Ziegler–Natta catalysts (mixture of metal alkyls and transition metal salts). Ziegler–Natta polymerization is quite common, and was developed by Ziegler and Natta who were awarded the Nobel Prize in chemistry in 1963 for this work (Tijssen et al., 2015). Because of its high energy demand, this polymerization process makes the total embodied energy of the finished plastics high (see Chapter 4).

A number of polymers are synthesized from intermediates rather than their chemical monomers. For example, polyethylene terephthalate (PET) is made through polycondensation of purified teraphthalic acid (PTA) with ethylene glycol (EG), followed by the esterification process to remove excess water, and pressurized prepolymerization and polymerization, while PTA is made through a reaction between the secondary petroleum product paraxylene (PX) and acetic acid (Hitachi, 2016). Therefore, the impacts of all intermediates should be considered when evaluating the final polymer's impact on human health and ecosystems, as well at the total energy used in its production.

During the manufacture of polymers other substances can be mixed into the main material. This tends to take two dominant forms: (1) intentional additives (stabilizers, antioxidants, fillers, etc.) used to adjust some of the properties of the given polymer; and (2) unintentional impurities such as catalysts used in the production of polymers that end up in the polymer itself. In some cases, specific combinations of catalysts and stabilizers are used. For example, when germanium is used as a catalyst in the production of PET phosphorus-based stabilizers are common (Kannan et al., 2016). As in the discussion of plasticizers in Chapter 7, any chemicals which are not part of the main polymer bond are more easily released once the material is in everyday use.

Stabilizers are used in polymers to adjust their properties in desirable ways, such as increasing heat resistance, decreasing oxidation, extending the period that the mixture can be worked on during manufacture, and increasing recyclability. Unfortunately, some of the substances used as stabilizers or fillers add toxicity to the polymers. For example, heavy metals such as lead and cadmium work well as stabilizers for many plastics. Their removal from the manufacturing process has been only recent. The deadline for voluntary removal of lead stabilizers from PVC in Europe was 2015, preceded in 2005 by the voluntary removal of lead in PVC piping for water supply (ECVM, 2016), and the 2001 voluntary phasing out of cadmium (Zhang et al., 2015). Other polymers are in similar situations. In 2016

Table 9.2 **Summary of common polymer plastics, fibers, paints, and varnishes**

Name	Chemical description	Common applications	Health impacts	Recycling numbers (if any)[a] and proportion of recycling
Polyethylene (PE) (previously polyolefin)	Polymer of ethylene: long chain of carbon atoms, with two hydrogen atoms attached to each carbon atom. It is vinyl polymer, made from monomer ethylene. Various densities use either free radical, Ziegler–Natta, or metallocene catalysis polymerization. Used as thermoplastics and fibers	The most popular plastic in the world. Used in a variety of forms (HDPE, LDPE, and LLDPE, see below) each with a significant share of the market. Used for shopping bags, shampoo bottles, toys, and much else. Industrial use commenced in the late 1930s	Appears to be safe, not known to leach	HDPE and LDPE recycled
High-density polyethylene (HDPE)	Long reasonably straight linear polyethylene. Tends to be made using Ziegler–Natta polymerization	Used for milk, water, and juice bottles, yogurt and margarine tubs, cereal box liners, detergent bottles, grocery bags. In construction used for: tubes, pipes, extrusions, tanks, electrical cables coverings, geomembranes, plastic lumber and wood-plastic lumber. In 2014, global revenue from HDPE was US$61.8 billion	Appears to be safe, not known to leach	No. 2 can be recycled About 30% recycled, with an increasing trend in recent years
Low-density polyethylene (LDPE)	Branched polyethylene. Usually free radical vinyl polymerization used to achieve branches	Mainly used in the packaging industry, for bags and sacks, bread and frozen food bags, squeezable bottles, plastic wraps, and packaging foam. In construction used as damp-proof membranes. In 2013, global revenue from LDPE was almost US$33 billion	Appears to be safe, not known to leach	No. 4 can be recycled but less than 6% is recycled
Linear low-density polyethylene (LDPE)	Harder to manufacture than LDPE but has better properties	Used for plastic bags, wraps, toys, covers, lids, pipes, flexible tubing, buckets, containers, electrical cable covering, and geomembranes. In 2013, global revenue of US$40 billion		Generally not accepted for recycling

(*Continued*)

Table 9.2 (Continued)

Name	Chemical description	Common applications	Health impacts	Recycling numbers (if any)[a] and proportion of recycling
Poly(ethylene terephthalate) (PET) or polyesters	Copolymer with ethylene and terephthalate groups. Has a hydrocarbon backbone which contains ester linkages. Made using transesterification from ethylene glycol and dimethyl terephthalate. Used as thermoplastics and fibers	Dominates the bottled drink market (water, soft drink, sport drinks bottles), also ketchup and salad dressings, peanut butter, and cosmetic containers. As fiber it is used as polyester in clothing which is the most common polymer used in clothing. In construction, it is used as polyester insulation, carpets, and as a substrate in thin-film solar cells. Developed in a series of patents during the 1940s and 1950s. It is the fourth most common polymer. In 2015 about 24 million metric tons were produced globally	Appears to be safe for short-term use —designed for single-use only, extended life can increase the risk of leaching and bacterial growth. Appears to leach under extreme conditions (see main text)	No. 1 can be recycled. About 31% is recycled in the US and 52% in EU, with an increasing trend in recent years
Polypropylene (PP)	Vinyl polymer, similar to PE but with methyl groups repeating. Used as thermoplastics and fibers	Higher melting temperature than PE, so used in dishwasher-safe plastics, ketchup bottles, yogurt and margarine tubs, medicine bottles, cereal liners, packing tape, drinking straws, and chips bags. Used in medical implants/meshes. As fiber used in thermal underwear and carpets. Also, used in plumbing supplies and outdoor carpeting, especially close to pools as it does not absorb water. In 2013, global market for PP was 55 million metric tons	Because of hazardousness of propylene, hazardous during production, but appears to be safe in use. Not known to leach harmful chemicals	No. 5 can be recycled About 3% recycled

Polycarbonate	Named after the carbonate groups in its backbone chain. Belongs to the family of polyesters. Often called polycarbonate of bisphenol A (BPA), because it is made from bisphenol A and phosgene using interfacial polymerization. Used as thermoplastic	Very clear plastic used to make lightweight eyeglass lenses. Used in water cooler bottles, baby bottles, microwave ovenware and eating utensils, plastic coating for metal cans. Recent removal from food storage has begun. Used to make car parts, especially shatterproof windows. In construction used as a transparent covering for indoor–outdoor areas, and other large transparent sheets. Used as one of the common components in two pot epoxy resins. Invented in the 1930s in the search for synthetic estrogen—thus not surprisingly a hormone disruptor. In 2010, revenue from BPA was around US$6 billion in the US only	Avoid—contains hormone disruptor with estrogenic activity; may cause chromosomal damage, can leach into food, especially as the product ages	Generally not accepted for recycling
Polystyrene	Long hydrocarbon chain, with phenyl group attached to every other carbon atom. Vinyl polymer made using either free radical chain or Ziegler–Natta polymerization of styrene. Used as thermoplastic	Hard, inexpensive and very common. Used for computer housing, foam packaging, car interiors knobs, CD and video cases, hairdryers, kitchen appliances, toys, clear plastic drinking cups, food containers. In construction used as expanded polystyrene insulation. Industrial use developed from the 1930s. In 2020, expected global revenue of about US$26 billion	Avoid. Benzene (used in production of styrene; see Table 8.2) is a known human carcinogen. May leach harmful substances (possible carcinogens and hormone disruptors). Persistent in the environment. Energy intensive	No. 6 can be recycled. Less than 1% is recycled
Poly(styrene-butadiene) SBR rubber and poly (styrene-butadiene-styrene) SBS rubber	These are replacements for natural rubber. SBR is a copolymer of styrene and butadiene, while SBS is the same but with long chain of polystyrene, polybutadiene, polystyrene in that order. SBR is more common.	SBR is used for the soles of shoes, car tyre treads, and carpet backing (see Chapter 8). Industrial manufacture of SBR started during WWII. In 2012, more than 5.4 million metric tons of SBR produced globally	SBR is harmful to aquatic life with long-lasting effects. It may cause an allergic skin reaction. It is manufactured using toxic chemicals	Generally not accepted for recycling

(Continued)

Table 9.2 **(Continued)**

Name	Chemical description	Common applications	Health impacts	Recycling numbers (if any)[a] and proportion of recycling
	As a combination of polystyrene (tough and hard plastic), and polybutadiene (rubbery), both have durable rubbery properties. These are thermoplastic elastomers, but also can be applied as an emulsion (E-SBR) or solution (S-SBR) to a surface			
Polyacrylonitrile (PAN) and copolymers known as acrylic	Vinyl polymer, made using free radical chain polymerization of acrylonitrile. Used as fibers	As a homopolymer it is used to make fibers for filtration systems, especially to make carbon fiber. As a copolymer it is used to make knitted clothing, tents		Generally not accepted for recycling
Poly(styrene-*co*-acrylonitrile) (SAN), and poly (acrylonitrile-*co*-butadiene-*co*-styrene) (ABS)	More solid acrylonitrile copolymers. SAN is a simple random copolymer of styrene and acrylonitrile. ABS is made by creating the same reaction in the presence of polybutadiene which then polymerizes, creating complex grafts	ABS is very strong and lightweight, stronger than polystyrene, and finds uses from kitchen appliances to car body parts, including plumbing fixtures and pipes and the faces of electric sockets. Developed during the 1930s and 1940s		Generally not accepted for recycling
Poly(vinyl chloride) (PVC)	Vinyl polymer, similar to PE but on every other carbon in the backbone chain, one hydrogen atom is replaced a chlorine atom. Produced using free radical polymerization of vinyl chloride. A thermoplastic	Some plastic wraps (meat, cheeses), food containers, blister packs, blood transfusion bags, garden hoses. Used in buildings for plumbing pipes and fixtures, vinyl flooring, window frames, electrical cable covering (see Chapter 7). Introduced into use after 1926 invention	Toxic in production and disposal. Plasticizers added for flexible applications can leach and be harmful (see Chapter 7)	No. 3 can be recycled. Less than 1% recycled. Releases dangerous gases during recycling, so this should be avoided

Poly(methyl methacrylate) (PMMA)	Polymer chain of methyl methacrylate. Vinyl polymer made by free radical vinyl polymerization from the monomer of methyl methacrylate. It is thermoplastic	Clear plastic (more transparent than glass) used for eyeglass lenses, dentures, shatterproof glass replacement, ice rink barriers. Known as Plexiglas, Perspex, Acrylite, Lucite. Used for, sinks, tubs, shower units. Also used in lubricating oils, and acrylic/latex paints which contain it suspended in water (it does not dissolve). When used as glass replacement similar to polycarbonate, but cheaper. Developed in 1928	Not classified as posing health issues	Generally not accepted for recycling
Poly(vinyl acetate) (PVA)	Vinyl polymer made by free radical polymerization of monomer vinyl acetate. Thermoplastic	Used as an adhesive for wood, paper, and textiles. Found in water-based acrylic paints, where PVA acts as latex (in form of poly(vinyl alcohol-*co*-vinyl acetate) which makes it possible for PMMA to be suspended in water. Discovered in 1912	Highly flammable liquid and vapor which may be harmful if swallowed and cause mild skin and serious eye irritation. Some indications of carcinogenesis in animals, but no data for humans. IARC Group 3	—
Nylons, aramids, and polyamides	A range of polymers with a chain of amide groups separated by phenylene groups (can be arranged in more than one way). Kevlar is made from terephthoyl chloride and p-diaminobenzene, Nomex from isophthoyl chloride and m-diaminobenzene, Nylon 6 from caprolactam, Nylon 6,6 from adipic acid and hexamethylene diamine. Currently nylon 6 and nylon 6,6 are most commonly used. Used as thermoplastics and as fibers	Around the 1940s nylon was first used for toothbrush bristles, women's stockings, parachutes, and ropes. Kevlar is used for bulletproof vests and puncture-resistant tyres. Used as fiber, molded shapes, and in films. In buildings it is found in carpets, fixings, and door furniture. Started with use during the 1940s		—

(*Continued*)

Table 9.2 **(Continued)**

Name	Chemical description	Common applications	Health impacts	Recycling numbers (if any)[a] and proportion of recycling
Polyurethanes (PU)	Family of polymers based on polymerization of urethane linkage, using polyols (polyester, polyether, etc.) and isocyanates (toluene diisocyanate, methylene diphenyl diisocyanate, isophorone diisocyanate, etc.). They can be elastomers in fibers and appear in paints and adhesives. Thermosetting or thermoplastic	The best-known polymer used to make foams. Very versatile family, used as fibers for clothing (Spandex/Lycra), wheels, tyres, hoses, shoes, and carpet underlay. In building used as adhesives, surface coatings, and sealants. Developed in 1938 for fibers. In 2014, in Europe 4.4 million metric tons of polyurethane were produced	Uses toxic chemicals in its manufacture, but as a thermoplastic is considered chemically inert when fully reacted. Adhesives, surface coatings, and sealants can lead to exposure to unreacted components	Generally not accepted for recycling

[a]Internationally standard recycling numbers go to number 6. Number 7 is often used to cover all types of plastics not covered by the specific numbers. Most municipalities do not recycle any of the number 7 plastics. In some parts of the world a greater range of numbered plastics and their recycling can be observed more recently.

Sources: PubChem Compound, 2016. Review of Chemicals by the National Center for Biotechnology Information (NCBI). Available from: www.ncbi.nlm.nih.gov (accessed October 2016.); ECHA, 2016. Chemicals Search from European Chemicals Agency (ECHA). Available from: echa.europa.eu (accessed October 2016.); Ceresana, 2016. Introductory Previews of Ceresana: Market Intelligence, Consulting reports. Available from: www.ceresana.com (accessed October 2016.); Polymer Science Learning Center, 2000. Polymer Science Learning Center, University of Southern Mississippi. Available from: www.pslc.ws (accessed October 2016.) (Polymer Science Learning Center, 2000); and other sources cited in this chapter.

however, the website of the European Council of Vinyl Manufacturers (ECVM) was still reporting on the long history, excellent properties, and cost-effectiveness of lead as a stabilizer of PVC (ECVM, 2016). Organic-based stabilizers (OBS) and 1,3-dimetyl-6-aminouracil (DAU) are the new replacement stabilizers, and outperform the older stabilizers in terms of heat resistance and increased recyclability (Asawakosinchai et al., 2015). Unfortunately, these replacement substances have not been thoroughly tested for their impact on human health. Detailed definitions of OBS are difficult to find, and DAU has only been used more intensely recently (PubChem Compound, 2016). The impacts of stabilizers and fillers should therefore be included in assessments of the total health risks posed by various polymers.

Toxic components can also be introduced into plastics through the use of copolymerization, aimed at enhancing the properties of common polymers. For example, PE or PVC could be exposed to chlorination through which some monomers in the base materials are substituted by others, creating a new copolymer (Akovali, 2012b). The physical and chemical properties of the copolymer can be significantly different than those of the initial polymer, and in some cases the extent of the replacement of monomers can further increase the range of different properties (Akovali, 2012b). Chlorination tends to result in a reduction in flammability and better performance at higher temperatures, and is used in chlorinated polyethylene (CPE) and chlorinated polyvinylchloride (CPVC) (Akovali, 2012b). Architectural applications of chlorinated copolymers include use in roofing and geomembranes, as liners of water tanks, and covering of electric cables. These materials have been reported as presenting health risks in manufacture, during their degradation, and potentially also while in use at higher temperatures (Akovali, 2012b). Many other copolymers and mixing of polymers and copolymers is common in plastics, and specific health features of each of the components should be assessed.

Residual catalysts present unintentional impurities in polymers and should also be evaluated for health impacts. A number of studies have observed release of traces of the catalyst antimony from PET bottles (Chapa-Martínez et al., 2016; Bach et al., 2014; Fan et al., 2014; Bittner et al., 2014). This is relevant given that more than 99% of beverage bottles now used in Europe are made of PET (Fan et al., 2014), and that antimony is a cumulative toxic element, with the level of toxicity depending on its state (Chapa-Martínez et al., 2016). Its release appears to be accelerated by higher temperatures and exposure to sun (Chapa-Martínez et al., 2016; Bach et al., 2014; Fan et al., 2014).

Catalyst residues are not the only issue with leaching of various substances from plastics. Recent studies have found that many kinds of plastics have a number of nonintentionally added substances leaching from them over time in certain conditions (Bach et al., 2014). For example, although bisphenol A (BPA) is not used in the manufacture of PET bottles, it has been found in water stored in them, leaving its origin unclear (Fan et al., 2014). Table 9.2 shows that BPA is a hormone disruptor with estrogenic activity which may cause chromosomal damage (see also Chapter 7). The study by Fan et al. (2014) found the release of BPA raised with prolonged water storage in PET bottles at a temperature of 70°C, concluding that storing PET bottles at high temperatures is not recommended. With an increase in the recognition of the estrogenic impacts of BPA, evaluations have been undertaken on related chemicals which might be released from polymer food containers. A 2014 study examined the release of estrogenic chemicals, such as BPA, from BPA-free rigid and reusable bottles still readily available on the market since the start of the campaign to eliminate BPA (Bittner et al., 2014). They found that polystyrene, Tritan polyester copolymer, polyethersulfone, and acrylic bottles released estrogenic chemicals, while glycol-modified polyethylene terephthalate (PETG), cyclic olefin polymer (COP), and copolymer (COC) did not (Bittner et al., 2014). Although complete explanations of these findings are still unavailable, these results suggest that many polymers could contain traces of other polymers and organic chemicals which were not part of their manufacture. Jointly, such findings suggest a need to better understand which nonintentional substances can be found in plastics and also poses the question as to whether the problem is inherently related to the nature of organic chemistry, where in certain conditions some substances react together.

Although leaching of chemicals out of food containers has little direct implications on construction materials, it does suggest that even when it comes to the long-term safety of food containers, there is currently insufficient understanding of the total implications of nonintentional chemical interactions on human health. The same is likely to be the case with building materials, which tend to be less researched. Heightened recognition of health concerns associated with BPA has led to some reduction in its use, and although in some countries regulations against BPA and polycarbonate (which is a polymer of BPA; see Table 9.2) in food packaging exist, these vary significantly from country to country (Metz, 2016). Polycarbonate is still widely used in many other applications. In architecture, e.g., its characteristics as a lightweight transparent plastic have meant it has been used as a glass replacement, especially in applications where high transparency and shatter resistance are important, such as in pergolas, dome lights, curved glazing, or noise barriers around busy roads. The argument for the use of polycarbonate for the windows of airplanes, trains, and cars has been made by emphasizing the reduced carbon emissions resulting from higher fuel efficiencies of the vehicles, attributed to the decreased weight of the plastic compared to glass (Metz, 2016). However, such assessments fail to consider the implications of the environmental release of BPA. Evaluations of environmental impacts have found that BPA can be readily recorded in the atmosphere throughout the world, including in the polar atmospheres (Metz, 2016). Similarly, BPA is released into waterways. Its release into seawater is accelerated compared to river waters (Sajiki and Yonekubo, 2003). At this point, full understanding of the ecosystem impact of BPA is still not available. What is known is that many impacts that are harmful for humans are also harmful for natural ecosystems, accepting that humans are part of living ecosystems. Therefore, the use of these materials in the built environment could lead to contamination during manufacture, through rainwater runoff if used in roofing, or when disposed of (the burning of discarded plastics is associated with the release of BPA into the environment (Metz, 2016)). Although elimination of such potentially dangerous substances from food and water containers should justifiably be seen as a higher priority than elimination from architectural applications, from the perspective of human and environmental health, a complete elimination of the use of BPA-containing materials is perhaps the safest solution. This is especially the case given that alternative materials exist (see Table 9.2). Issues with the use of BPA in building materials also appear in Section 9.5.2 when discussing epoxy.

9.4 Disposal of plastics

When considering the use of plastics in sustainable construction, issues related to the disposal of polymers should be acknowledged. Because of the close relation to fossil fuels, many plastics can easily be used as a fuel or incinerated, and for the majority of the seven key hydrocarbons controlled burning is the only safe disposal method (see Table 9.1). However, for some polymers issues have already been raised about the possible toxic implications of such practices. Combustion of PVC, e.g., comes with the risk of release of dioxins, which are toxic and environmentally persistent organic pollutants (Zhang et al., 2015). However, research has found that while this appears to be the case for many house, backyard, and landfill fires, if combustion is controlled and at high enough temperatures, such issues are mitigated (Zhang et al., 2015). This work did not report on the impact of the release of the stabilizers (possibly lead) or plasticizers during the incineration of PVC.

Biodegradation of polymers takes place in two steps: (1) fragmentation of the polymer (due to oxidation, photodegradation, or hydrolysis or degradation by microorganisms), and (2) bioassimilation of the polymer fragments by microorganisms and their mineralization (Vroman and Tighzert, 2009). However, the biodegradability of polymers depends on their chemical structure and the environmental conditions (Vroman and Tighzert, 2009). Even through such processes, many polymers take a very long time to biodegrade. For example, polyurethane degrades slowly with some estimating that it takes hundreds of years for polyurethane molecules to return to the environment (Álvarez-Berragán et al., 2016).

While much research has been done more recently to understand which microorganisms might be able to accelerate the biodegradation of polymers, many of the promising findings so far require highly specialized species and conditions for degradation to occur faster (Álvarez-Berragán et al., 2016).

9.5 Film-forming finishes: paints, varnishes, and oils

9.5.1 *Acrylic paints*

The diverse group of finishing and decorating paints, varnishes, and oils make a special group of polymers used in the built environment. Historically, various natural materials have been used for similar purposes (see Part I). Archeological evidence suggests that casein curd from milk was used as a glue in Ancient Egyptian times (Lambuth, 2007). Casein has also historically been used as a paint stabilizer, a binding medium, and an additive to increase the water resistance of water-based paints. It was included in a popular house paint which created a very matte surface, was easy to make, and emitted no harmful fumes or odors (Ren et al., 2016). In fact, a wide range of natural substances were used to treat the surfaces of materials (see Chapter 10).

Immediately after WWII, there was a period of rapid development and increased demand for acrylic or latex paints (then also called "rubber-based paints") (Scholz, 1953). Many of the older paints relied heavily on the use of organic solvents (see Chapter 8). The main difference between plastics and these film-forming surface coverings is that the latter tend to polymerize on the site of their installation, through the process normally described as drying. This implies that the chemical reaction takes place on the site and could impact on air quality, depending on the chemistry of these processes. This broad group includes several families of chemicals: acrylic polymer emulsions, polyvinyl acetate emulsions, and the reaction polymers discussed later in this section. While solvents played an important role in the development of these paints, today the market is dominated by formulations with little or no solvents (Du, 2007).

Acrylic polymer emulsions are the dominant technology in paints. Emulsions are designed to prevent polymerization before application by using water-soluble surfactants to physically separate small monomer droplets from each other (Lombardi and Gasper, 2007; and explanation in Table 9.2). Therefore, the emulsion contains the monomer droplets, surfactants, and water-soluble initiators of the process. The majority of acrylic paint emulsions are designed to polymerize at temperatures between 10°C and 40°C, which makes mixtures of methyl acrylate and methyl methacrylate most suitable, given these polymerize at 8°C and 106°C, respectively (Lombardi and Gasper, 2007). Acrylic monomers suitable for the development of paints are esters of acrylic and methacrylic acid and include: methyl, ethyl, isobutyl, *n*-butyl, 2-ethylhexyl, octyl, lauryl, and stearyl (Lombardi and Gasper, 2007). To create acrylic paints, mixtures of these monomers are chosen for their desired properties. For example, butyl acrylate or 2-ethyl hexylacrylate give acrylic paint adhesive strength, while methyl methacrylate and methyl acrylate give it cohesive strength (Lombardi and Gasper, 2007). Unfortunately, each of these four examples is classified by the ECHA as a serious irritant for humans and also harmful to aquatic life (ECHA, 2016). This suggests that many of the ingredients used pose serious questions in terms of both human and ecosystem health. In addition, melamine and epoxy curing agents may also be added to achieve crosslinking (Lombardi and Gasper, 2007). Issues with these substances are discussed later in this section.

Unfortunately, the majority of manufacturers of such products currently disclose very little or no specific information about the chemical mixtures of their products. For example, a data sheet might only list that the paint is 100% acrylic. Currently, the only sources of more detailed information are initiated by organizations working on increasing the level of available information, such as the program Declare by the International Living Future Institute (International Living Future Institute, 2016). In addition to the full breakdown of ingredients, Declare also makes it clear how far each of the components has traveled. Unfortunately, to date few materials are covered in such a way (see Chapter 18 for a discussion of issues arising when trying to obtain more complete information for Declare).

Despite the uncertainty over the exact composition of paints, with increased recognition of the issues associated with VOCs, recently many manufacturers have voluntarily started to decrease VOCs in their products, and include this in their product information. These disclosures take the form of listing the VOC content of the paints in grams per liter, along with the total VOC (TVOC) content. When reading these disclosures, it is important to remember that TVOCs have been shown to differentiate poorly between more harmful and safer VOCs (see Chapter 8). A study that measured which VOCs were released while four different low-VOC acrylic paints were drying, reported the presence of benzene, toluene, xylene, formaldehyde, and other aldehydes (Zhao et al., 2016). The authors observed the results were influenced by the specific formulations of the paints and the substrate. Lower VOCs were measured from paints that were applied on more porous surfaces (Zhao et al., 2016). Therefore, the key problem is that VOCs were found in all samples they tested, suggesting that the real level of harmfulness of these might be higher than indicated by their TVOC count.

9.5.2 Reaction polymers: epoxies, phenolics, and polyurethanes

Another large group of surface treatment materials is reaction polymers, which include epoxies, phenolics, and polyurethanes.

Epoxy resin systems are sold in two separate pots. While there are a number of different combinations of components epoxy systems could contain, generally they contain base resin (epoxy monomer), curing agents, reactive diluents, solvents, plasticizers, pigments, reinforcing fibers, and fillers (Bray, 1999). Epoxy can be used for adhesives, hardwearing coatings, and in construction of electronics circuit boards but also for composite materials such as fiberglass. However, any unreacted chemicals can pose significant health risks, and this is especially the case because 75%–90% of epoxy resins are based on epichlorohydrin and BPA, which react to form diglycidylether of bisphenol A (DGEBA) (Prodi et al., 2015). Some work suggests that DGEBA is a sensitizer for 60%–80% of patients who develop allergic dermatitis to unhardened epoxy resins (Prodi et al., 2015). Conversely, fully reacted epoxy resins (which are not always achieved) are neither irritants nor sensitizing. Construction workers and painters were found to be at higher risk of sensitization to epoxy resins (Prodi et al., 2015).

Reaction polymers include the previously discussed urea-formaldehyde (see Chapter 7), and melamine-formaldehyde and phenol-formaldehyde (phenolic). In both melamine- and phenol-formaldehyde, formaldehyde has a stronger chemical bond than urea-formaldehyde and consequently is less likely to be released than from urea-formaldehyde. As a consequence, both related polymers are stronger and have found uses since their development early in the 20th century (see Chapter 3 and Chapter 7). These are also used as hardeners of epoxy, and can be used in durable paints (Bourlier, 2007). Phenol-formaldehyde is also used in the manufacture of laminate products, such as formica, which are manufactured by impregnating several layers of kraft paper with phenol-formaldehyde resin, applying compression, and then curing with heat (Formica, 2016; see Chapter 3). Health issues associated with formaldehyde are discussed in Chapter 7.

Melamine is toxic if swallowed, and may be harmful if inhaled or absorbed through skin. It is an irritant to skin, eyes, and mucous membranes, can cause kidney damage, and when heated emits highly toxic fumes (PubChem Compound, 2016). Phenol is toxic if swallowed, inhaled, or in contact with skin, is suspected of causing genetic defects and in certain cases damage to organs if there is prolonged or repeated exposure. It is also harmful to aquatic life with long-lasting effects (PubChem Compound, 2016). Therefore, all products from this family of chemicals should be considered as toxic in their manufacture and when used in construction. They are, however, currently considered to be stable when fully reacted.

Polyurethanes are used in a variety of forms, with coating varnish systems being one application in buildings. Urethanes are known for chemical, water, and abrasion resistance, with high tensile and impact strength (Merlatti et al., 2008). As Table 9.2 explains polyurethanes are made using a variety of polyols and isocyanates. The isocyanate components are especially problematic when assessing the health impact of polyurethanes. Basic health issues associated with the most commonly used isocyanates are as follows:

- Toluene-2,6-diisocyanate is fatal if directly inhaled. It is an irritant, may cause allergic reactions, is suspected of causing cancer (IARC Group 2B), and is harmful to aquatic life with long-lasting effects (PubChem Compound, 2016).
- 4,4′-Diphenylmethane diisocyanate (MDI) is fatal if directly inhaled, is an irritant, may cause allergic reactions, and causes damage to organs through prolonged or repeated exposure (PubChem Compound, 2016).
- 1,6-Diisocyanatohexane is known as a severe irritant which can cause chronic lung problems after long-term exposure. No information is currently available on its long-term health impacts (PubChem Compound, 2016).
- Isophorone diisocyanate (IPDI) is toxic when inhaled (fatal if directly inhaled), toxic when absorbed through the skin, is a strong skin irritant, and is toxic to aquatic life with long-lasting effects (PubChem Compound, 2016).

Given the high level of toxicity of these chemicals it is not surprising that, as discussed earlier, some have been involved in serious industrial accidents resulting in many deaths (Akovali, 2012b). Clearly, any use of such toxic components should be avoided.

However, the problems with polyurethane do not end there. More traditional polyurethane coatings have relied on organic solvents to prevent the mixture from reacting. Once a coat was applied, this solvent had to evaporate before the curing chemical reaction would commence. For such applications, polyester and polyether were common polyols (Álvarez-Berragán et al., 2016; Pourreau and Smyth, 2005).

In recent years, demand for lower VOCs has led to development of a number of water-based alternative urethane products. In order to develop varnishes without solvents, polyurethane-polyacrylate special blends have been developed, and also acrylic/urethane hybrid polymers (Derby et al., 1995). One such combination uses isophorone diisocyanate and allyl-based acrylic polyols to formulate moisture-curable coatings which prepared acrylated urethane acrylic oligomers for coatings (Pourreau and Smyth, 2005). When combining acrylics with polyurethanes, the variety of possible water-based polyurethanes increases dramatically. This makes it hard to anticipate likely formulations given manufacturers' poor disclosure of the ingredients. What is relevant to note is that the chemistry of water-based varnishes appears to be evolving towards decreasing the polyurethane content and increasing the content from acrylic paint systems. Some of the water-based acrylic systems currently disclosed on the Declare program appear to have no isocyanate content, and instead use common acrylic paint chemistry (International Living Future Institute, 2016).

9.5.3 Natural oil varnishes

Finding sustainable painting and varnishing solutions has always been possible. Throughout the period of popularity of solvent-based polyurethanes and the development of their water-based alternatives, natural wood oil systems persisted. For a period many used significant solvent content to accelerate curing processes. More traditional wood oil systems relied on oils such as linseed and tung oil in order to cure through a reaction of their polyunsaturated oils with oxygen in the air. This meant that, depending on the ambient temperature and humidity, these oil finishes might need days, weeks, or even months to fully cure. Interest in more sustainable alternative materials has inspired the rejuvenation of these industries and some such products can be found through the Declare program (International Living Future Institute, 2016). Many of the paint formulations based on traditional natural paints have also remained available from many small companies around the world.

Film-forming materials are not free of the other issues observed with polymers. Many are made using toxic chemicals, which can find their way into installation sites, including homes. Some are dangerous in use and although much smaller in volume, they have the same issues in the disposal stage. The fact that a polyurethane finish can be sanded off only increases the potential exposure of humans to these toxic substances. Smaller particles still take hundreds of years to chemically decompose.

9.6 Biopolymers

In recent years, more work has been done on the development of biopolymers, and there is more than one approach to such innovations. For example, because of the great similarities in organic chemicals extracted from crude oil or from grown materials, some efforts have focused on developing systems for extracting the same seven key hydrocarbon building blocks used in petroleum-based polymers from grown bioresources. In principle at least, the same key chemicals can be derived using: ethanol C_2H_5OH, glycerol $C_3H_8O_3$, xylose $C_5H_{10}O_5$, fructose $C_6H_{12}O_6$, and glucose $C_6H_{12}O_6$ (Christensen et al., 2008). One of the great advantages of this way of developing hydrocarbons is that instead of energy-intensive steam cracking of fossil fuels, grown carbohydrates can be transformed into simple sugars and fermented into ethanol and glycerol, and from these ethylene and butadiene can be produced (Christensen et al., 2008). This approach could lead to the production of the same polymers that are already in use, but sourced from bioresources. Other research efforts are seeking to find ways to develop new polymers which are better suited to the chemistry of bioresources. Specific alternative technologies include the production of olefins using bioethanol dehydration, methanol-to-olefins, catalytic fast pyrolysis of lignocellulosic biomass, and bio-oil upgrading (Pyl et al., 2012). Unfortunately, to date manufacturing polymers using these processes has not arrived to the point of financial viability (Pyl et al., 2012), which is understandable, given that research focusing on the development of these chemicals has only begun reasonably recently. Ethanol is one chemical that can be readily obtained from biomass by fermentation, and in recent years its production has been increasing (Rass-Hansen et al., 2007).

One innovative approach uses fluid catalyst cracking of hydrodeoxygenated vegetable oils derived from Norwegian spruce pulping (Pyl et al., 2012). The intention of the initial research was to examine if similar processes commonly used to derive hydrocarbons from petroleum could be effective for breaking down naturally occurring oils into hydrocarbons. The resulting chemicals were closely comparable to those in steam cracking of naphtha and were especially successful in deriving ethane and propene (Pyl et al., 2012). All approaches to develop hydrocarbons from bioresources are still in the innovation phase, when it is still hard to anticipate which of the solutions will prove to be most useful.

Using bioresources, which are presumably renewable, to develop petrochemicals already in use would be an improvement to relying on finite fossil fuel resources. Another approach to the development of new polymers is to develop new chemistry based around the chemicals commonly found in bioresources without mimicking the existing petrochemical industry. Poly(lactic acid) (PLA) which has entered the market as an alternative to PET and polystyrene is one such product. PLA is a recyclable, biodegradable, and compostable alternative for many packaging and more durable plastics (Christensen et al., 2008). In PLA, the monomer, lactic acid, is produced by bacterial fermentation of corn starch or sugar (Christensen et al., 2008). Wood dust can be added to PLA as a natural stabilizer and filler, without any reduction to recyclability and biodegradability of the biopolymer. However, the disadvantage of PLA is that like many other polymers it does require very specific conditions for biodegradation, and to date municipal collection of PLA for recycling does not exist in many places.

While there is great promise and reason for excitement with the development of biopolymers, it should also be acknowledged that such practices are likely to mean competition arising between land to grow food crops and land to grow crops for the production of biopolymers. In fact, in late 2006, the price of tortilla flour in Mexico doubled, because of the rise in the US corn prices due to US interest in the manufacture of biofuels from corn (Runge and Senauer, 2007). This provoked a significant public outcry because about half of Mexico's population lives in poverty and relies on tortillas as a main food source (Runge and Senauer, 2007). This aspect of considering the benefits and disadvantages of using more biopolymers has to be carefully considered. In this example, the efforts to increase production of biofuels in the developed world led to hunger of poor in less developed countries. Biopolymers could present similar costs.

9.7 Radon

9.7.1 Radon in building materials

Natural materials can also present naturally occurring hazards. After all, both lead and asbestos (see Chapter 5) are naturally occurring. However, another potential health hazard, radon gas, might be harder to observe.

Radon is a radioactive, colorless, and odorless gas which naturally occurs in some soils and in excavated building materials (Kayakökü et al., 2016). Radon leads to exposure to ionizing radiation which has been associated with genome damage and cancer development, but appears to also increase health risk of cardiovascular and immunological disturbances (Fucici, 2012). The first regulative intervention laying down the basic standards for protection against ionizing radiation was set in 1959 for the EU, and has had many subsequent adjustments (Council Directive 2013/59/Euratom, 2014). The 2013 European Directive provided stricter limits for radon and signaled a need to establish national reference levels for indoor radon concentrations, and to encourage appropriate actions where the levels are high (Council Directive 2013/59/Euratom, 2014; Smetsers et al., 2016). Higher prevalence of radon is associated with certain soil types, thus some regions have a greater problem with radon (Hahn et al., 2015; Collignan et al., 2016; Smetsers et al., 2016). Many parts of the US and Continental Europe are known for high radon prevalence and relevant national bodies provide information on specific differences within countries. However, because of the contemporary use of nuclear energy (either for peaceful or potential war purposes), modern societies have greater exposure to ionizing radiation than in the past (Fucici, 2012). Indoor exposure is especially important because radon can build up indoors in much higher concentrations than outdoors. Exposure to radon indoors has been correlated with an increased risk of lung cancer, especially for exposure levels from 100 Bq/m^3 (Fucici, 2012). Therefore, careful design of the built environment in areas of risk, such as installing radon barriers, should be paramount.

Three key factors influence the level of radon found indoors: the level of radon in the local soil, the presence of building materials which might emit radon, and the type of ventilation present. Radon can also be introduced through the water supply and indoor cigarette smoking (cigarettes contain polonium and radioactive lead (Desideri et al., 2007)). Any materials or components of materials excavated from areas naturally high in radon could also introduce radon into built indoor environments. Because of the natural variability in radon levels in the ground, this is difficult to predict and requires systematic examination of many samples. Any use of building materials such as earth, bricks, ceramic tiles, stone, gravel, sand concrete (because of its aggregate), and even gypsum presents the risk of radon being present. Existing data suggest that adobe/earth construction, sand, and gravel present lower risks than materials exposed to heat during their processing, such as bricks (Szabó et al., 2013; Gharbi et al., 2012; Kayakökü et al., 2016). Nevertheless, some stones, such as granite and perlite have been associated with very high indoor radon concentrations (Collignan et al., 2016; Kayakökü et al., 2016).

Bricks can be a problem, because heat treatment during production causes changes to the internal structure of the material and therefore can influence subsequent emissions of radon and thoron, technically called "exhalation" (Szabó et al., 2013; Kayakökü et al., 2016). This is why bricks can have elevated levels of background radioctivity. The temperature at which they were fired can influence this. A study which compared three different types of brick found that some bricks had several times the radioactive level of others (Kayakökü et al., 2016), although other studies have found more consistent and less alarming results for bricks (Gharbi et al., 2012). Because of the total variability with bricks both in terms of the source material and the temperature treatment used in manufacture, it is worth noting that radiation content should be regularly considered as a possible hazard associated with the use of bricks made in certain parts of the world.

Although this section explains the risks associated with radon in many excavated materials, it is important to emphasize that bricks and stone should not be seen as unsustainable materials because some samples have been found to have high radon levels. Rather, the natural variability of these materials should be supported with more complete assessments of the material for radioactive content. These materials present many other advantages in sustainable construction.

9.7.2 Sustainable approaches to construction which can increase risks from radon

Because concrete is not set by heat, it tends to have a lower radioactive content and thus compares favorably with bricks (Szabó et al., 2013). However, more recently, coal slag and fly ash have become common additives to concrete, and these tend to have a higher radioactive content (Szabó et al., 2013; Fucici, 2012). The radioactivity of such additives depends on the natural radioactivity of the source coal itself, which varies as with all other excavated materials (Szabó et al., 2013). After burning coal, the remaining slag or ash tends to be more radioactive because it gets enriched with radionuclides during the process (Szabó et al., 2013). Cement can be replaced by slag or fly ash in proportions ranging from 10% to 80%, which can be seen as an affordable or arguably sustainable use of a waste product, but unfortunately using it requires special waste management procedures due to its radioactive content (Fucici, 2012). Some studies have found samples of concrete using slag or fly ash to have an ionizing radiation content above the regulated levels (Szabó et al., 2013). The use of slag or fly ash in concrete is a good example of an intervention intended to elevate the sustainability of a product by using a waste from another process, that could actually be working against the long-term health of building inhabitants.

The increased use of ceramic tiles for flooring is another choice associated with sustainable construction which could pose radioactive risks. Ceramic tiles are similar to fired brick and the majority of health issues found with the latter are likely to be found with ceramic tiles. In addition, some glazes can increase the radioactive exhalation from the tiles. Zircon sand has been studied and concentrations of radionuclides in the glaze which have exceeded the regulated limits have been found

(Yahong et al., 2002). Traditionally tiles were mainly used in kitchens and bathrooms, but, more recently, they are increasingly used in living and other rooms as people reject carpet on health grounds, thus increasing the potential for exposure to radiation (Fucici, 2012). This is especially problematic for young children as radon is heavier than air and tends to accumulate close to the floor surfaces (Fucici, 2012).

Another situation when the problem has increased in recent years is houses constructed using concrete slab-on-grade, which is a classic way of making a passive solar house. This form of construction tends to accumulate higher radon concentrations (Collignan et al., 2016). This can present a double risk if the concrete slab-on-grade is finished with ceramic tiles, which is common.

Finally, the recent focus on energy efficiency has led to many houses being retrofitted to reduce air infiltration. A study of houses converted to be energy efficient but without the installation of mechanical ventilation, has found that these had higher average radon concentrations (Collignan et al., 2016; Francisco et al., 2016). The research team explained that this was due to the increase in achieved airtightness. While airtightness is sometimes desirable for efficient energy use, it can create unhealthy indoor environments, which is why it should be accompanied with proper ventilation systems.

Based on this discussion it is clear that there is sufficient acknowledgment of the health risks associated with radon, and from that perspective radon should be in the Stage 3 of recognition. However, because it is impossible to eliminate such a naturally occurring gas, what this overview suggests is that radon is likely to be in a permanent Stage 2 of recognition, needing additional clarifications and the passing of more strict regulations. The natural variability of radon makes regular assessments of potential risk necessary.

9.8 Titanium dioxide (TiO_2)

Titanium dioxide (TiO_2), or titanium white, is a common additive to acrylic and other paints, which adds whiteness and reflective properties and helps to stabilize these chemically. In many ways, titanium dioxide is a replacement for lead in paints. It is also found in many other products from food, medicine, and toothpastes, to plastics, papers, and inks. In 1966 the United States and in 1969 the EU approved food-grade titanium dioxide as an inactive ingredient in human food and labeled it E171 (Jovanović, 2014; Proquin et al., 2016). Between 1916 and 2011, an estimated total of 165 million metric tons of titanium dioxide was produced globally (Jovanović, 2014). This prolific use of titanium dioxide, at both micro- and nanoparticle sizes (both are found in E171), has only recently inspired research on its impacts on human health. Pele et al. (2015) demonstrated that a proportion of ingested titanium dioxide is absorbed into the bloodstream, while others have investigated the toxicity of such particles (Lappas, 2015; Proquin et al., 2016; Wormington et al., 2016). Titanium dioxide appears to cause inflammation, pulmonary damage, fibrosis, and lung tumors. It also appears to be genotoxic and possibly carcinogenic to humans (Chen et al., 2014). Currently reported health issues tend to focus more on nanoparticles than microparticles of titanium dioxide (Jovanović, 2014; Lappas, 2015; Proquin et al., 2016; Wormington et al., 2016), leaving the assessments still incomplete. While titanium dioxide in paints is unlikely to be absorbed as directly as when added to food, the patterns of the poor recognition of the potential risks of this commonly used substance echo the patterns observed with lead (see Chapter 6). Titanium dioxide can be seen as making transition into Stage 2 of recognition with some regulative action likely to follow the recent increase in research in this area. Unfortunately, currently it is impossible to avoid products containing titanium dioxide, as for many alternatives do not exist.

9.9 Linoleum—a natural composite

Linoleum is a good example of a natural composite. Without trying to breakdown bioresources into their chemical constituents as biopolymers do, linoleum simply combines natural materials to make a new composite. Furthermore, this discussion of linoleum should be contrasted with the earlier discussion of PVC-based vinyl flooring (see Chapter 7), given that both provide almost identical flooring solutions.

Linoleum was invented in 1860 by Frederick Walton, and was the first form of resilient flooring to replace more traditional bare wood or dirt floors (Lent et al., 2009). It is made by mixing oxidized linseed oil with resins from pine trees, wood flour, cork, and limestone fillers, with added pigments, which are then pressed onto a cloth backing to make sheet linoleum (Lent et al., 2009). These are natural and renewable materials (Akovali, 2012a). This general formula has remained largely unchanged since linoleum was first made (Lent et al., 2009). However, during the 1960s, vinyl flooring rapidly overtook the use of linoleum as the predominant resilient floor covering (Lent et al., 2010). The last US linoleum plant closed in 1975 (Lent et al., 2009, p. 14) and by the mid-1980s only three linoleum producers were left in the world (BBC, 2014). The most enduring and largest by volume manufacturer of linoleum, Fabro (previously Nairn in Kirkcaldy, Scotland) manufactures linoleum under the brand name Marmoleum (Anon., 2013).

More recently, with the recognition of the health and environmental concerns associated with vinyl there has been an upsurge of interest in linoleum (Lent et al., 2009). In 2009, Lent, Silas, and Vallette conducted a comprehensive review of the available resilient floor coverings for the Healthy Building Network, and recommended linoleum as either already better than all other alternatives or most able to be improved. Their research and similar comparisons show some possible concerns associated with linoleum. Because oxidation of linseed oil is at the core of the process, while the flooring is new, this oxidation can continue, leading to persistent odors (Lent et al., 2009). Recently, the VOC emissions of linseed-based paints have also been investigated, showing that a number of aldehydes are produced in the early hours and days post application (Fjällström et al., 2002). In one State of California study, two linoleum products failed Californian strict standards due to high emissions of acetaldehyde (Lent et al., 2009). However, this is not an unavoidable problem, but rather indicative of the relative neglect of research into improving linoleum production during the second half of the 20th century. Not all samples of linoleum suffer from the same problems and by developing new combinations of linseed and other oils this chemistry can be significantly improved (Fjällström et al., 2002; Lent et al., 2009). Similarly, farming of flax (linseed) as a crop has been associated with the use of environmentally persistent pesticides (such as trifluran) and eutrophication, both of which are avoidable through improvements in farming practices (Lent et al., 2009). Linoleum is also biodegradable.

The conclusion of this analysis is that linoleum should be reasonably healthy for users and neutral in impact for the environment, although it is important to research specific linoleum products, as these vary on both levels. In returning to the unavoidable comparison with PVC-based vinyl flooring (see Chapter 7), it is clear that a more natural and healthier alternative has always existed. While vinyl dominated the market because of its lower price and greater compositional predictability, and quickly worked its way into some regulations, leading to commonly perceived advantages in using it (Petrović et al., 2016), these benefits are in part negated because of the toxic chemicals used in its manufacture, toxicity release while in use, and the still only partly understood toxicity at the disposal stage (Section 9.4 and Chapter 7). Seen in that light, linoleum offers advantages well beyond those of vinyl flooring.

9.10 Timber

A wide range of natural materials can be viewed as having few health issues in terms of indoor air quality. Many grown materials, including timber may not have such issues and have sustainability advantages already discussed in Parts I and II of this book. However, grown materials can present problems in contemporary construction because these are prone to aging and can be both dimensionally and visually harder to control. In response, a number of strategies have become common to increase the predictability of grown materials. Some of these strategies mean the material becomes a less healthy option. In the context of this research, the most important example of this is timber.

When used in interior settings, issues associated with the use of timber include risk of fire, wear and tear, color changes, staining, and discoloration (Feist, 2007). Air temperature, humidity, the species of the wood, and the cut of the timber all influence the ways in which timber elements shrink and swell. For surfaces that are to be painted, the most desirable situation is no dimensional variation because most paint polymers struggle to accommodate such changes without cracking (Feist, 2007).

In terms of building construction, the most problematic feature of wood is that it is sensitive to biological (fungal (mold) and insect) attacks (Schiopu and Tiruta-Barna, 2012; Feist, 2007). Fungal development not only structurally weakens timber elements, but can also release odors and spores indoors, which are harmful to human health. In order to prevent this, a number of treatments have been developed over the years. Chapter 3 and Chapter 10, explain the traditional approaches to achieve this which ranged from strategies as diverse as using vinegar or wax, or charring the timber, to using the resin found in particular woods. More recently, a variety of biocides have been used for protection against decay (Schiopu and Tiruta-Barna, 2012). Contemporary chemical treatments of timber are designed "to destroy, deter, render harmless, prevent the action of, or otherwise exert a controlling effect on any harmful organism" (Schiopu and Tiruta-Barna, 2012). Thus, not surprisingly, these are not healthy for living creatures. This leads to the question: which timber treatments are more or less harmful in terms of human health? Although there are good reasons to consider some treatments as more and others as less harmful, and there is research that makes such comparisons (Schiopu and Tiruta-Barna, 2012), the question of greater importance is: could contemporary construction develop approaches which require no or lower reliance on human-made chemical treatments? Chapter 17 discusses some recent moves towards reducing the regulatory requirements for the chemical treatment of timber in the EU. This can be construed as a very positive sign.

In the absence of fungi, left outdoors completely untreated (with no chemical treatment or paint applied), timber naturally weathers, and this process gradually changes the color and texture of wood, often rendering wood gray. Once this change has taken place, timber stays almost unchanged for a long time, slowly wearing away in a process called erosion (Feist, 2007). Most softwoods erode at an average rate of 6.35 mm per century, while dense hardwoods only erode at a rate of 3.18 mm (Feist, 2007). Very low-density softwoods, such as western red cedar, may erode at a rate as high as 12.7 mm per century, although in cold climates this might drop to 0.79 mm per century (Feist, 2007). The slow progress of erosion of timber shows that timber can be a very robust and enduring building material providing it is kept protected from fungi and insects.

In the past, many timber constructions were designed to be regularly maintained and replacement of the elements most vulnerable to biological attack was planned. Chapter 3 and Chapter 10, review some of these traditional strategies. Unfortunately, contemporary construction seldom has forward-thinking long-term solutions for the replacement of components, but rather relies on regular maintenance using paints, or chemical treatments to make timber more durable. Paint can be used to create a waterproof film on timber, however, regardless of the number of coats of paint used, this approach never creates an entirely moisture-proof layer. Instead, the natural shrinking and swelling of wood tends to crack such films (Feist, 2007). Once moisture enters into the painted timber elements it is very difficult for it to dry and consequently good conditions for potential rotting occur (Feist, 2007). Therefore, using paint to protect timber from rotting can lead to the opposite. Design solutions which allow for the easy replacement of the most exposed elements could present better long-term solutions. This is not to say that exterior painting should be avoided, rather contemporary reliance on it for waterproofing is too great.

Another issue which more directly impacts indoor air quality is strong smells from some timber resins. For example, pine species are abundant in monoterpenes,

α-pinene, limonene, and 3-carene, which have been related to irritation of human respiratory systems in dwellings, although the same compounds are used as a cure for asthma (Manninen et al., 2002). Two studies compared emissions from air-dried and heat-treated wood, observing a significant decrease in TVOCs, but also a change in which VOCs were present in heat-treated samples (Manninen et al., 2002; Esteves et al., 2008). During heat treatment, the chemical structure of timber altered as the cellulose thermally degraded through the process (Manninen et al., 2002). Because of this, although the TVOC emissions were several times lower after heat treatment (and that was especially true for the terpenes), the components of other VOCs increased including some that are potentially toxic (Manninen et al., 2002). Furthermore, because tree resin acts as a natural biocide to protect wood, the heat-treated timber tested as less resistant to certain fungal species (Stratev et al., 2011). This is an area where more research is needed to fully understand the best ways to dry timber for construction use, especially given that heat treating adds to the embodied energy of timber (see Chapter 4).

What this section shows is that although some indoor health issues can occur with timber, it performs best when its natural properties are accepted and consciously included in design. The appearance of timber will change over time, and accepting this is the best approach, and avoids the use of paints and varnishes. Leaving timber as it is allows for its natural biocides to protect the wood without introducing unnecessary, and potentially health-damaging human-made biocides into indoor environments. Although each tree species and its timbers require a separate analysis, the same general trends are likely to apply. In other words, grown materials can be excellent for creating good indoor air quality, as long as their gradual decay is accepted and included in the design.

9.11 Conclusion

When applying the principles of the three stages of the recognition of hazards as presented throughout Part 3 of this book to the materials discussed in this chapter, new types of issues become apparent. As this chapter shows, very hazardous substances are used in the production of petroleum-based polymers. The key ingredients of petroleum-based polymers should be classified in either Stage 2 or 3 of recognition. It is well known that these are hazardous generally both for people and natural ecosystems, yet their use under controlled manufacturing conditions continues. Some of the polymers appear to be safe while in use, while others are known to pose health problems. The majority are single-use products, or are used a small number of times before disposal. The full environmental and human health cost of polymer degradation in the environment is still unknown. Therefore, while there are many advantages to using polymers and life without these is hard to imagine, it is also important to reassess the way we think about polymers. When considering the totality of their adverse effects, they should perhaps be treated in a similar way to precious metals. The latter are in everyday use, but are used sparingly and for very specific situations. A significant change in the way contemporary culture uses polymers is probably urgently needed.

Buildings predate the development of polymers and the petrochemical industry and, therefore, should be a good place for the easy elimination of polymers. However, in recent years, polymers have gone through an intense expansion in use in construction, from their irreplaceable utilization as damp-proof membranes and for rubber in engineered window systems, to avoidable replacements for nails (as glues), plastic extrusion coverings, and polymer fillers for surface imperfections. In addition, the majority of indoor finishes and fittings include polymers of some kind, such as paints, varnishes, vinyl, formica laminate products, acrylic, plywood, particleboard, MDF, and carpets and their backings. Polymers are also increasingly used to replace tried and tested materials; e.g., glass is with polycarbonates, and decking timber with polymer timber. Nevertheless, in the building industry it should be possible to return to traditional nonpolymer construction techniques, and to start to actively resist this increase in polymers and other human-made materials in construction. Due to the comparatively large size of buildings in relation to other consumer goods, any changes in the construction industry could have significant impacts on the total global consumption of such materials.

This chapter shows that although less artificial materials could generally be safer for human health, they still come with risks. A number of natural materials can present health risks whether from their use of heavy metals, the introduction of radon into interior environments, or high emission of VOCs. Yet, it is the improvement and development of these materials that show great promise to provide solutions which address criteria discussed in this book. However, manufacturing technology and research data that can confirm the healthy and sustainable outcomes of such materials still requires much development.

References

Akovali, G., 2012a. Plastic materials: polyvinyl chloride (PVC). In: Pacheco-Torgal, F., Jalali, S., Fucic, A. (Eds.), Toxicity of Building Materials. Woodhead Publishing, Cambridge, pp. 23–53.

Akovali, G., 2012b. Plastic materials: chlorinated polyethylene (CPE), chlorinates polyvinylchloride (CPVC), chlorosulfonated polyethylene (CSPE) and polychloroprene rubber (CR). In: Pacheco-Torgal, F., Jalali, S., Fucic, A. (Eds.), Toxicity of Building Materials. Woodhead Publishing, Cambridge, pp. 54–75.

Alvarez, R., Barriocanal, C., Díez, J.L., Cimadevilla, G., Casal, M.D., Canga, C.S., 2004. Recycling of hazardous waste materials in the coking process. Environ. Sci. Technol. 38, 1611–1615.

Álvarez-Berragán, J., Domínguez-Malfavón, L., Vargas-Suárez, M., González-Hernández, R., Aguilar-Osorio, G., Loza-Tavera, H., 2016. Biodegradable activities of selected environmental fungi on a polyester polyurethane varnish and polyether polyurethane foams. Appl. Environ. Microbiol. 82 (17), 5225–5235.

Anon., 2005. Victims sue Formosa Plastics Corporation, Texas, due to injuries sustainable on October 6, 2005 explosion and fires at their Point Comfort, Texas plant. PR Newswire12th December 2006.

Anon, 2013. Marvelling at marmoleum. Contr. Flooring J.32–34, February 2013.

Asawakosinchai, A., Jubslip, C., Rimdusit, S., 2015. Organic based heat stabilizers for PVC: a safer and more environmentally friendly alternatives. Key Eng. Mater. 659, 321–326.

Avis, S., Hutton, C., 1993. Acute benzene poisoning: a report of three fatalities. J. Foren. Sci. 38 (3), 599–602.

Bach, C., Dauchy, X., Severin, I., Munoz, J.-F., Etienne, S., Chagnon, M.-C., 2014. Effect of sunlight exposure on the release of intentionally and/or non-intentionally added substances from polyethylene terephthalate (PET) bottles into water: chemical analysis and *in vitro* toxicity. Food. Chem. 162, 63–71.

BBC, 2014. Scotland's Landscape: Kirkcaldy (Fife). Available from: http://www.bbc.co.uk/scotland/landscapes/kirkcaldy/ (accessed June 2014.).

Biscaldi, G.P., Mingardi, M., Pollini, G., Moglia, A., Bossi, M.C., 1981. Acute toluene poisoning: electroneurophysiological and vestibular investigations. Toxicol. Europ. Res. 3 (6), 271–273.

Bittner, G.D., Yang, C.Z., Stoner, M.A., 2014. Estrogenic chemicals often leach from BPA-free plastic products that are replacements for BPA-containing polycarbonate products. Environ. Health. 13 (41), 1–14.

Bourlier, K., 2007. Phenolic resins. In: Tracton, A.A. (Ed.), Coating Materials and Surface Coatings. CRC Press, Boca Raton, FL, pp. 8-1–8-5.

Bray, P.G., 1999. Epoxy resins. Occup. Med. (Chic. Ill). 14 (4), 743–758.

Broughton, E., 2005. The Bhopal disaster and its aftermath: a review. Environ. Health Global Access Sci. Sour. 4, 6.

Ceresana, 2016. Introductory Previews of Ceresana: Market Intelligence, Consulting reports. Available from: www.ceresana.com (accessed October 2016.).

Chapa-Martínez, C.A., Hinojosa-Reyes, L., Hernández-Ramírez, A., Ruiz-Ruiz, E., Maya-Treviño, L., Guzmán-Mar, J.L., 2016. An evaluation of the migration of antimony from polyethylene terephthalate (PET) plastic used for bottled drinking water. Sci. Total Environ. 565, 511–518.

Chemical Safety Board US (CSB), 2006. Fire at Formosa Plastics Corporation: Evaluating Process Hazards: Case Study. Available from: http://www.csb.gov/ (accessed October 2016.).

Chen, T., Yan, J., Li, Y., 2014. Genotoxicity of titanium dioxide nanoparticles. J. Food Drug Anal. 22, 94–104.

Christensen, C.H., Hansen, J.R., Madsen, C.C., Taarning, E., Egeblad, K., 2008. The renewable chemicals industry. ChemSusChem. 1 (2008), 283–289.

Collignan, B., Le Ponner, E., Mandin, C., 2016. Relationship between indoor radon concentrations, thermal retrofit and dwelling characteristics. J. Environ. Radioact. 165, 124–130.

Council directive 2013/59/Euratom, 2014. Council directive 2013/59/Euratom of 5 December 2013 laying down basic safety standards for protection against the dangers arising from exposure to ionising radiation, and repealing directives 89/618/Euratom, 9o/641/Euratom, 96/29/Euratom, 97/43/Euratom, 2003/122/Euratom. Official Journal of the European Union 17.01.2014: L 13/1-73.

Derby, R., Gruber, B., Klein, R.J., 1995. Performance attributes of acrylic urethane hybrid polymers in aqueous wood lacquers. Am. Paint Coatings J.57, 79.21 (30th Jan 1995).

Desideri, D., Meli, M.A., Feduci, L., Rpselli, C., 2007. 210Pb and 210 Pb inhalation by cigarette smoking in Italy. Health. Phys. 92 (1), 58–63.

Du, J.W., 2007. Surfactants, dispersants, and defoamers for the coatings, inks, and adhesive industries. In: Tracton, A.A. (Ed.), Coating Materials and Surface Coatings. CRC Press, Boca Raton, FL, pp. 30-1–30-13.

ECHA, 2016. Chemicals Search from European Chemicals Agency (ECHA). Available from: echa.europa.eu (accessed October 2016.).

Environment Protection Agency US (EPA), 2016b. Health Effects of Petroleum Coke. Available from: https://www.epa.gov/petroleum-coke-chicago/health-effects-petroleum-coke#whatis (accessed October 2016.).

Esteves, B., Graça, J., Pereira, H., 2008. Extractive composition and summative chemical analysis of thermally treated eucalypt wood. Holzforschung. 62 (3), 344–351.

European Council of Vinyl Manufacturers (ECVM), 2016. Lead Stabilisers. Available from: www.pvc.org (accessed October 2016.).

Fan, Y.-Y., Zheng, J.-L., Ren, J.-H., Luo, J., Cui, Z.-Y., Ma, L.Q., 2014. Effects of storage temperature and duration on release of antimony and bisphenol A from polyethylene terephthalate drinking water bottles in China. Environ. Pollut. 192, 113–120.

Feist, W.C., 2007. Exterior wood finishes. In: Tracton, A.A. (Ed.), Coating Materials and Surface Coatings. CRC Press, Boca Raton, FL, pp. 111-1–111-12.

Fjällström, P., Andersson, B., Nilsson, C., Andersson, K., 2002. Drying of linseed oil paints: a laboratory study of aldehyde emissions. Ind. Crops Prod. 16 (3), 173–184.

Formica, 2016. Formica Guide Specification. Available from: http://www.formica.com/ (accessed October 2016.).

Francisco, P.W., Jacobs, D.E., Targos, L., Dixon, S.L., Breysse, J., Rose, W., et al., 2016. Ventilation, indoor air quality, and health in homes undergoing weatherization. Indoor. Air. 2016, 1–15.

Fucici, A., 2012. The main health hazards from building materials. In: Pacheco-Torgal, F., Jalali, S., Fucic, A. (Eds.), Toxicity of Building Materials. Woodhead Publishing, Cambridge.

Gharbi, F., Oueslati, M., Abdelli, W., Samaali, M., Ben Tekaya, M., 2012. Exposure to radiation from the natural radioactivity in Tunisian building materials. Radiat. Prot. Dosimetry. 152 (4), 418–422.

Guercio, V.J., 2010. US producing, exporting more slurry oil. Oil Gas J. Available from: http://www.ogj.com/articles/print/volume-108/issue-37/processing/us-producing-exporting-more-slurry-oil.html (Accessed October 2016.).

Hahn, E.J., Gokun, Y., Andrews, W.M., Overfield, B.L., Robertson, H., Wiggins, A., et al., 2015. Radon potential, geological formations, and lung cancer risk. Preven. Med. Rep. 2 (2015), 342–346.

Hitachi, 2016. Production Process for Polyethylene Terephthalate (PET). Available from: www.hitachi.com (accessed October 2016.).

International Living Future Institute, 2016. Declare. Available from: www.living-future.org (accessed October 2016.).

Jovanović, B., 2014. Critical review of public health regulations of titanium dioxide, a human food additive. Integr. Environ. Assess. Manag. 11 (1), 10–20.

Kannan, G., Grieshaber, S.E., Zhao, W., 2016. Thermoplastic polyesters. In: Obabisi, O., Adewale, K. (Eds.), Handbook of Thermoplastics, *second ed.* CRC Press, Boca Raton, FL (2016).

Kayakökü, H., Karatepe, Ş., Doğru, M., 2016. Measurements of radioactivity and dose assessments in some building materials in Bitlis, Turkey. Appl. Radiat. Isotopes. 115, 172–179.

Pele, L.C., Thoree, V., Bruggraber, S.F.A., Koller, D., Thompson, R.P.H., Lomer, M.C., et al., 2015. Pharmaceutical/food grade titanium dioxide particles are absorbed into the bloodstream of human volunteers. Part Fibre Toxicol. 12, 26.

Lambuth, A., 2007. Soybean, blood and casein glues. In: Tracton, A.A. (Ed.), Coating Materials and Surface Coatings. CRC Press, Boca Raton, FL, pp. 19-1–19-11.

Lappas, C.M., 2015. The immunomodulatory effects of titanium dioxide and silver nanoparticles. Food Chem. Toxicol. 85, 78–83.

Lent, T., Silas, J., Vallette, J., 2010. Chemical hazards analysis of resilient flooring for healthcare. HERD: Health Environ. Res. Design J. 3 (2), 97–117.

Lent, T., Silas, J. and Vallette, J. (Healthy Building Network), 2009. Resilient Flooring and Chemical Hazards: A Comparative Analysis of Vinyl and Other Alternatives for Health Care: Report for Health Care Research Collaboration. Available from: http://www.healthybuilding.net/docs/HBN-ResilientFlooring&ChemicalHazards-Report.pdf. (accessed June 2014.).

Levshina, S.I., Efimov, N.N., Bazarkin, V.N., 2009. Assessment of the Amur River ecosystem pollution with benzene and its derivatives caused by an accident at the chemical plant in Jilin City, China. Bull. Environ. Contam. Toxicol. 83, 776.

Lombardi, R.A., Gasper, J.D., 2007. Acrylic polymers. In: Tracton, A.A. (Ed.), Coating Materials and Surface Coatings. CRC Press, Boca Raton, FL, pp. 1-1–1-10.

Luo, J.C., Nelsen, K.G., Fischbein, A., 1990. Persistent reactive airway dysfunction syndrome after exposure to toluene diisocyanate. Br. J. Ind. Med. 47, 239–241.

Manninen, A.-M., Pasanen, P., Holopainen, J.K., 2002. Comparing the VOC emissions between air-dried and heat-threated Scots pine wood. Atmos. Environ. 36, 1763–1768.

Merlatti, C., Perrin, F.X., Aragon, E., Margaillan, A., 2008. Natural and artificial weathering characteristics of stabilized acrylic-urethane paints. Pol. Degrad. Stab. 93, 896–903.

Metz, C.M., 2016. Bisphenol A: understanding the controversy. Workplace Health Safety. 64 (1), 28–36.

Petrović, E.K., Vale, B., Wilson, M.S., 2016. Vinyl and linoleum flooring: health issues as perceived by lay people and architects. J. Green Build. 11 (1), 159–177.

Polymer Science Learning Center, 2000. Polymer Science Learning Center. University of Southern Mississippi. Available from: www.pslc.ws (accessed October 2016.).

Pourreau, D.B., Smyth, S., 2005. New low-viscosity acrylic-urethane prepolymers and their acrylated oligomers for moisture and UV-*curable* coatings. JCT Coatings Tech. 2005, 42–48.

Prodi, A., Rui, F., Fortina, A.B., Corradin, M.T., Filon, F.L., 2015. Occupational sensitization to epoxy resins in Northeastern Italy (1996–2010). Int. J. Occup. Environ. Health. 21 (1), 82–87.

Proquin, H., Rodríguez-Ibarra, C., Moonen, C.G.J., Urrutia Ortega, I.M., Briedé, J.J., de Kok, T.M., et al., 2016. Titanium dioxide food additive (E171) induces ROS formation and genotoxicity: contribution of micro and nano-sized fractions. Mutage. Adv. 2016, 1–11.

PubChem Compound, 2016. Review of Chemicals by the National Center for Biotechnology Information (NCBI). Available from: www.ncbi.nlm.nih.gov (accessed October 2016.).

Pyl, S.P., Dijkmans, T., Antonykutty, J.M., Reyniers, M.-F., Harlin, A., Van Geem, K.M., et al., 2012. Wood-derived olefins by steam cracking of hydrodeoxygenated tall oils. Bioresour. Technol. 126, 48–55.

Rass-Hansen, J., Falsig, H., Jørgensen, B., Christensen, C.H., 2007. Perspective bioethanol: fuel or feedstock? J. Chem. Technol. Biotechnol. 82, 329–333.

Ren, F., Atlasevich, N., Baade, B., Loike, J., Arslanoglu, J., 2016. Influence of pigments and protein aging on protein identification in historically representative casein-based paints using enzyme-linked immunosorbent assay. Analy. Bioanalyt. Chem. 408, 203–215.

Runge, C.F., Senauer, B., 2007. How biofuels could starve the poor. The New York Times May. 7, 2007. Available from: http://www.nytimes.com/cfr/world/20070501faessay_v86n3_runge_senauer.html?pagewanted=print&_r=0 (accessed October 2016.).

Sajiki, J., Yonekubo, J., 2003. Leaching of bisphenol A (BPA) to seawater from polycarbonate plastic and its degradation by reactive oxygen species. Chemosphere. 51, 55–62.
Schiopu, N., Tiruta-Barna, L., 2012. Wood preservatives. In: Pacheco-Torgal, F., Jalali, S., Fucic, A. (Eds.), Toxicity of Building Materials. Woodhead Publishing, Cambridge, pp. 138–165.
Scholz, H.A., 1953. History of water-thinner paints. Industrial Eng. Chem. 45 (4), 709–711.
Smetsers, R.C.G.M., Blaauboer, R.O., Dekkers, S.A.J., 2016. Ingredients for a Dutch radon action plan, based on a national survey in more than 2500 dwellings. J. Environ. Radioact. 165, 93–102.
Stratev, D., Gradinger, C., Ters, T., Kackler, K., Kuncinger, T., Srebotink, E., 2011. Fingal pretreatment of pine wood to reduce the emissions of volatile organic compounds. Holzforschung. 65 (4), 461–465.
Szabó, Z., Völgyesi, P., Nagy, H.É., Szabó, C., Kis, Z., Csorba, O., 2013. Radioactivity of natural and artificial building materials – a comparative study. J. Environ. Radioact. 118, 64–74.
Tijssen, K.C.H., Blaakmeer, E.S.M., Kentgens, A.P.M., 2015. Solid-state NMR studies of Ziegler-Natta and metallocene catalysts. Solid. State. Nucl. Magn. Reson. 68-69, 37–56.
United States Steel Corporation (USS), 2016. Safety Data Sheets for Steelmaking By-Products. Available from: www.ussteel.com/uss/portal/home/aboutus/safety/SDS-Labels (accessed October 2016.).
Unwin, J., Coldwell, M.R., Keen, C., Mcaliden, J.J., 2013. Airborne emissions of carcinogens and respiratory sensitizers during thermal processing of plastics. Ann. Occupat. Hygiene. 57 (3), 399–406.
Vogt, E.T.C., Weckhuysen, B.M., 2015. Fluid catalytic cracking: recent developments on the grand old lade of zeolite catalysis. Chem. Soc. Rev. 44 (20), 7342–7370.
Vroman, I., Tighzert, L., 2009. Biodegradable polymers. Materials. 2 (2), 307–344.
White, W.C., 2007. Butadiene production process overview. Chem. Biol. Interact. 166, 10–14.
Wormington, A.M., Coral, J., Alloy, M.M., Damare, C.L., Mansfield, C.M., Klaine, S.J., et al., 2016. Effect of natural organic matter on the photo-induced toxicity of titatnium dioxide (TiO_2) nanoparticles. Environ. Toxicol. Chem. Available online as accepted December 2016.
Yahong, M., Yigang, L., Guang, Z., Xiaolei, H., 2002. The studies on radiological limits of color-glazed tiles used in home decoration. Health. Phys. 82 (4), 510–512.
Zhang, M., Buekens, A., Jiang, X., Li, X., 2015. Dioxins and polyvinylchloride in combustion and fires. Waste Manage. Res. 33 (7), 630–643.
Zhao, P., Cheng, Y.-H., Lin, C.-C., Cheng, Y.-L., 2016. Effect of resin content and substrate on the emission of BTEX and carbonyls from low-VOC water-based wall paint. Environ. Sci. Pollut. Res. 23, 3799–3808.

Part IV

Case Studies

Sustainability and the material aspect of traditional residential buildings in Serbia

10

Ana Radivojević, Mirjana Roter-Blagojević and Ljiljana Đukanović
University of Belgrade, Belgrade, Serbia

Because of the complex geographical and historical circumstances, and a reasonably wide range of climatic conditions, Serbian traditional architecture developed a diversity of types and used a range of materials, thus providing good examples of typical vernacular development in many parts of not only the Balkans, but also, Central Europe and Asia Minor. This case study shows that the traditional solutions tended to be good examples of a relationship related to sustainability on many levels. They were sensitive to the site and context, predominately used local materials which were only modestly altered from their natural state, made use of durable materials that were reused, and the more perishable materials returned to nature at the end of their lifespan. The diversity of building types, construction techniques, and the materials selected can be considered one of the central values of vernacular architecture in Serbia. The suitability of such approaches for sustainable practices is recognized in good integration of the vernacular architecture with the natural environment. This is achieved through the use of appropriate siting, which takes into account the local bioclimatic features, optimization and reduction of resources used for the construction, and contribution to the creation of a healthy living space (Correia et al., 2015).

10.1 Historical, cultural, and social context

Set on the Balkan Peninsula, Serbia has records of settlements since prehistoric times, with one of the early archeological sites, the *Lepenski vir* settlement on the Danube River, established during the Neolithic period, in 7–6 millennia BC (Pejić, 1998). Due to main roads connecting Europe with Asia Minor passing through this region, it has witnessed the influences of many cultures from both East and West. In early medieval times, the late 6th and early 7th centuries, the most significant influence on the current culture was the migrations of Slavic tribes from northeast Europe to the Pannonia Plan and Balkan region. The newly arrived Slavic tribes lived in dispersed rural settlements in one-room semidugout or aboveground houses, in solid wood construction (log cabins) or timber-framed construction, with adobe infill (Milošević, 1997). During the 12–15th centuries, residential buildings

Materials for a Healthy, Ecological and Sustainable Built Environment: Principles for Evaluation.
DOI: http://dx.doi.org/10.1016/B978-0-08-100707-5.00010-1

were built in fortified urban, or in unfortified rural settlements. The houses of Serbian feudal rulers were typically built of stone and solid wood construction, but the greatest number of houses and utilitarian buildings were log cabins or timber-framed houses (Milošević, 1997).

In the postmedieval period, from the mid-15th century, the region was under the domination of the Ottoman Empire. Initially this had little impact on the existing vernacular architecture in rural settlements, but over time its influence increased in urban settlements. Between the 16th and 19th centuries, rural houses in the central and southern parts of contemporary Serbia, and in other parts of the Balkans (Greece, Macedonia, and Bulgaria), did not change much. Changes were visible in urban areas where eastern (Asia Minor) types of houses were dominant (Pavlović et al., 1987). However, the Ottoman influence did not spread across all of contemporary Serbia, and from the late 17th to the early 20th centuries the northern part, set in the Pannonia Plain, Vojvodina, was part of the Habsburg Monarchy (later Austro-Hungarian Monarchy) and developed under a Central European influence (Samardžić et al., 2014). Massive brick construction systems dominated in urban buildings, while Pannonia Plain houses in rural settlements tended to be predominantly built of rammed earth (Pavlović et al., 1987).

From the early 19th century Ottoman domination started to weaken in the Balkan region and Serbia moved towards political independence, giving way to a Western, mainly Central European, influence. During the same period, the political, social, economic, and cultural situation was improved (Nestorović, 2006). These changes influenced a gradual transformation of urban settlements and their residential architecture. Two specific influences on the new forms emerged: changes to the building and engineering regulations and a general improvement of education in the field of civil engineering and architecture. These influenced the development of new spatial principles and ensured their acceptance, and brought control and organization of the whole construction process, from the design to the finishing of buildings, but only in the main towns (Roter-Blagojević, 2006). However, rural architecture during the same period had limited change, retaining many of its already established and traditional characteristics (Kojić, 1949; Deroko, 1968).

10.2 Basic types of traditional rural and urban residential buildings

It is possible to say that the key characteristics of the traditional vernacular architecture of Serbia did not change much from medieval times until the early 19th century.

In rural settlements, the natural and climatic conditions had a primary impact on the house characteristics with the readily available materials, such as wood, earth, and stone, dominating vernacular construction (Table 10.1). The presence of these materials, their use, and combinations depended on the particular geographical area, and these differences reflected variations in the local climate and soil. Regardless of the particular place and conditions, construction of traditional rural houses has

Table 10.1 A review of applied building materials in traditional rural and urban houses

	Rural house			Urban house	
	Log cabin	**Timber-framed house**	**Rammed earth house**	**Oriental house**	**Central European house**
Foundation	Stone	Stone	Rammed earth	Stone, brick	Stone, brick
Wall structure	Wood	Wood	Rammed earth, Branches, stems	Wood	Brick
Wall infill		Wood, adobe, wattle		Adobe	
Ceiling structure	Wood	Wood	Wood and loam	Wood	Wood
Roof structure	Wood	Wood	Wood	Wood	Wood
Roof covering	Shingle, straw, bark, slate	Barrel-tile, flat-tile, slate	Straw, reed, flat-tile	Barrel-tile	Flat-tile
Floor	Earth, wood	Earth, wood	Earth, wood	Earth, wood	
Wall finishing		Mud and straw	Mud and straw	Mud and straw	Lime mortar

always been a collective act. It was also a way for master-builders to exchange and improve their acquired building knowledge and skills and to transfer these to the next generation of builders.

The climate conditions strongly influenced some key characteristics, such as the spatial characteristics and the roof pitch of houses. In the warmer regions, roofs were low pitched with deep eaves and houses were more open towards the countryside, having large porches, while in the regions with long, snowy winters, the roofs were steeply pitched (to prevent piling of snow), and the houses only had small openings. This shows an organic relationship between the natural environment and the corresponding traditional house forms. Synchronicity with the natural environment and preservation of the ecological balance, which are clearly seen here, are also specific universal characteristics that connect different vernacular architectures in the world stemming from similar natural conditions (Cvijić, 2000).

The variety found in Serbian vernacular architecture can be organized into three distinct rural and three distinct urban house types (Fig. 10.1). Each type can be found in the surrounding geographical areas, beyond the contemporary political borders of Serbia.

10.2.1 The rural types

Three main construction techniques dominated rural houses: log cabins, timber-framed houses (*bondruk-style* house), and rammed earth houses (Fig. 10.1A–C).

Figure 10.1 Traditional rural and urban houses types: (A) log cabins; (B) timber-framed house; (C) rammed earth house; (D) Oriental house; (E) Transitional house; (F) Central European house.

10.2.1.1 Log cabins

In the mountain areas of central and west Serbia houses were built of quality timber in the form of log cabins. The topography of these mountainous sheep-grazing areas was the main reason for the dispersed settlements, with separated groups of houses, one for each family cooperative (in these areas many people lived in extended family groups or cooperatives). A number of separate houses, for cooking and dining, gathering, or the use of smaller groups of people, like young married couples and guests, were placed freely in the courtyard, surrounded with work-related buildings, gardens, and orchards. These timber houses were very small, with a rectangular plan and simple spatial organization. They were either a one-room structure (room with hearth called *kuća*, meaning a "*home*") or a two-room structure with "home" (*kuća*) and an additional room for gathering and guests. Rarely, a porch was part of a log cabin. The same type is found in the hilly areas of the western Balkans (the so-called *Dinara*-type of log cabins).

The basic materials used for the walls and roofing of log cabins were massive logs of circular cross-section or, more often, rectangular beams (Fig. 10.2A). They were notched at their ends to form corner joints, usually being saddle notched (Fig. 10.2B). At first, walls were made of horizontally laid unhewn logs, from which only the bark might have been stripped. More complex processing of logs occurred later when the logs were trimmed both on the inner and outer sides in order to obtain a straight wall surface. Assembling of wooden elements was initially done by simple contact between two logs, but since such connection was not resistant to deformations from drying of the material and uneven settlement, voids had to be filled with moss and grass to ensure better air-tightness of the wall structure and improve protection against weathering. Over time the joint was improved by connecting logs using wooden pegs made of hard woods, such as oak. Some houses were built of a combination of logs for the walls of the "home" (*kuća*) and timber-framed walls covered with lath, clay, and mud plaster (so-called *čatma*).

In the earliest stages of the formation of the structure, foundation walls did not exist. Rather, four tree trunks were laid directly on the ground, notched at the corners and interconnected. Other logs were placed over them and joints were sealed with moss, clay, or cow manure. Later, houses had shallow foundations of broken stone, and if they were built on sloping terrain, they usually had a stone-walled cellar under one part of the house, where livestock and other necessities were kept. Due to the climate conditions, the roof was steeply pitched and the rafters usually covered with wooden shingles (*šindra*). In regions abundant in shale-stone, shale-stone slates were used for roofing. There are examples of log cabins that were thatched with straw. The area under the roof was conjoint with the space of the *home* (*kuća*), with an open hearth. The *home* had an earthen floor and, when present, the floor and ceiling of the additional room were made of wood.

10.2.1.2 Timber-framed houses

In the flatter and agriculturally richer valleys of the Morava River and eastern Serbia, houses were made of a timber-framed construction (the *bondruk*-style house) (Fig. 10.2C). These were freely placed in rural courtyards, similar to log cabins, and groups of buildings, both residential and work-related, were created in dispersed or semicompact settlements. These houses were small, with a one-space (*kuća* with fireplace) or two-space plan, and each house was for a single family as part of a rural cooperative. The form and structure of the houses were similar to the

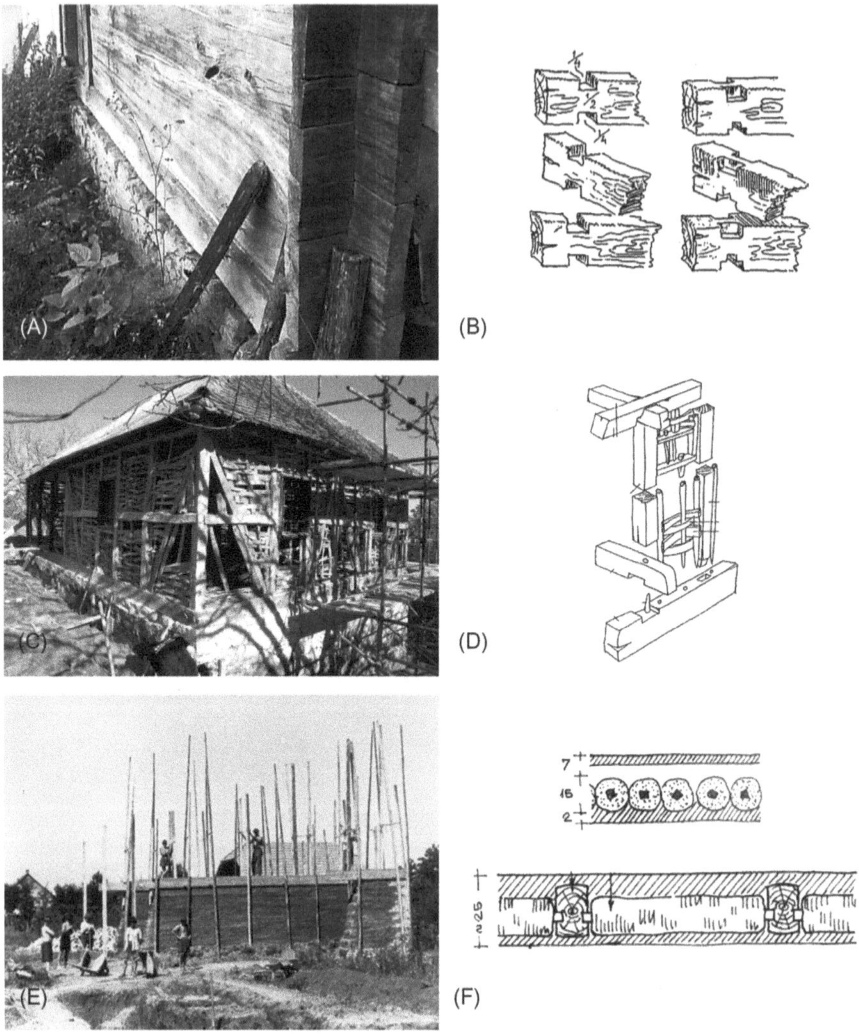

Figure 10.2 Typical structures of traditional rural houses: (A) log cabin wall structure; (B) corner joints; (C) timber-frame wall structure; (D) wattle; (E) building of a rammed earth wall; (F) ceiling structure of a rammed earth house with a prefabricated cylindrical infill.

log cabin, but due to a milder climate, such houses usually had an open porch (*ajat, hajat*) at their entrance. The porch was used for various farming chores, but it had a separate part for sitting and resting (*doksat*). A house could be built on flat terrain or on a slope when it then had a stone cellar (*izba*) for storing farm produce and tools or for keeping livestock. A similar type can be seen in the eastern and southern Balkans, in Macedonia, Bulgaria, and Greece, and in Turkey (Deroko, 1968; Pavlović et al., 1987).

The basic timber frame of the house consisted of a post and beam structure with trusses or braces supporting the corner. This type of timber framing was widely used, as it allowed the houses to be built relatively quickly and the timber did not have to be of top quality. It also had the advantage of being quite safe in seismic conditions. The horizontal sill timbers were placed on top of the low stone foundation, the latter built by dry stacking the stones or with the use of mud plaster and then, at about 1 m spacing, vertical posts were placed and connected to the roof structure with horizontal tie beams. The timber-framed walls were filled with different types of material, but the simplest way was wattle (*pleter*)—stakes woven with branches and daubed with a combination of mud and straw (Fig. 10.2D). Another type was horizontal wattle anchored to the supporting vertical posts or slats nailed to the vertical posts, filled with clay, crushed stone, and straw, and covered with mud plaster (*čatma*). Adobe bricks (*čerpić*) could also be used in combination with mud mortar. Such walls were covered with two layers of plaster, which was a combination of loam and cut straw, and painted with lime. These houses usually had shallow-pitched roofs, formed over a square base, on a structure of rafters and battens. The shallower roof slope induced a change of roof covering and barrel tiles were used. The greater weight of these contributed to the complexity of the roof structure. Typical for this type of roof was the creation of deep eaves that protected the walls from rain and excessive overheating during summer.

10.2.1.3 Rammed earth houses

In Vojvodina, in the large Pannonia Plain, rammed earth houses dominated, because earth was the most available material (Fig. 10.2E). Such houses have been built since the early 18th century in regulated compact settlements and placed facing the street with gardens and auxiliary buildings positioned to the back of a narrow plot. The narrow side of the houses faced the street, with the longer side running back into the plot. All houses had a lateral yard porch (*kong, gang*) with columns, directly connected with the street entrance, which was used as an open corridor for accessing the rooms and for various farming jobs. The house had a small entrance hall that led into the semiopen area of the kitchen with a vaulted hearth (built with brick) and into a room that looked towards the street. The additional rooms were built on the yard side along the porch (Kojić, 1949; Deroko, 1968; Pavlović et al., 1987).

Typically, the low thick walls of these rammed earth houses were built by compressing a semidry mixture of earth into wooden molds, packing layer upon layer. Corners and joints were reinforced with tree or bush branches inserted every 20–30 cm from the base of the wall; reinforcing could also be done with bricks. Often reeds, branches, or vines were inserted every 20–30 cm, providing a degree of seismic reinforcement to the walls. The foundations were also built of rammed earth, thus creating a serious problem with capillary suction and moisture in the walls.

The ceiling structure of all three types of rural house was of round logs or hewn beams with different kinds of infill. Wooden laths or planks were most often used for this purpose, in combination with reed and mud mortar, with a layer of mud above (Fig. 10.2F). The exterior and interior walls were painted in white with lime. The houses usually had timber gable roofs, covered with straw or reed, but later on, from the start of the 20th century, with flat ceramic roof tiles (*crep*).

10.2.2 The urban types

From the early 19th century urban houses followed the same construction principles as the timber-framed rural ones and the Eastern type of urban houses (*bondruk*-style) were developed by the Ottomans in Asia Minor (Kuban, 1993) and the Balkan region (Deroko, 1968; Roter-Blagojević, 2006). Changes refer to transformation of building types and structures imported from Central Europe in the mid-19th century, as well as to the use of new building materials which were not necessarily natural in their origin but were the subject of various technological processes, such as the use of brick, cement, concrete, terracotta, gypsum, artificial stone, and terrazzo (Table 10.1).

The three distinct types trace the integration of these Eastern and Western influences on house development (Fig. 10.1D–F).

10.2.2.1 Oriental house

The timber-framed structure was deeply rooted in the Balkan region from the medieval period, for both urban and rural houses, as in other European and Asia Minor areas (Deroko, 1968). Until the early 19th century, the traditional structural systems and building techniques of residential architecture were similar throughout the entire Balkan region (modern Bosnia and Herzegovina, Serbia, Macedonia, Greece, Bulgaria, and Romania) and in Asia Minor (modern Turkey), because all these regions were part of the Ottoman Empire. This is the reason why this house type is known in the literature variously as the Turkish House, Balkan House, Levantine House, and Oriental House (Kojić, 1949; Deroko, 1968; Findrik, 1986; Pavlović et al., 1987; Roter-Blagojević, 2006). This chapter uses the term Oriental house, because this covers a wider area where this type was developed. Concepts typical for the Balkan towns, combined with traditional forms of rural dwellings that were transferred and adapted to the urban surroundings, prevailed in Serbian towns (Fig. 10.1D). Houses were located in separate residential areas (called *mahalas*) with steep, narrow, winding streets, and built as free-standing one or two-storey buildings surrounded by gardens and high walls (Kojić, 1949).

The Oriental house had a traditional timber-framed structure with different types of infill, made from wattle and daub (*čatma*), or adobe bricks (*čerpić*). The basic structural elements were the same as for similar rural houses, but the craftsmanship was of better quality, especially when the house was built for a wealthy owner.

10.2.2.2 Transitional house

In the period from 1820 to the 1850s, the Oriental urban house was transformed into a Transitional type (Vujović, 1986). These buildings retained the characteristics of the Oriental architecture regarding space organization, but the traditional timber-framed structure was combined with a new material, fired ceramic brick (Fig. 10.3A,B), and the European baroque and classical style (Fig. 10.1E). In some cases, buildings kept a traditional structure while applying new building techniques, such as massive stone, and brick construction (Fig. 10.3C,D).

The upper parts of the Transitional type of urban house consisted of timber frame and brick infill, which was possibly a reflection of the fact local craftsmen still had limited experience of the new construction technique of pur masonry brickwork. The outer and inner walls were stiffened with wooden columns, beam, and struts, although the wall infill of baked bricks was thicker, more massive, and structurally stable (Fig. 10.3A,B). It is obvious that, at that time, masonry walls without

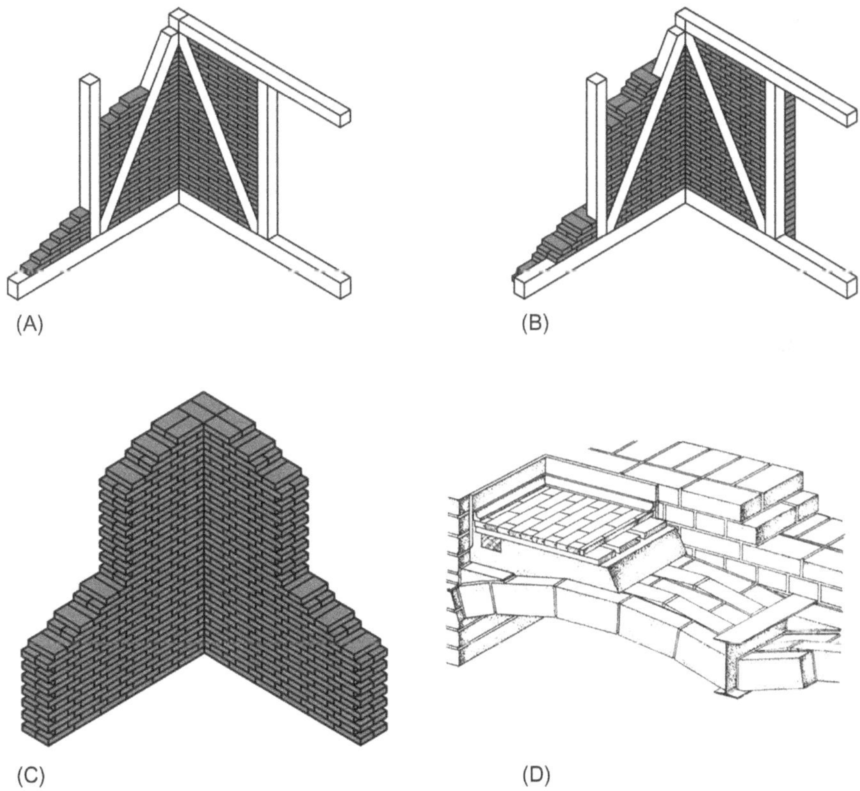

Figure 10.3 Typical structures of traditional urban houses: (A) wall structure of a timber-framed construction with brick or adobe infill; (B) combined solid brick and timber-framed construction; (C) solid brick wall; (D) Prussian vault.

wooden reinforcements represented a partial mystery, while the timber-framed structure with thicker masonry infill seemed to be a safer solution and an easier choice for skilled local builders.

10.2.2.3 Central European house

From the early 18th century, a Central European house type became common for urban settlements in Vojvodina, which was a part of the Austrian Monarchy. In the mid-1830s massive masonry structures appeared in the central part of Serbia that was still within the Ottoman Empire. Since the use of brickwork was expensive for the time, it was used for a few public buildings or houses for the upper classes. The larger transfer to the Central European dwelling type was realized in the 1860s and 1870s, but the buildings were still only one or two storeys (Fig. 10.1F). More complex structures, such a multifamily multistorey "residential palaces," appeared in the last two decades of the 19th century and the early 20th century (Nestorović, 2006; Roter-Blagojević, 2006).

The massive masonry houses were built of stone and brick, laid in lime mortar (Fig. 10.3C). Floor structures were either "*Prussian-vaults*" (a combination of shallow brick vaults and transverse iron beams) above the basement and ground floor (Fig. 10.3D), or wooden structures with rough wooden planks or, in better houses decorated with carvings, for the other storeys. Local builders were familiar with such structures, but when it was necessary to set up the *stucco mortar* (called "*štokator*"), masters from Vojvodina had to be engaged. The roof structure was timber covered with flat ceramic roof tiles (*crep*). Facades were decorated in a modest but classical style, most often using lime plaster. Basement walls and plinths were made of rustic stone.

Use of innovative materials and structural elements, that were typical of residential architecture in the Austro-Hungarian Empire and Germany, became dominant in Serbia in the last decade of the 19th and the first decade of the 20th centuries (Roter-Blagojević, 1997). Foreign and locally trained architects and civil engineers, educated in European schools (Budapest, Vienna, Munich, Berlin, Karlsruhe, Zurich, and Achen), replaced traditional builders, and residential urban houses change their spatial organization, construction, and materials following European standards and concepts (Roter-Blagojević, 2006). Cast-iron supports were used for elements of floor constructions, galleries, and balconies (Đurić-Zamolo, 1980). Finishing works on the facades of houses were made using artificial stone, prefabricated stylistic elements and details in terracotta, cement, and concrete, as well as using the *sgraffito* technique. In the interior, terracotta and terrazzo tiles were used on staircases as cladding. Stone stairs are also found and prefabricated gypsum decorated elements. These elements were mostly obtained from abroad until the end of the 19th century when the first factories for production of different ornamental elements in cement, gypsum, and concrete were opened in Serbia (Nestorović, 2006).

Regarding the town morphology the transitional and European types of house were built along the streets, being terraced to form an urban block. At this time, lots became smaller, narrower, and deeper, creating small inner courtyards and light wells in the urban block. This situation was undesirable for general housing conditions, especially for insulation and the ventilation of buildings and urban blocks. In general, traditional rural and urban buildings were modest in size and the way they were furnished. At first, the size of a building was closely related to the limitations of the materials used and the skills and knowledge of builders. Over time, the growing needs of the population have affected the size of the houses which have become larger and more comfortable. In rural areas, such tendencies have contributed to expansion of timber-framed houses, which has been manifested in reduced dimensions in terms of the spatial characteristics of the houses, since in the case of a log cabin size was directly related to the length of the timber. Another reason for the gradual replacement of log cabins with timber-framed houses was because during the 19th century there was an intensive settling of Serbian territory which induced serious deforestation of oak forests and, consequently, led to a reduction in the amount of available timber (Cvijić, 2000).

10.3 Environmental features of building materials and structures

This case study clearly shows that the selection of materials for the construction of a house can significantly affect the extent to which it would be characterized as a sustainable or healthy home (Sassi, 2006). The use of materials can be analyzed with respect to their environmental profile in terms of embodied energy and pollution, waste generation, and recycling possibilities, as well as their contribution to energy conservation and energy efficiency (Kim and Righton, 1998).

The material basis of the traditional residential architecture of Serbia reveals a natural cycle for these buildings in that many organic materials will biodegrade at the end of the building life or return to the earth in the case of earth-based materials. Additionally, the closer a material is to its natural state (e.g., less treated), the less dangerous it becomes to the environment, suggesting that the use of local, natural, nontoxic, and biodegradable materials, as in the case of traditional houses in Serbia, produces both quality of life and lower environmental impact during the whole lifecycle of the building.

Typically, traditional houses used materials that were obtained locally and in the case of rammed earth houses, even from the building site itself. Such practice decreased the transport component of the embodied energy of materials which already had a low embodied energy. The impact that buildings have on energy resources and their potential depletion is highest during the useful life of a building. In the so-called building phase of a house (Kim and Righton, 1998), comfort is most often identified with the experience of thermal comfort as a result of the insulating characteristics of materials and structures, their potential for passive solar heating and cooling, as well as behavior regarding water vapor movement through the structure and the type of heating regime used. Structures of traditional buildings were permeable and as such contributed to the indoor air quality of a house. Earth structures also proved to be good regulators of humidity. In spite of the fact that the insulating properties of the thermal envelope of traditional houses were considerably poorer in comparison to the requirements of today's buildings (Radivojević et al., 2012, 2014), and heating systems were not very efficient (in rural houses the use of wood burned in the hearth, and in urban houses the use of wood-burning stoves), when combined with the use of thermal zoning and adequate orientation, they provided comfort that was good enough for the time, although whether still sufficient to meet modern standards, especially in urban areas, is the critical point. Traditional buildings implytraditional ways of using them. Harmonization with environmental conditions is especially significant for traditional log cabins which, due to the harsh climatic conditions of mountainous areas where they were built, had a true need to harmonize with the environment, starting with the choice of locations for their erection. As a rule, they were placed on sunny slopes that were well protected from dominant winds, having openings (windows and doors) on their sunny sides, while on the north they had very small or no windows, covered with wooden shutters.

Longevity of buildings is a relevant factor for efficient and economical use of resources. A long building life can be achieved by the means of different measures, such as the use of durable materials, adequate maintenance of the component parts of the building, possible reuse of materials, or recycling of components and materials at the end of the building life.

Until more recent times, durable building materials, such as stone and bricks, were seldom used for the construction of traditional residential houses in Serbia, usually being reserved for the lower parts of buildings (for foundations and storage basements), because they were hard to handle, needed specific skills of builders and craftsmen and were unaffordable for modest rural or urban houses. However, durable building materials were available from the late 19th century for the urban houses of wealthier citizens that were typically built with masonr brickwork. Other types of traditional houses relied predominantly on degradable, less durable materials. In the case of wood which is less durable compared to stone or brick, traditional builders made careful selection of wood types, based on their availability and durability. Consequently, for timber structural elements they used hard and durable oak. When wood was used as the roof covering of a log cabin, its selection was predetermined by the availability of certain wood types. Therefore, shingles might have been made from oak, although pine was even more convenient since it contains a lot of resin which makes it less susceptible to rotting. Additionally, in summer, resin comes out from the pine wood, and while it was still soft, builders used to spread it over the roof covering and sprinkle it with rye flour, creating in this way a protective waterproof layer. In order to extend the life of timber structural elements, traditional builders typically used different nontoxic and ecologically acceptable protective measures and treatments from natural resources, such as vinegar, linseed oil, wax, and tar, to protect wood from drying out and from fungi and insects (Tomovska and Radivojević, 2015). Another typical method was to protect the surface of wood in the timber structure of a log cabin with a layer of soot (carbon) provided by burning the wood in a fire in an open hearth. In this way, the fungi that cause rotting were destroyed and this extended the life of the wood. The intention to provide a longer life for some less durable materials can also be seen in the use of straw as a roof covering for traditional log cabins. When used for this purpose, the straw was dipped in a solution of lime which was a technique that improved its resistance to fire (Deroko, 1968).

Maintenance was of particular importance for traditional houses, especially in the case of potentially loose earthen structures that needed special attention. For instance, traditional rammed earth houses were sporadically coated with mud mortar which was applied in two layers (Višekruna, 2010). Such plaster was made of earth, with the addition of quicklime and water, but it might also be made of earth with the addition of sand, water, and chaff or chopped straw.

Building materials and elements (such as wooden elements, clay tiles, slates, etc.) were used in different ways enabling easy separation and repair. With respect to this, especially interesting is the evolution of the wall structure of a traditional log cabin. Although joints between the wooden elements of traditional houses evolved over time, they were always thought of in a way that enabled dismantling. However, in the case of the earliest types of log cabins which were laid directly on the ground, since the logs were directly exposed to water in the soil, which led to their rotting, a complete system for raising the house and replacing rotted logs was designed (Fig. 10.4B). Dismantling of the elements of a wooden structure was a principle that was also used in timber-framed structures which were typically connected with dowel-type joints, while secondary elements, like laths, were typically nailed.

Construction methods and types of joints used by traditional builders for houses contributed to the house being demountable and as such easily transferable to other locations. Smaller buildings were transported in their entirety (Fig. 10.4A,B), while larger ones were disassembled, transported, and then reassembled at the new location. In the past, the advantages of such wooden structures have proved to be very useful in rural areas, especially during the frequent turbulent historical circumstances, or due to certain economic reasons (for instance, division or reorganization of a property) when the peasants had to be relocated from their original place of

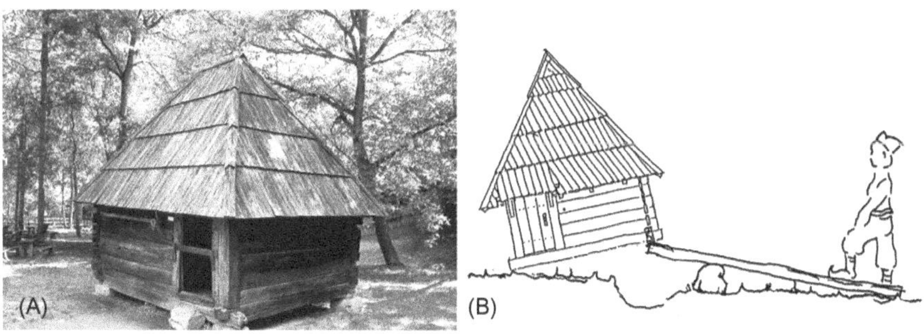

Figure 10.4 A small log cabin suitable for transportation: (A) appearance; (B) reconstruction of the process of raising the house.

residence (Findrik, 1986). Today this fact enables easy removal of traditional houses, especially log cabins, to another location which is often used as a method for preservation of this type of built heritage.

Rammed earth houses might have had a type of prefabricated element as an infill between the wooden beams of the ceiling structure. Around the laths whose length matched the distance between the wooden beams straw mixed with mud mortar was wound. These cylindrical elements that had uncoated laths at their end were placed into the grooves of wooden beams and completely filled the space between. This primitive method of prefabrication relied on locally available building materials and was used to speed up the building process.

Although the massive parts of traditional houses were not predisposed to disassembly, during their demolition in the 19th and early 20th centuries traditional builders would separate and clean bricks and stone blocks for their reuse in another masonry structure. The separation could be done relatively easily with a couple of strong shakes, since the binders used were either mud or lime-based mortars which did not create an especially strong bond between the masonry elements.

It should be noted that, as a result of the nature of building materials and techniques, the building process was very economical and did not produce much construction waste. Even, waste materials such as bark, wood chips, cut straw, earth, etc., tended to find a secondary use, either as a roof cover, in the case of bark, or as a constituent of elements of earthen structures, mortars, and plasters.

10.4 Conclusion

Because of the diversity of geographic and natural potentials that exists in Serbia, traditional architecture has been shaped by the natural environment. The influence of natural factors is identified in all segments of the house: use of natural materials from the immediate environment, dimensional adjustment of the house to fit the available natural materials, as well as adjustment to the terrain, which led to forming a functional space in accordance with climatic conditions (porch and open galleries), and specific development of certain elements of the house (eaves, roof slope).

The principles of sustainable architecture are recognized in many ways, such as the biodegradability, recycling, and reuse potentials of materials and structures, as well as reduction of the negative environmental impacts in all phases of the building lifecycle. By relying on natural building materials, traditional builders ensured their houses would be healthy living spaces.

However, inevitable changes that happened over time first influenced urban architecture, resulting in the use of new building materials which were more durable, but which carried a larger environmental burden in comparison to natural ones. In this way, modernization has contributed to the improvement of certain aspects of living comfort, and increased the longevity of buildings, but the change in the main construction techniques and materials has also meant moving away from the traditional links with the natural environment and its cycles.

Acknowledgments

This chapter was produced as a part of a research project "*Housing development in Serbia in the context of globalization and integrations*" (no. 36034) financed by the Ministry of Education, Science and Technologic Development of Republic of Serbia.

References

Correia, M., Dipasquale, L., Mecca, S. (Eds.), 2015. Versus. Heritage for Tomorrow. Vernacular Knowledge for Sustainable Architecture. DIDA Dipartimento di Architettura Università degli Studi di Firenze, Firenze.

Cvijić, J., 2000. Sabrana dela: Balkansko poluostrvo. third ed. SANU i Zavod za udžbenike i nastavna sredstva, Beograd.

Deroko, A., 1968. Narodno neimarstvo, I i II. Spomenik SANU, CXVIII, Beograd, SANU.

Đukanović, Lj, 2015. Tipologija i valorizacija gradjevinske strukture stambenih zgrada Beograda sa stanovišta komfora stanovanja (doctoral dissertation). Arhitektonski fakultet, Beograd.

Đurić-Zamolo, D., 1980. Beograd 1898 – 1914, Iz arhive Građevinskog odbora. Muzej grada Beograda, Beograd.

Findrik, R., 1986. Prenosive gradjevine u narodnoj arhitekturi. Glasnik Etnografskog muzeja u Beogradu. 50, 101–111.

Ginder, E., 1973. *Vojvođanske seoske kuće od naboja. Materijal, konstrukcija i izvođenje* (habilitation work). Arhitektonski fakultet, Beograd.

Kim, J.-J., Righton, B., 1998. Sustainable Architecture Module: Qualities, Use, and Examples of Sustainable Building Materials. National Pollution Prevention Center for Higher Education, College of Architecture and Urban Planning, The University of Michigan, Michigan. Available from: http://www.umich.edu/~nppcpub/resources/compendia/ARCHpdfs/ARCHsbmIntro.pdf (accessed 10.12.15.).

Kojić, B., 1949. Stara gradska i seoska arhitektura u Srbiji. Prosveta, Beograd.

Kuban, D., 1993. The Turkish Hayat House. Istanbul.

Milošević, G., 1997. Stanovanje u srednjovekovnoj Srbiji. SANU, Beograd.

Nestorović, B., 2006. Arhitektura Srbije u XIX veku. Art Press, Beograd.

Pavlović. St., D., Angelova, R., Mucopulos, N.K., Stojka, Ž., Sezgin, H., 1987. Narodno graditeljstvo na Balkanu. RZZSKS, SANU, DKS, Beograd.

Pejić, S. (Ed.), 1998. Spomeničko nasleđe Srbije, nepokretna kulturna dobra od izuzetnog i velikog značaja. RZZSK, Beograd.

Radivojević, A., Roter-Blagojević, M., Rajčić, A., 2012. Preservation of vernacular architecture in Serbia – authenticity versus thermal comfort issues. In: Proceedings of the International Conference on Structural Analysis of Historical Constructions, Volume III, SAHC 2012, 15–17 October 2012, Wroclaw, Poland, 2750–2759.

Radivojević, A., Roter-Blagojević, M., Rajčić, A., 2014. The issue of thermal performance and protection and modernization of traditional half-timbered (*bondruk*) style houses in Serbia. J. Arch. Conserv. 20-3, 209–225.

Roter-Blagojević, M., 2006. Stambena arhitektura Beograda u 19. i početkom 20. Veka. Arhitektonski fakultet & Orion art, Beograd.

Roter-Blagojević, M., 1997. Nastava arhitekture na višim i visokoškolskim ustanovama u Beogradu tokom 19. i početkom 20. veka, uticaj stranih i domaćih graditelja. Godišnjak grada Beograda. 44, 125–168.

Samardžić, N., Samardžić, R., Roter-Blagojević, M., 2014. Kulturna istorija Beograda, XVIII vek. Hesperia-edu, Beograd.

Sassi, P., 2006. Strategies for Sustainable Architecture. Taylor & Francis, Oxon. Available from: http://library.uniteddiversity.coop/Ecological_Building/Strategies_for_Sustainable_Architecture.pdf (accessed 15.01.16.).

Tomovska, R., Radivojević, A., 2015. Environmental features of building materials of traditional Ohrid house and their contribution to its human design. In: International Academic Conference Places and Technologies (2; 2015; Nova Gorica) Keeping up with technologies to make healthy places Book of conference proceedings, 86–92.

Vujović, B., 1986. Umetnost obnovljene Srbije 1791-1848. Prosveta, Beograd.

Višekruna, D., 2010. Održavanje kao vid čuvanja vojvođanske kuće. Glasnik Društava konzervatora Srbije. 34, 254–257.

Palm thatched building in Mexico

11

Jaime Rios-Calleja
BUAP, Benemérita Universidad Autónoma de Puebla, Puebla, Mexico

11.1 Introduction

This case study looks at the vernacular building tradition of using palm as a thatching material in the states of Colima, Jalisco, Nayarit (west coast of Mexico), and Puebla (highlands of central Mexico). It focuses on quantifying the impact of these natural materials on the environment using the ecological footprint measure of global hectares (gha) (Wackernagel et al., 2005), where a global hectare is an area of land of average global productivity.

In the last few decades palm thatching has become popular in urban and suburban areas in these areas of Mexico. Sometimes these uses are for tourist purposes by deliberately creating buildings that showcase Mexican building traditions and sometimes people use thatch just because they like its appearance. The aim of this case study is to compare the impact of using thatch in a traditional way for making rural buildings and its use in these more urban settings. The information on which the impact assessment has been based has involved many site visits and interviews with thatchers and building owners and a full description of this work is available elsewhere (Rios, 2015).

11.1.1 The natural environment and palms used for thatching

Palms are a very diverse species and are found in many different climates around the world. Generally, west coast Mexico falls into the category of deciduous forest and along the coastal lands dry to subhumid tropical forests dominate the landscape. Such a climate is very suitable for palms leading to a huge variety of palm species in the region. These sometimes occur in plantations but other palms that self-set in mixed arable lands are also harvested for thatching. Palms have economic value in these regions wherever optimum conditions for growth occur and due to the many products and byproducts derived from them. Products range from food such as seeds, fruits, or young shoots called "palmettos," to utensils like brooms and a wide variety of baskets, to wearable products like hats or rain capes, and to building items such as mats or thatching for roofs and walls. Other considerations include being easy to access and harvest, as well as being fast-growing and easily regenerated or propagated.

Despite the wide variety of palm species found in the middle west coast of Mexico, only three are popularly used for thatching. Of these, two are found in

Materials for a Healthy, Ecological and Sustainable Built Environment: Principles for Evaluation.
DOI: http://dx.doi.org/10.1016/B978-0-08-100707-5.00011-3

more coastal locations and one in the highlands. One of the more coastal species is *Attalea guacuyule* (Liebm. ex Mart.) Zona, (Missouri Botanical Garden, 2014; Quero, 1992; Zona, 2002) traditionally known as "coyul" palm, which is an oily palm mostly found in the tropical coastal lowlands along the Pacific coast of Mexico. This grows well as a solitary palm but is also found in patches as a secondary forest mixed with other plants. It is tall, reaching up to 20 m in height when fully mature and can live for over a century. Although leaves for thatching can be harvested from very young (7 years, 3 m tall) to very old trees (100 years, 20 m), thatchers prefer young palms because the older, taller palms present more risk when harvesting the leaves. *A. guacuyule* palm was once highly valued and planted because of its oily nut, although it has now been replaced by the coconut palm. Like many other palms, *A. guacuyule* is highly resistant to fire in its natural environment (Tejero-Díez et al., 2008).

The second popular thatching palm in the west coast region is *Sabal rosei* (O.F. Cook) Beccari. This palm belongs to a genus proper to the Americas, but is limited to the northern hemisphere within the Caribbean, Mexico, and Central America (Uhl and Dransfield, 1987), and grows well along the Pacific coast of Mexico. Within the core of this area in the states of Colima and Michoacán a similar palm of the same family *Sabal pumos* (Kunth) Burret is also found (Quero, 1992). There is almost no difference to the untrained eye between these two palms (information given personally by Dr. Hermilo Quero in Rios-Calleja, 2015). These palms are widespread in dry tropical forest areas in the west coast of Colima, Jalisco, and Nayarit, predominantly within a band of 0–500 m above sea level and up to 300 km inland.

The *Sabal rosei* and *Sabal pumos* palms are locally known as "palapa" palm, which is also the name given to a thatched building or palm thatched roof in the same region. *Sabal* palm is a medium-height tree reaching no more than 7 m and adapts well to mixed forests. It lives for 30–50 years and the leaves are valued for thatching and the mature trunks are used as structural posts. Harvesting of leaves can begin when the palm is 7–10 years old. *Sabal* ssp. can be the dominant species in an area but also occurs as a secondary species in other types of forest, and very dense secondary palm groves are found when the forests are destroyed, mainly by fire to make grasslands for cattle farming (Quero, 1992). It is rarely cultivated.

The third common palm for thatching, *Brahea dulcis* (Kunth) Mart, belongs to the highland area. Today, this palm is widely used for a range of products from diverse arts and crafts artefacts, to utensils such as baskets, and to roof coverings (Ugent, 2000). It mainly prefers limestone soil and is found in semidesert colonies from north Veracruz State, to San Luis Potosi and Guatemala (Quero, 1994). It is a dwarf species and grows well in clusters as secondary forest. The palm buildings in this case study are found in the Tentzo protected natural reserve in the highlands of the state of Puebla, 1500 m above sea level. Here settlements formed almost entirely of palm thatched houses are found.

Although the thatching technology and the resulting buildings in these regions are iconic in their use of palm, a wide range of other materials are normally involved. These will be introduced when discussing the technological aspects of the buildings before making the environmental assessment.

11.1.2 Vernacular palm thatching traditions in selected regions of Mexico

The thatched building tradition can be traced back to pre-Columbian times, and archeological, ethnographical, and ceramic records give a hint of the way early Mesoamerican houses were built. The first solid evidence belongs to the Formative period, which occurred between 1500 and 500 BC (Flannery and Sabloff, 2009), "It was at the start of this period that true, permanent villages of pole-and-thatch (wattle-and-daub) houses first became widespread in Mesoamerica" (Flannery and Sabloff, 2009). There is also evidence of clay model houses of thatched appearance from the Classic period up to AD 500, and these houses are sometimes arranged in a circle around circular pyramids (Kelly, 1980), thus resembling the "Guachimonton" architectural tradition. Reliable ceramic figurines of thatched houses come from the Postclassic (AD 800–1500) periods, with clear evidence of thatched pitched roofs in the region from these times (Von Winning et al., 1996). The bifold books (codex) from the highlands of Puebla and Oaxaca also show illustrations with a striking similarity to modern palm thatched houses (see Fig. 11.1).

The Mexican palm thatching tradition has also been influenced by migrations. Although there are few official records, the migrations of diverse ethnic groups during the colonial period in the 16th, 17th, and 18th centuries reinforced the thatching traditions. Those from the Philippines were mostly illegal migrants looking for a better life (Bruman, 1945; Cook and Borah, 1971), while Africans were brought as slaves (Cook and Borah, 1971; Pi-Sunyer, 1957; Vasquez, 2010). Today the regions that contain the largest populations of Asiatic and African races are on the west coast from Nayarit to Oaxaca and Veracruz. What is relevant for this case study is the syncretism of vernacular architectural traditions related to the use of palms. What evolved was a subtle application of building ideas, suitable for the local environment and available materials, including palm (Edwards, 1983), particularly as these groups adapted to the rural environment (Fuller, 2008). The coconut palm *Cocos nucifera* L. clearly points to the Philippines as its point of introduction into the Americas (Bruman, 1945; Clement et al., 2013), and some words can be traced to an Asiatic origin. The word "palapa" is particularly relevant to this investigation as it is widely used in Mexico to describe a woven thatched roof. However, the word "palapa" is from the Malayo language of the Philippines, where it means a pulpy leaf with fibrous veins, referring to the leaves of both bananas and palms (Gómez-Azpetia, 2006; Manuel, 1958). Polynesian types of building are also a good match for their west coast Mexican counterparts. They are made of triangular framing with roping and binding as a flexible technique more suitable for areas exposed to hurricanes and earthquakes than the rigid framing of the Spanish found in many other regions of Mexico.

11.1.3 Building features, materials, and land management

Today there are few places where a building can be achieved solely out of natural vegetative material, and it is clear that some of the last places where this type of

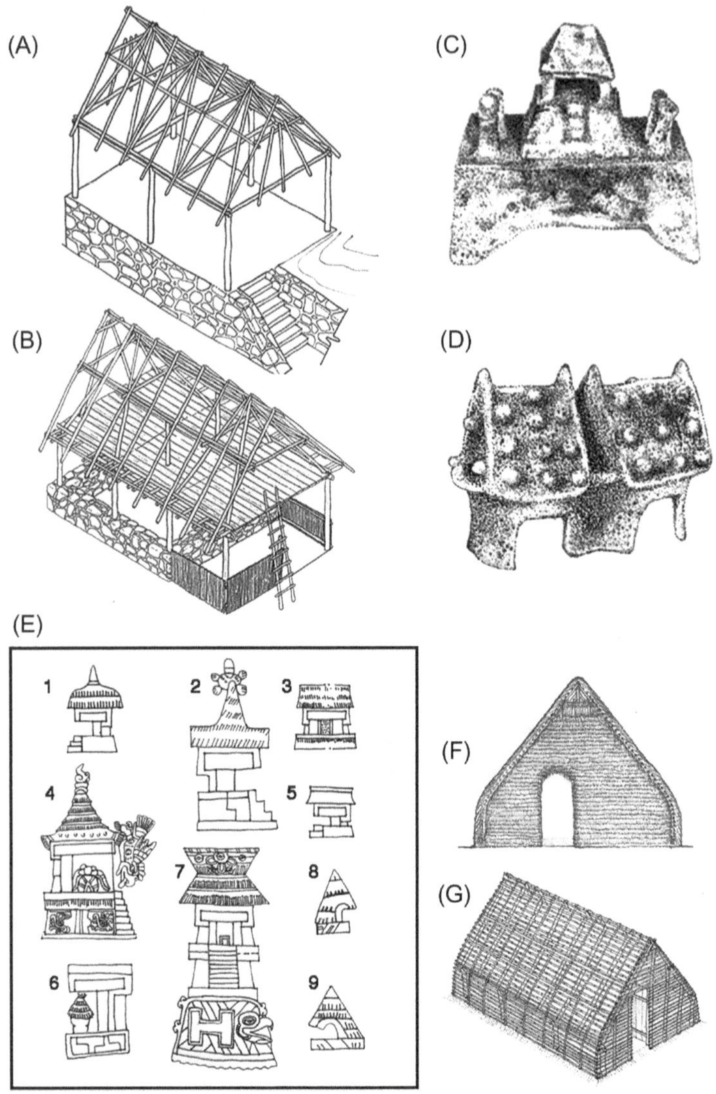

Figure 11. 1 Inventory of building materials related to palm thatched buildings in Mexico. (A) Isometric of case study in rural location in the west coast using *A. guacuyule* palm, (B) Isometric of case study in remote location in the west coast using *A. guacuyule* palm, (C, D) Clay models of temples from EI Chanal, culture in Colima, Postclassic, after Olay and Reyes 2001, (E) (1) Codex Fejérváry-Meyer, Plate 30; (2) Codex Borgia, Plate 19; (3) Codex Borgia, Plate 65; (4) Codex Zouche-Nuttall, Plate 9; (5) Codex Laud, Plate 13(12); (6) Codex Egerton-Sánchez, Plate 27; (7) Codex Zouche-Nuttall, Plate 45, (8, 9) Codex Vaticanus B, Plate 71, (F) Front view of case study in rural location in the highlands of Puebla using *B. dulcis* palm, (G) Isometric of case study in rural location in the highlands of Puebla using *B. dulcis* palm. All drawings by Jaime Rios-Calleja.

building is found are in isolated Pacific Austronesian cultures. It is thus important that some palm thatched building structures in west coast Mexico made without the use of a single nail still exist, although more commonly iron components are used to attach certain building elements together. Contemporary thatched structures also use an extensive inventory of natural vegetative materials, including hard and semi-hard woods, bamboos, reeds, vines, fibrous plants for roping, and branches for wattle and daub work. Knowledge is needed to make the best use of each natural material. It is also important to note that conventional architects, planners, or builders seem unfamiliar with this wide inventory of plants and that in these areas of Mexico palm thatching knowledge is acquired by being apprenticed to other thatchers rather than from formal academic training.

Today's palm thatching typologies resemble those built thousands of years ago (see Fig. 11.2) with consistent features such as the use of open walls, the angle of pitched roofs, the elevated foundations of the buildings, and the size of the structure and the use of bioclimatic features such as top opening windows in the thatch to provide ventilation. However, there are differences in the adaptation of thatched roofs and structures to more modern building expressions, in terms of building forms, materials, and technologies. The main differences are the use of conventional materials such as concrete posts and beams, and metal beams, rafters, ties, struts, ridge poles, and joists, as well as the use of metal wire or a reinforcing bar to connect modern conventional building materials with vegetative ones. Finally, the use of manmade walling materials such as fired bricks and concrete blocks, glass (sometimes framed in aluminum), together with cement and gypsum plaster finishes can all be found in modern thatched buildings.

The case studies discussed below reveal such mixing of materials and adaptations of traditional techniques and these mostly occur in buildings in urban and semiurban areas rather than rural, remote ones. It was also observed that decisions about the use of materials in these cases were made mostly by the clients or architects that employed the thatchers for the roof structure. These modern thatched buildings are residences and commercial buildings such as hotels, restaurants, bars, or stores. Often in such buildings use is also made of precious hardwoods prohibited by the Mexican organization formed to control the use of resources in particular environments. Another obvious variable is the short life of many commercial buildings using these expensive natural and conventional materials due to business failures, and bad design decisions such as choosing an inappropriate roof pitch for a thatched building, leading to quicker deterioration or more frequent maintenance.

In contrast the survey of palm thatching techniques in rural and remote areas highlighted how the selection of materials, size of the building, form of the structure, and purpose of the building were decided based on an awareness of the proper use of local resources and the environmental and economic cost involved in the process, often leading to simpler designs and enhanced building performance. In such cases the local people contract directly with a thatcher instead of going through an architect or designer. In some cases the building owners spoken to during this research undertook the construction process from the gathering and treatment of materials, to their application in the building in order to reduce the labor cost and

Figure 11.2 The six case study palm thatched buildings.

to obtain a higher standard of the construction. Even in such cases the important decisions are mostly made by the thatcher, such as where to harvest the materials, the selection of forest species for the structure according to the location and availability of appropriate timber species, and the season of harvesting and number of items (e.g., leaves, vines) taken per area. In fact, the wide variety of plants involved shows a great understanding of the local ecosystems, and their diversity and cycles (see Table 11.1). This is a clear example of how local traditional knowledge has a lot to do with having sustainable management of land and resources when it comes to making buildings. In west coast and highland Mexico tropical environment ecosystems with their high diversity of flora and fauna, like the deciduous dry and subhumid tropical forest, are particularly complicated to manage on a sustainable basis. However, issues such as maintaining biodiversity and minimizing local natural ecosystem disturbance seem to be taken into account by local rural land owners and thatchers through use of traditional knowledge that has been handed down, whereas such issues are mostly ignored by foreign architects, developers, and designers. For the latter often the use of a natural material is considered enough to be sustainable, without sufficient thought to how and where that material is obtained. Only natural materials that are produced on a sustainable basis can be considered part of sustainable buildings.

11.2 Palm thatch building assessment

The surveys undertaken in rural areas showed many houses lasted for decades without rethatching, as maintenance patches were regularly applied rather than waiting until it was necessary to rethatch the whole structure. Houses are sometimes lived in for long periods, sometimes over 100 years, and a structure will typically last at least 50 years.

However, once a building is abandoned most of the useful items are dismantled for reutilization, and in fact some houses surveyed used third-generation posts over 200 years old. Also, from interviews with thatchers and surveys of their working spaces, it was noticed that nothing left over was thrown away, and even small pieces of wood were preserved for future use. Moreover, the profound knowledge of the local environment becomes evident when interviewing the thatchers, as from their discussion it was clear they knew that not all natural building materials are also healthy materials. Even when some types of timber seem appropriate in terms of hardness, strength, and weather resistance, they are not used because they are not healthy for humans, as some are prone to producing allergic reactions (e.g., Lechozo, the common name, which could possibly be either *Brosimum alicastrum* Sw. or *Vatairea lundellii* (Standl.) Killip ex Record).

The main purpose of such a large research project into vernacular palm thatch building technology (Rios, 2015) was to address the lack of quantitative studies of the architectural use of vegetative and forest materials and the many issues involved in sustainable building construction. During the investigation many challenges emerged, such as the identification of species. This required several trips to where the natural materials were available to collect seeds, fruits, leaves, and make a photographic record for proper identification later. Not all the plants used and named by locals could be successfully identified. Another issue was to establish inventories of materials used, the distance covered from where resources are harvested to the

Table 11.1 **Inventory of building materials related to palm species and regions in Mexico**

Common names	Scientific name	Author	Published in
Hardwood species used in the west coast region (*A. guacuyule* and *S. rosei* palms)			
Barcino	*Cordia elaeagnoides* DC	Candolle, Augustin Pyramus de	Prodromus Systematis Naturalis Regni Vegetabilis 9: 474. 1845. (1 Jan 1845)
Botoncillo	*Cordia alliodora* (Ruiz & Pav.) Oken	Oken, Lorenz	Allgemeine Naturgeschichte 3(2): 1098. 1841.
Balsamo	*Myroxylon balsamum* (L.) Harms	Harms, Hermann August Theodor	Notizblatt des Königlichen botanischen Gartens und Museums zu Berlin 5(43): 94. 1908. (1 Sept 1908)
Llora Sangre	*Apoplanesia paniculata* C. Presl	Presl, Carl Boøivoj	Symbolae Botanicae, sive, Descriptiones et icones plantarum novarum aut minus cognitarum 1: 63, pi. 41. 1831. (September–December 1831)
Tabachin	*Caesalpinia mexicana* A. Gray	Gray, Asa	Proceedings of the American Academy of Arts and Sciences 5: 157. 1862.
Guayabillo	*Psidium sartorianum* (O. Berg) Nied.	Niedenzu, Franz Josef	Die Natürlichen Pflanzenfamilien 3(7): 69. 1893.
Guayabillo borcelano	*Piranhea mexicana* (Standl.) Radcl.-Sm.	Radcliffe-Smith, Alan	Kew Bulletin 51: 546. 1996.
Granadillo	*Dalbergia retusa* Hemsl.	Hemsley, William Botting	Diagnoses Plantarum Novarum . . . Mexicanarum 1: 8. 1878. (Jul 1878)
Granadillo; Sangualican	*Dalbergia granadillo* Pittier	Pittier, Henri Francois	Journal of the Washington Academy of Sciences 12(3): 62–63. 1922.

Verdecillo; Amapa amahlla Mapilla	*Tabebuia chrysantha* (Jacq.) G. Nicholson	Nicholson, George	The Illustrated Dictionary of Gardening, . . . 4: 1. 1887.
Amapa rosa; Amapa morada; Rosa Morada	*Tabebuia impetiginosa* (Mart, ex DC.) Standl.	Standley, Paul Carpenter	Publications of the Field Museum of Natural History, Botanical Series 11(5): 176. 1936.
Habillo; Habilla; Jabillo	*Hura crepitans* L.	Linnaeus, Carl von	Species Plantarum 2: 1008. 1753. (1 May 1753)
Palo Fierro	*Olneya tesota* A. Gray	Gray, Asa	Plantae Novae Thurberianae 328. 1854.
Tampisira; Sangualican	*Dalbergia granadillo* Pittier	Pittier, Henri Francois	Journal of the Washington Academy of Sciences 12(3): 62–63. 1922.
Lechozo	**Brosimum alicastrum* Sw. **Vatairea lundellii* (Standl.) Killip ex Record	Swartz, Olof (Peter) Killip, Ellsworth Paine	Nova Genera et Species Plantarum seu Prodromus 12. 1788. Tropical Woods 63: 5. 1940.
Palo de Brasil	*Haematoxylum brasiletto* H. Karst.	Karsten, Gustav Karl Wilhelm Hermann	Florae Columbiae terraumque adjacentium specimina selecta in peregrinatione duodecim annorum observata delineavit et descripsit 2: 27–28, pl. 114. 1862.
Coconut Palm	*Cocos nucifera* L.	Linnaeus, Carl von	Species Plantarum 2: 1188. 1753. (1 May 1753)
Tintilagua	Species not recognized		
Comengalo	Species not recognized		
Chinillo	Species not recognized		
Softwood species used in the west coast region (*A. guacuyule* and *S. rosei* palms)			
Otate	*Otatea aztecorum* (McClure & E.W. Sm.) C.E. Calderón & Soderstr.	Calderón, Cléofe E. Soderstrom, Thomas Robert	Smithsonian Contributions to Botany 44: 21. 1980.
Bamboo	*Bambusa* Schreb.	Schreber, Johann Christian Daniel von	Genera Plantarum 1: 236. 1789.

(*Continued*)

Table 11.1 **(Continued)**

Common names	Scientific name	Author	Published in
Species used as binding materials in the west coast region (*A. guacuyule* and *S. rosei* palms)			
*Bejuco Corralero	*Hippocratea volubilis* L.	Linnaeus, Carl von	Species Plantarum 2: 1191. 1753. (1 May 1753) (Sp. Pl.)
*Bejuco de Agua	*Davilla nitida* (Vahl) Kubitzki	Kubitzki, Klaus	Mitteilungen der Botanischen Staatssammlung München 9: 95–99. 1971. (31 Aug 1971) (Mitt. Bot. Staatssamml. München)
Sisal	*Agave sisalana* Perrine ex Engelm.	Perrine, Henry	Transactions of the Academy of Science of St. Louis 3: 305, 316, pl. 2–4. 1875.
Lechugilla	*Agave lecheguilla* Torr.	Torrey, John	Report on the United States and Mexican Boundary . . . Botany 2(1): 213–214. 1859.
Henequen	*Agave fourcroydes* Lem.	Lemaire, (Antoine) Charles	L'illustration horticole 11: 66. 1864. (III. Hort.)
Raphia	*Raphia nicaraguensis Oerst.*	Oersted, Anders Sandoe	Videnskabelige Meddelelser fra Dansk Naturhistorisk Forening i Kjøbenhavn 1858(1–4): 52. 1859.
Cuichi or garrobo	Species not recognized		
Hardwood species used in highland region Puebla (*B. dulcis* palm)			
Sabino	*Taxodium mucronatum* Ten.	Tenore, Michele	Annales des Sciences Naturelles; Botanique, sér. 3 19: 355. 1853.
Yucca	*Yucca aloifolia* L.	Linnaeus, Carl von	Species Plantarum 1: 319. 1753. (1 May 1753)
Encino	*Quercus alba* L.	Linnaeus, Carl von	Species Plantarum 1: 319. 1753. (1 May 1753)
*Cuitlapil	*Cestrum nocturnum* L.	Linnaeus, Carl von	Species Plantarum 1: 319. 1753. (1 May 1753)
Temacate de petaca	*Pithecoctenium crucigerum* (L.) A.H. Gentry	Gentry, Alwyn Howard	Taxon 24(1): 123. 1975. (Feb 1975)

Cuilote	*Montanoa grandiflora* DC.	Candolle, Augustin Pyramus de	Prodromus Systematis Naturalis Regni Vegetabilis 5: 565. 1836. (1-10 Oct 1836)
Pino	*Pinus montezumae* Lamb.	Lambert, Aylmer Bourke	A Description of the Genus Pinus, ed. 3 1: 39, t. 22. 1832. (1 May 1753)
Softwood species used in highland region Puebla (*B. dulcis* palm)			
Otate	*Otatea aztecorum* (McClure & E.W. Sm.) C.E. Calderón & S oderstr.	Calderón, Cléofe E.	Smithsonian Contributions to Botany 44: 21. 1980.
		Soderstrom, Thomas Robert	
Carrizo	*Arundo donax* L.	Linnaeus, Carl von	Species Plantarum 1: 81. 1753. (1 May 1753)
Guadua	*Guadua aculeata* Rupr. ex E. Fourn.	Ruprecht, Franz Josef (Ivanovich)	Mexicanas Plantas 2: 130. 1886.
Species used as binding materials in highland region Puebla (*B. dulcis* palm)			
Sisal	*Agave sisalana* Perrine ex Engelm.	Perrine, Henry ex author Engelmann, Georg (George)	Transactions of the Academy of Science of St. Louis 3: 305, 316, pi. 2–4. 1875.
Lechugilla	*Agave lecheguilla* Torr.	Torrey, John	Report on the United States and Mexican Boundary . . . Botany 2(1): 213–214. 1859.
Henequen	*Agave fourcroydes* Lem.	Lemaire, (Antoine) Charles	L'illustration horticole 11: 66. 1864.
Espadin	*Agave angustifolia* Haw.	Haworth, Adrian Hardy	Synopsis plantarum succulentarum . . . 72. 1812.

References: http://www.tropicos.org (search by scientific name).
*Possible species.

building, and yield factors. In fact, the vast range of natural materials used in these Mexican buildings only came to light during the field trips. An initial step was to identify thatchers and make a survey of buildings in order to select a range of case studies based on their locations. Structured interviews were then carried out with expert thatchers and trips made in different seasons to where the natural materials grow to collect plant material and to assess palm populations, their productivity, and to plot the various useful species within a given area to create yield values. Other data collected included information on materials other than palm, the quantity of hand labor used, fuels used and means of transport and distances non palm materials traveled, to acquire a better understanding of the energy involved in the lifecycle of the building (construction, maintenance, renovation, and demolition). Operating energy was not taken into account as this can vary significantly depending on the use of the building and the habits of the users, and thus this study only looks at lifecycle embodied energy. Since the type of building better fits a process line lifecycle methodology, this was selected and adapted to the vernacular context. Other issues related to evaluating the building were measuring the carbon content of vegetative materials such as hardwoods and whether carbon capture should be taken into account, and the problem of the embodied energy of mineral materials such as rocks used in retaining walls in particular case studies, as there is a lack of place-specific information about these. As a result, figures from other case studies were applied, but it should be noted this may have inflated the energy figures as these were more related to industrial rather than vernacular society buildings. In the latter local stone is usually carried by hand with little or no shaping prior to construction.

The use of fossil fuels for transporting materials was also assumed in some instances as the type of fuel, engine capacity, distance, and load capacity could not be found for all the case studies (Rios, 2015). It is important to highlight the difficulties in applying models developed for industrial building processes to vernacular ones.

During the research several buildings were assessed but based on the selection criteria explained above the study focused on the six case studies where the fewest assumptions were made and these are the ones evaluated here. All six buildings were reasonably similar in size. Three buildings were rural, one was in a remote location, one was suburban, and the other urban. Each was evaluated using an inventory of materials subdivided into organic materials such as hardwoods, softwoods, and fibrous materials, and inorganic materials such as mineral or industrial materials, considering processes from harvesting, treatment to make thatching elements and application in situ, or industrial manufacture of materials. Also included were distances from where materials were harvested or from where they were imported, and means of transport, fuel type, and load capacities. All evaluation was for a 50-year building life. Once all the information was assembled quantities of vegetative and mineral materials were converted to weight based on densities, and carbon and ecological footprints were then calculated for the buildings and their elements, and as a value per square metre (see Table 11.2).

Table 11.2 **Carbon footprint, embodied energy, and ecological footprint per m^2 of the six case studies as first constructed**

Final comparison of the impact assessment of six case studies of palm thatched buildings in Mexico as first constructed per m^2						
	***A. guacuyule* case study remote building**	***A. guacuyule* case study rural building**	***B. dulcis* case study rural building**	***B. dulcis* case study suburban building**	***S. rosei* case study rural building**	***S. rosei* case study urban building**
Carbon footprint (CT)	0.23	0.152	0.076	0.195	0.115	0.195
Embodied energy (GJ)	11.079	10.029	2.093	2.453	3.256	5.365
Ecological footprint (gha/year)	0.11	0.102	0.021	0.025	0.033	0.054

11.2.1 Results

Looking at the six case studies of palm thatching it seems that not all vernacular buildings are equally sustainable in terms of their impact on the environment. Table 11.2 gives the ecological footprints in gha of the six case studies, by material group and per m^2 when first constructed.

Surprisingly, the *Attalea* palm thatched buildings had higher ecological footprints than the other four, even though they were situated in the middle of forest lands. This shows that some building methods have to be used carefully. Here the imported industrial materials and the extensive use of rock to create retaining walls and, for A-R1, a modern-style deck, makes a higher environmental impact than the rural vernacular uses of *Brahea* and *Sabal* which seem to adhere to a more traditional way of building that has remained little changed for centuries. In some sense this shows that traditional vernacular architecture is more sustainable than the current idea of what vernacular architecture is, especially when it is imported into the urban scenario as for *Brahea* B-SU and *Sabal* S-U.

Table 11.2 shows the breakdown of the carbon, energy, and ecological footprints of all six buildings per m^2 when first constructed to allow for small differences in their sizes.

The two smallest carbon footprints were for the *Brahea* rural and *Sabal* rural buildings, which used all natural materials with earth instead of concrete floors, but the *Brahea* suburban also has a small per m^2 footprint. For those buildings using industrial materials the carbon footprint is larger because of the CO_2 emissions from their production through the use of fossil fuels.

11.3 Conclusion

The more traditional buildings, both in rural areas and that look very similar to those found in historical records as described earlier, have the lowest environmental impacts. This suggests that choosing natural materials will not lead to more sustainable buildings unless these materials are used in traditional ways to make traditional buildings. In other words, using natural materials can lead to sustainable buildings where vernacular processes are followed, but the materials themselves will not necessarily produce such buildings.

References

Bruman, H.J., 1945. Early coconut culture in western Mexico. Hispan. Am. Historic. Rev. 25, 212–223.

Clement, C., Zizumbo-Villarreal, D., Brown, C., Ward, R.G., Alves-Pereira, A., Harries, H., 2013. Coconuts in the Americas. Botanic. Rev. 79, 342–370.

Cook, S.F., Borah, W.W., 1971. Essays in Population History: Mexico and the Caribbean. University of California Press, Berkeley.

Edwards, J.D., 1983. The first comparative studies of Caribbean architecture. Nieuwe West-Indische Gids/New West Indian Guide. 57, 173–200.
Flannery, K.V., Sabloff, J.A., 2009. The Early Mesoamerican Village. Left Coast Press, Walnut Creek, CA.
Fuller, S., 2008. Creating and contesting boundaries: exploring the dynamics of conflict and classification. Sociol. Forum. 18 (1), 3–30.
Gómez-Azpeitia, G., 2006. Importance of being palapa. Palapa. 1 (1), 3–4.
Kelly, I.T., 1980. Ceramic Sequence in Colima: Capacha, an Early Phase. University of Arizona Press, Tucson.
Manuel, E.A., 1958. Tayabas Tagalog Awit fragments from Quezon Province. Musical score. Reprinted from Folklore Studies: Vol XVII. National Library of the Philippines: Manila.
Missouri Botanical Garden, S.L., 2014. Tropicos.org [Online]. Saint Louis, Missouri. Available from http://www.tropicos.org/Name/50225119 (accessed 10.03.14.).
Pi-Sunyer, O., 1957. Historical background to the Negro in Mexico. J. Negro History. 42, 237–246.
Quero, H., 1992. Current status of Mexican palms. Principes. 36, 203–216.
Quero, H., 1994. Flora del Valle de Tehuacan-Cuicatlán. Fasc. 7: Arecaceae. Instituto de Biología. Universidad Autónoma de México, México.
Rios Calleja, J., 2015. Environmental Impact Assessment of Vernacular Thatch Building Tradition in Mexico: Case Studies of Three Palm Species and Related Technology along the Trans-Mexican Volcanic Belt (Ph.D. thesis). Victoria University of Wellington. Available at: http://hdl.handle.net/10063/4675.
Tejero-Díez, J.D., Ledesma-Corral, J.C., Torres-Díaz, A.N., 2008. El palmar de Orbignya guacuyule al sur de Nayarit, México. Polibotánica. 26, 67–100.
Ugent, D., 2000. The master basket weavers of the Toluca market region (México). Econ. Botany. 54, 256–266.
Uhl, N.W., Dransfield, J., 1987. Genera Palmarum: A Classification of Palms Based on the Work of Harold E. Moore, Jr. Allen Press, Lawrence Kansas.
Vasquez, I.A., 2010. The longue durée of Africans in Mexico: the historiography of racialization, acculturation, and Afro-Mexican subjectivity. J. Afr. Am. History. 95, 183–201.
Von Winning, H., Weigand, P.C., Williams, E., 1996. El arte prehispánico del occidente de México. Michoacan, El Colegio de Michoacán AC.
Wackernagel, M., Monfreda, C., Moran, D., Wermer, P., Goldfinger, S., Deumling, D., et al., 2005. National Footprint and Biocapacity Accounts 2005—The Underlying Calculation Method. Global Footprint Network, Oakland, CA.
Zona, S., 2002. Name changes in Attalea. Palms. 46, 132–133.

The effect of global trade on the New Zealand house

12

Nigel Isaacs
Victoria University of Wellington, Wellington, New Zealand

12.1 Introduction

In order to calculate environmental impact through measuring embodied energy, greenhouse gas (GHG) emissions or resource use, it is necessary to understand the material from its origin to it final use. Modern analysis techniques and software can provide values to a high precision, but the manufacturing origin as well as the details of methodology and data accuracy can play a critical role. For example, lifecycle analysis of modern, multicrystalline silicon photovoltaic cells finds GHG emissions varying from 5 to 1690 gCO_{2e}/kWh and the energy payback time from 0.2 to 4.7 years, depending on the analysis method, the manufacturing location, and the associated processing energy and origins of the raw materials (Yao et al., 2014).

This chapter explores the international origins of a small number of selected materials and asks whether it has ever been possible to prepare a standardized analysis of the environmental impacts. Even something as apparently simple as the timber-framed house with a corrugated iron roof may use materials sourced locally, nationally, and internationally. Extraction of the raw material(s), processing into the appropriate product, transport from the manufacturing location to the site of use, and the construction into the completed building can each have a wide range of paths. This chapter focuses on the import and manufacture of selected products, exploring the ways this has changed over time and the lessons this gives for modern analysis.

12.2 Background

From the earliest arrival of humans, Aotearoa/New Zealand has been part of international trade. With the signing in 1840 of the Treaty of Waitangi between the United Kingdom and the indigenous Maori, trade links expanded as the new colony became part of the greater British Commonwealth. Sailing ships carried people and a wide range of building materials from Britain, Europe, and the Americas. Building supplies were largely imported with local processing of indigenous materials (such as tree to timber) while technologies and skills were imported in finished or kitset form (including nails, sawmills, and carpenters). An increasing population combined with an improved understanding of indigenous materials soon led to the promotion of local industrialization.

Materials for a Healthy, Ecological and Sustainable Built Environment: Principles for Evaluation.
DOI: http://dx.doi.org/10.1016/B978-0-08-100707-5.00012-5

Table 12.1 Development periods

Period	Time	Techniques	Source	Skills	Notes
Importation	1790s–1870s	Import+limited manufacture	Import	Import	
Industrial development	1880s–1930s	Import+ manufacture	Import+ local	Training + limited import	Focus on increasing employment
Industrialization	1940s–1980s	Limited import + manufacture	Import + local	Training + limited import	Increased specialization of materials

Table 12.1 sets out a time structure for the evolution of building techniques based on the techniques, sources, availability of skills, and related environmental changes (Isaacs, 2015, p. 268). The earliest European settlements saw most construction materials imported, with limited local manufacture. From the 1880s to the end of the 1930s a national focus on increasing employment and industrial development led to the emergence of a range of manufacturing industries. Technologies were developed to suit the unique needs of the country while improved education and training was implemented. Building controls changed in response to the extensive damage of the 1931 Napier earthquake, although the first national code dated back to 1924.

From the 1940s to the 1980s, increased industrialization helped deal with the ongoing aftermath of WWI and the devastation of WWII. By the 1970s it was possible to build a complete house using New Zealand-made materials, standards, and labor, although less than 20 years later there was a return to international sourcing of building materials.

12.3 Material selection and data sources

Table 12.1 provides a basis on which to explore the materials used in the creation of houses. Import, export, and manufacturing statistics coupled with newspapers and other contemporary publications have been used to provide data to explore the different construction materials. Even from the earliest locally published newspapers, construction imports were listed (Isaacs, 2009, p. 85).

In this chapter, the value, quantities, and origins of seven building materials are examined: asbestos cement sheet; cement; window glass; galvanized iron; nails; gypsum plaster; and roofing slates. The first five were at some time made in New Zealand, the sixth (gypsum plaster) is a critical raw material for plasterboard and the last (slate) had a full lifecycle within the period of interest.

Since 1841 summary statistics on merchandise external trade (imports and exports) have been published by the Customs Authorities or Statistics New Zealand and its predecessors. Over this period, these statistics have become increasingly detailed (Statistics New Zealand, 2004, pp. 153–158), which coupled with changes in collection and categorization makes longitudinal research difficult beyond 1967. Data have been extracted from the relevant annual import statistics reports for 1868 and 1869 and then on a 5-yearly cycle from 1870. Based on detailed analysis of annual nail import statistics, 5-yearly data (1880, 1885, 1890 ...) provide an acceptable time series. Where necessary, import statistics for additional years were investigated. A detailed discussion of the process of selecting and analyzing the import statistics is provided in Isaacs (2015, pp. 47–55).

The replacement of imports by local manufacture was actively encouraged from the 1860s. Bounties for the supply of locally made product for Government use (e.g., New Zealand Parliament, 1880) as well as the pride generated through local industrial exhibitions (e.g., New Zealand Industrial Exhibition, 1886) provided positive encouragement to investors and the industry.

Unfortunately, the manufacturing statistics are not as detailed as import statistics, limiting historic analysis. For example, nail import statistics are available from 1871 but manufacturing statistics are only available from 1938, even though manufacture commenced about 50 years earlier (Isaacs, 2009, p. 93).

12.3.1 Cement

From relatively early European settlement there was an interest in the use of cement and concrete. Fig. 12.1 shows that until 1920, the large majority (over 90%) of cement imports were from the United Kingdom. In 1920 UK imports reduced to 23%, with the majority from Canada (33%) and Australia (39%). In the following years, the UK continued its dominance, although it was not until 1965 that it again provided such a high proportion of imports.

The 1880 "Royal Commission into Local Industries" was in favor of encouraging the cement industry which "should become very valuable." They recommended the Government "should offer to purchase 100 tons of cement from any bona fide local manufacturer, subject to proper tests, and the price to be the then market price of the best [imported] Portland cement" (Evening Post, p. 2, 13 August 1880; New Zealand Parliament, 1880, pp. 11–12). The following years saw cement manufacture commence in Northland and Otago, as well as attempts in Fiordland and Golden Bay. By 1895, the Public Works Office supply contract for "Cement and Lime" was met from New Zealand-manufactured cement.

Fig. 12.2 gives total cement use by decade by origin, whether imported or New Zealand made. The number of cement and lime plants are given in brackets. Until 1875 only a combined *Plaster of Paris and Cement* category was provided in the import data, but then each was separately enumerated. However, for manufacturing, until 1937 lime and cement data were combined, so it is not until the 1940s that the data are solely for cement manufacture. Where data are available for the 1930s, 1940s, and 1950s, cement averages 60% of the total value of cement and lime production. For the 1960s this increases to 83%. The number of manufacturing plants increases through to the 1940s, when there is apparently a sizeable decline but this probably relates to the separation of lime and cement manufacture.

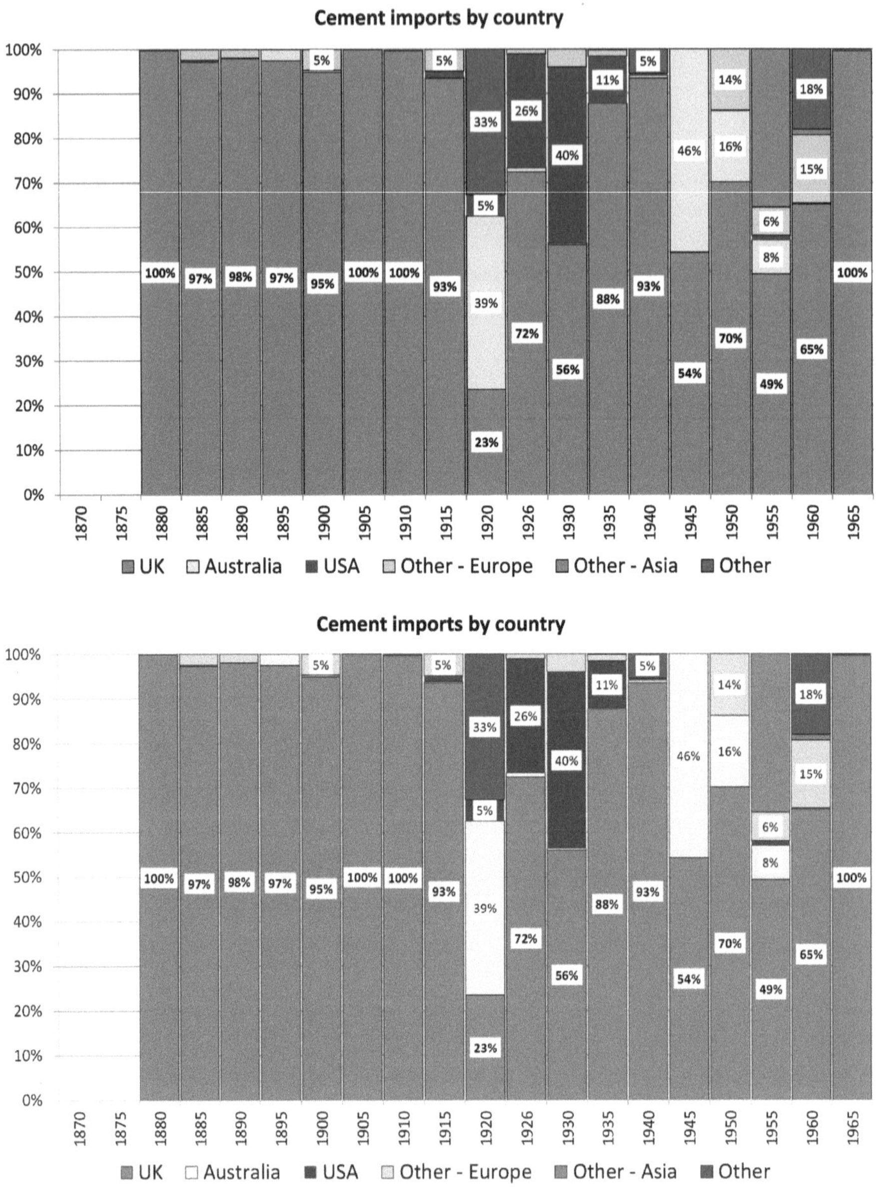

Figure 12.1 Cement imports by country 1870–1965 (value).

For example, in the 1953 year there were 85 lime plants, most of which employed less than six people, while there were only three cement works with an average of 129 employees.

Fig. 12.2 also shows that the development of the local industry in the early 1900s resulted in a steadily falling per capita demand for imported product,

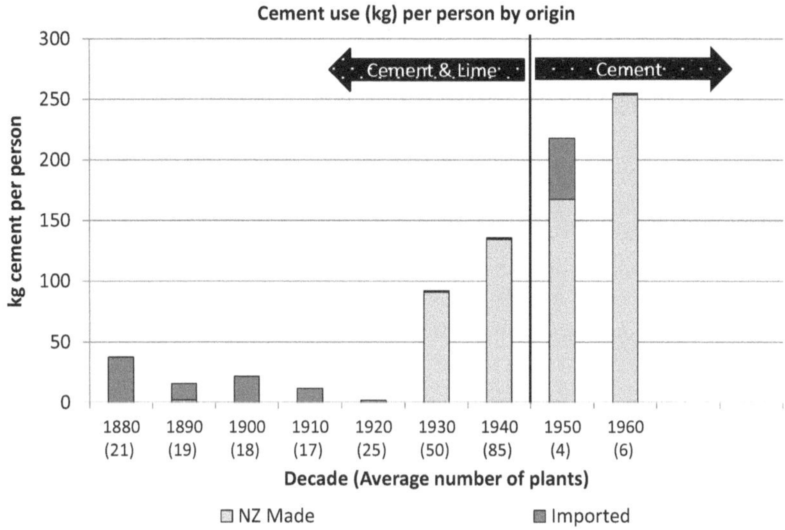

Figure 12.2 Total cement use per person by origin (weight).

which by the 1920s was at low levels. The large increases in imports per capita immediately following WWII suggest the local industry was unable to meet the demand, but it soon expanded as can be seen by the low import levels in the 1960s.

12.3.2 Asbestos cement boards

As discussed in Chapter 6, asbestos has been used by humans for thousands of years and by the later part of the 19th century health risks associated with asbestos were increasingly recognized. The modern asbestos cement board was invented in 1900 by Ludwig Hatschek of Austria (The Monopolies Commission (Great Britain), 1973, p. 26), and patented in 1900 as UK patents 22,139 and 6455 (The Times, p. 5, 12 December 1908). The process was patented in the United States in 1904 (Hatschek, 1904) and reissued in 1907 (Hatschek, 1907; Wilson and Snodgrass, 2008, p. 4).

Asbestos cement products were first advertised in New Zealand 4 years after the British patent. Over the next two decades many different products and trade names were advertised. Analysis of asbestos sheet or tile newspaper advertisements from 1904 to 1923 found a total of 15 products: five of UK origin, two from Australia, two from America, with one each from Germany, Austria, Canada, and New Zealand, and two which could not be matched to a country.

The first time asbestos sheeting (as opposed to asbestos fiber) was listed in the import statistics was 1915, with most imports coming from the United Kingdom and Italy. Belgium first appears in the 1920 statistics. With the development of the asbestos board industry across the Tasman, Australian manufactured product gradually took over, except for the WWII-impacted years, 1945 and 1950, when the United Kingdom was again the main source. Later, 1955 and 1960 show a growth in Belgium imports.

Although the Australian company James Hardie Ltd. were selling French-manufactured, Hatschek process, "fibro-cement" in Australia in 1903, English companies also provided supplies (Carroll, 1987, pp. 49–51). In 1917 Australian manufacture commenced in New South Wales, with other plants opening in Perth (1921), Brooklyn, Victoria (1927), Brisbane (1934), and Adelaide (1947) (Carroll, 1987, pp. 67, 72, 73, 75, 107). The first NZ asbestos fiber-cement sheet manufacturing plant was opened by James Hardie Ltd. at Penrose, Auckland in 1940. It continued to manufacture sheet until September 1983 when treated cellulose replaced the asbestos, with corrugated sheet manufacture centralized on the Perth plant (Carroll, 1987, pp. 79, 218, 220, 230). A second asbestos sheet plant was operating in Christchurch from 1943 to 1974 (National Asbestos Medical Panel, 2006, pp. 35–36). As Chapter 6, Persisting issues with the most recognized building materials health risks: lead and asbestos, discussed, from the 1970s recognition of health risks associated with asbestos brought a gradual decrease to its use.

12.3.3 Window glass

In the 1840s whalers lived in raupo reed houses where "two square holes in the wall serve as windows, with wooden shutters for the night" (Wakefield, 1908, p. 239). In 1892 Alice Clark felt she was the envy of the other wives in the remote bush camp for she was "the first to have glass windows put in the shanty" (Reed, 1964, p. 88).

Window glass was brought out from Europe, protected from the rigors of sea transport by straw and crates. In April 1840 the sale of window glass was advertised in Wellington (New Zealand Gazette and Wellington Spectator, p. 1, 25 April 1840). The glass was from the New Zealand Company's ship Glenbervie that had left London on 2 October 1839 and arrived safely 5 months later. Included in its cargo were 78 boxes of glass and 24 pairs of windows (New Zealand Gazette and Wellington Spectator, p. 3, 2 May 1840). Windows (and doors) were more easily purchased than made in the bush (Leys, 1883, p. 723), and after their use in temporary raupo housing or similar they could be shifted to a permanent house (Salmond, 1989, p. 68).

Although window glass was obtainable from the 1840s, little information is available on its value or origin until 1870 and not until 1890 for consistent physical units. Fig. 12.3 shows that until 1900, the majority of window glass imports by value were from the United Kingdom, although a small proportion came from Europe, mainly Belgium and Germany. Imports from Germany ceased with WWI and did not resume until 1955. The majority of European glass in this period came from Belgium. American glass is only a significant proportion in the 1915 year.

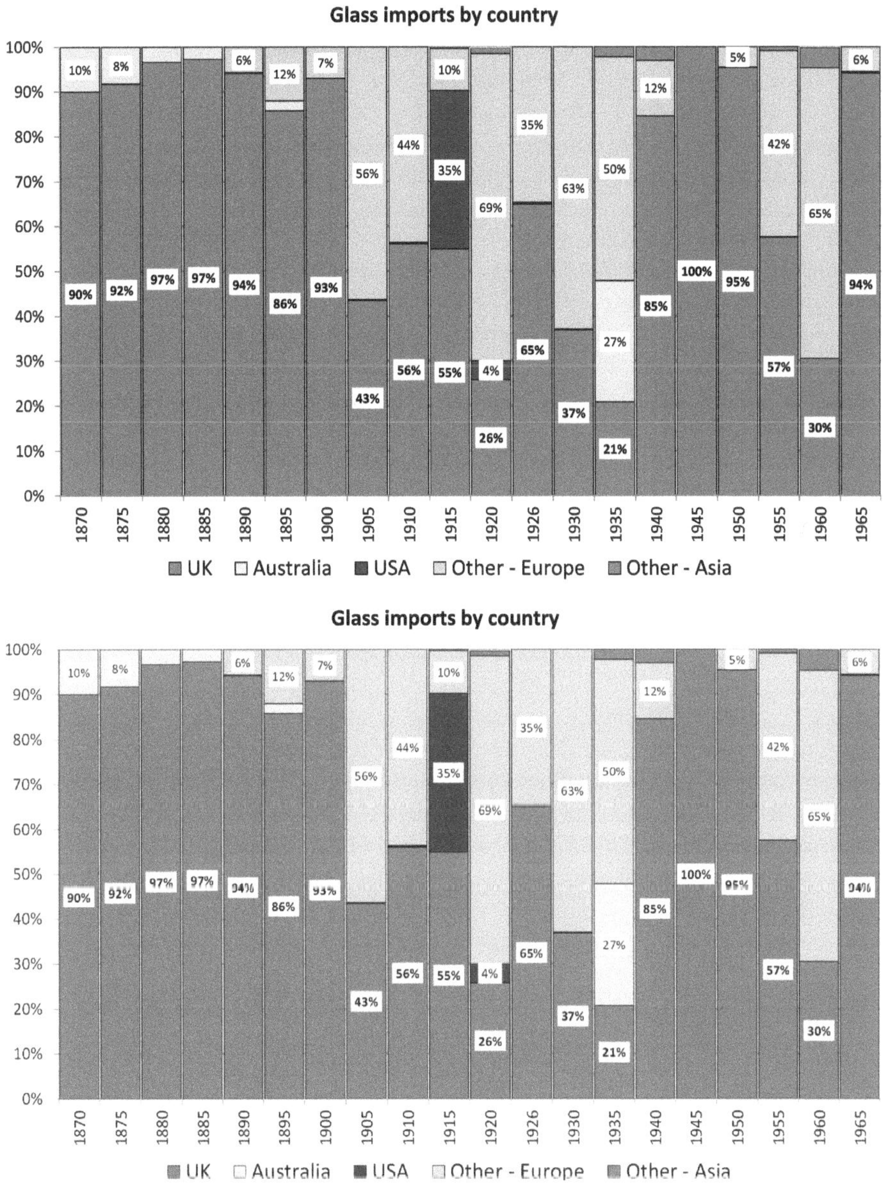

Figure 12.3 Glass imports by country 1870–1965 (value).

Until 1932, there was no Australian glass industry (Cowan, 1998, p. 53), so the glass reported as originating there from 1870 to 1890 was most likely reexported.

In the late 1950s there was a growing New Zealand glass market serviced entirely by imports. A local company saw an opportunity to manufacture window glass, and established the McKendrick Glass Manufacturing Company Limited in

Whangarei, building a major plant around a glass tank feeding two Foucault machines (see McGrath, 1961, pp. 49–51, 79). Trials started on 11 July 1962 but were a failure. By April 1964 the plant had been converted to the Pittsburgh Plate Glass process (see McGrath, 1961, p. 52), and commenced commercial production. Some 30,000 tonnes of sheet glass were produced each year, in a range of thicknesses from 2 to 6 mm (Callan, 1989, p. 74; New Zealand Window Glass Ltd., 1982).

The raw materials came from around the world, the bulkiest from nearby. In 1987 the plant used: 24,000 tonnes of silica sand from Parengarenga, NZ; 6000 tonnes of soda ash from United States; 6000 tonnes of dolomite from Golden Bay, NZ; 500 tonnes of limestone from Te Kuiti, NZ; 200 tonnes of feldspar from Norway or Canada; and 20 tonnes of coal dust from Australia. By 1988, the plant employed 215 people, with a sales force of eight based in Auckland. It produced about 600 tonnes/week or 80,000 m^2 of glass, and exported 40% of its production to Australia, Japan, and the Pacific (Callan, 1989, pp. 68–74). Glass was supplied in four grades (Horticultural, Standard, Selected, and Special Selected) and as quality increased, so did the price (New Zealand Window Glass Ltd., 1982).

In glass making, the price of raw materials was important, but none more so than energy. Initially the cost of energy was so low that wastage was not a problem. The 1981 regular tank rebuild saw a significant investment in thermal insulation to reduce tank heat loss and hence energy needs. By 1983 natural gas had replaced oil, halving the fuel bill as well as providing a cleaner fuel (New Zealand Window Glass Ltd., 1981) and bringing the pipeline to Whangarei for the rest of the community. Unfortunately the float glass process, patented by Pilkington Brothers PLC in 1959, revolutionized the manufacture of window glass. The first float plant in the Southern Hemisphere was opened in Dandenong, Victoria in 1974, producing up to 6000 tonnes a week (CSR Viridian New World Glass, 2010), 10 times as much as Whangarei at its peak. The Whangarei plant closed on 14 February 1991 (Northern Advocate, 2 May 1991).

12.3.4 Corrugated iron

In a country where corrugated iron could be used not only for roofing but also as a wall cladding, a chimney, and even in boats, imports played a critical role from the beginning of European settlement. By 1961 houses with iron roofs accounted for 70% of dwellings falling to 61% in 1981 as new roofing types, notably pressed metal tiles, came into use (Isaacs, 2015, p. 43).

From 1870 to 1940, the UK provided the large majority of imports, with product processed in the Australian colonies, principally Victoria, providing significant imports up to 1890. Although corrugated iron was imported into Australia from the 1830s, sheet steel was not made until 1914 (Cowan, 1998, p. 61; Mornement and Holloway, 2007, p. 159). WWII halted UK imports, replacing then with product from Australia and America. Following the war, UK imports resumed but from the 1960s were largely replaced by Australian product.

12.3.5 Nails

Even from the earliest days of European settlement, the majority of dwellings were wooden framed and wooden clad, although framing technology changes: from the 1790s to the 1890s braced framing was in use, from the 1860s to the 1940s balloon framing and from the 1880s to present day platform framing (Isaacs, 2015, pp. 132–136). Balloon and platform framing required increasing quantities of nails, the product that holds the timber together.

It could be expected that local nail manufacture would have developed early to support the extensive use of timber construction. Although different authors have postulated the existence of an indigenous nail manufacturing industry in the mid to late1800s (Salmond, 1989, p. 105; Cottrell, 2006, p. 429; Thomson, 2005, p. 107), the first nail machine was only imported in 1887 (Isaacs, 2009, p. 93).

Prior to 1862 no nail import statistics are available but newspaper advertisements and shipping reports provide qualitative information showing the availability from the earliest European settlement of iron bar for the local forging of nails, as well as machine-made wrought and cut nails. In the 1860s French wire nails and American cut nails were being sold, with English wire nails being advertised in the 1870s (Isaacs, 2009, p. 85). The last advertisement found for nail rod iron, necessary for hand forging of nails, was in 1863 (Lyttelton Times, p. 4, 19 August 1863), although this may only mean that it was so common that advertising was not required. The availability of imported nails coupled with improved transport probably made it preferable to use imported factory-made, rather than locally handmade nails.

Fig. 12.4 shows over time the dominant country supplier of nails varied widely. Until 1893 the main source of imported nails is clearly British, whether from the United Kingdom or its Australasian colonies. From 1893 the sources changed from Britain firstly to Europe, then the United States, and then Canada. Nails from Canada were not subject to preferential rates of duty, so it would appear unlikely that American nails were being trans-shipped for export. The large-scale entrance of Canadian suppliers from 1905 ultimately supplanted those of the United States.

There are only limited published statistics for the local nail manufacturing industry. From 1892 to 1910 only the number of works are available, while from 1938 to 1965 the number of works, value of production, and from 1951 production weight are reported.

The local nail manufacturing industry became active from the 1890s and is probably in part responsible for the reduced weight of nail imports per capita. Fig. 12.5 shows that from 1870 to 1885 the demand for nail imports by weight per capita (kg/person) fell, possibly as the initial need for permanent house construction was meet and the local manufacturing industry developed. The import demand rose again in the 1900s before beginning a steady decline to almost zero in the 1940s. However, the local manufacturing data (only available from 1951) show that consumption by weight of nails per capita ranges from 4.9 kg/person (1955) to 4.5 kg/person (1965), which compares with the import high of 4.4 kg/person in 1910.

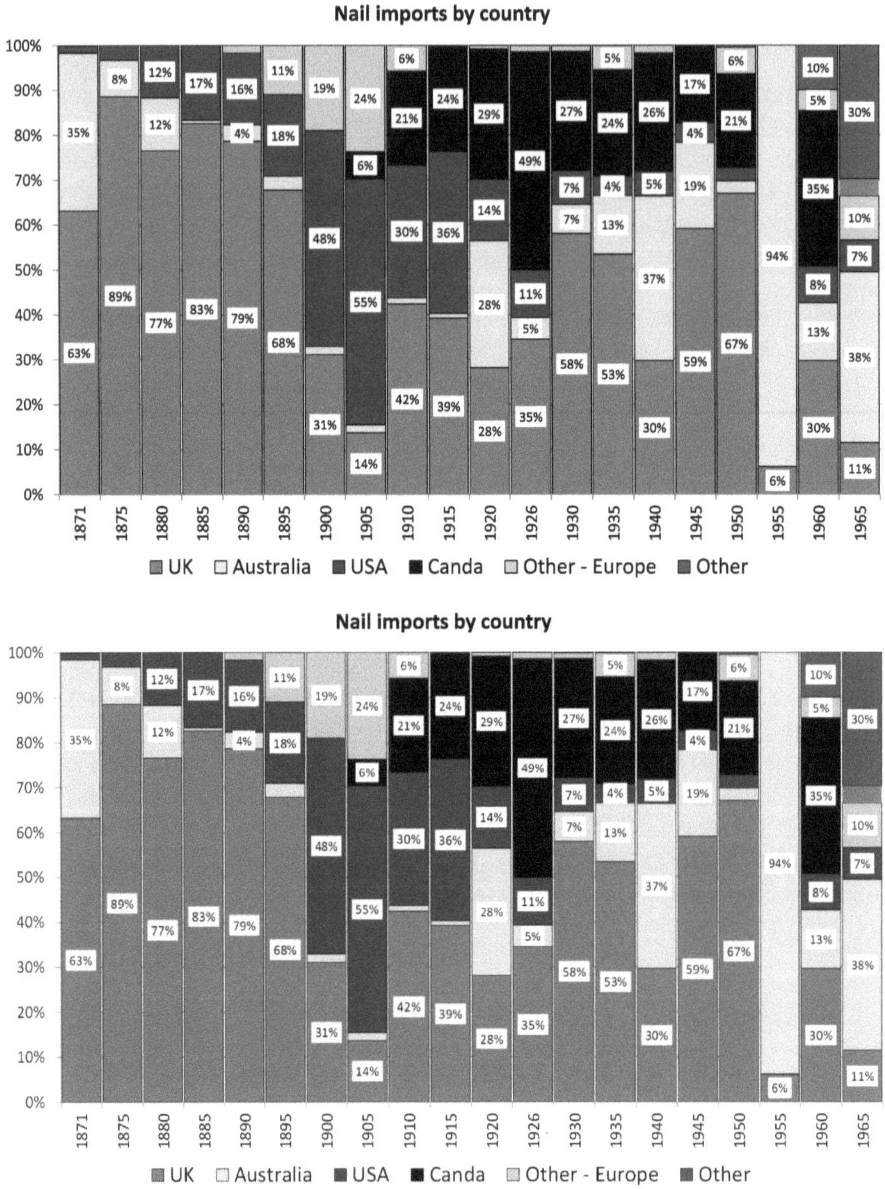

Figure 12.4 Sources of nail imports 1871–1965 (value).

12.3.6 Fibrous plaster and plasterboard

Plaster of Paris is formed by heating natural gypsum rock and grinding it to a fine powder (Torres et al., 2006, p. 22). When mixed with water the plaster could be shaped, soon setting to a hard, but brittle, solid. It had many uses, ranging from holding broken bones to creating precast decorative features for ceilings and cornices.

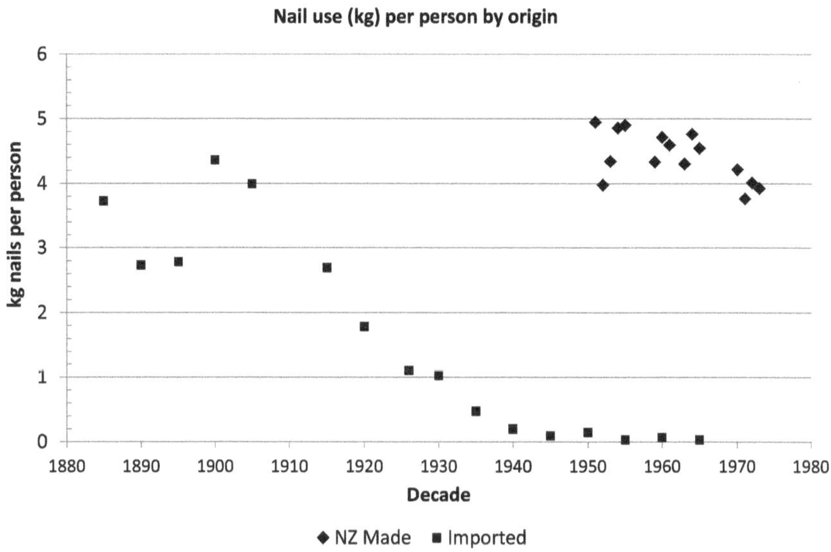

Figure 12.5 Nail imports and local production per capita (weight).

Up to 1905, the majority of gypsum imports were from the United States, but from 1910 Australia played an increasingly important role. However, gypsum imports were very low until the 1920s, when the first plant manufacturing plasterboard commenced operation. Following WWII as the use of plasterboard linings increased (replacing timber and timber products) imports started to increase, notably in the 1960s.

No commercially significant gypsum deposits are found in New Zealand (Williams and McKee, 1974, pp. 220–221), so it is necessary to import this product. As well as it use in construction, gypsum is also used in agriculture as a soil conditioner. Gypsum imports were less than 0.01 kg/person until 1910, increasing to nearly 2 kg/person in 1915, and then increasing each year until reaching 41 kg/person in 1965.

Direct use included fibrous plaster products which were being advertised by R. Wardrop in Dunedin from 1901 (Otago Daily Times, p. 1, 2 December 1901). In Wellington the NZ Carrara Fibrous Plaster Company was advertising product in May 1903 (Evening Post, p. 6, 11 May 1903). They used *Stuccolin* (Progress, p. 89, 2 January 1907), a patented product of Ernst and Cie (1906). Cast fibrous plaster sheeting was being made and sold in Wellington by 1923 (Evening Post, p. 5, 26 October 1923), and a plant was operating in Auckland by November 1925 (Auckland Star, p. 1, 21 November 1925).

In May 1890 Augustine Sackett, a New York engineer, patented a type of plasterboard which developed into the form widely used today. By 1917 plasterboard was being produced in Great Britain (Torres et al., 2006, p. 78). In New Zealand, plasterboard was initially imported from North America (Simpson, 1965, p. 161), although the quantities must have been small as it was not until 1936 that plasterboard imports were separately listed (Census and Statistics Office, 1937, p. 346). *Sackett* plasterboard arrived on the New Zealand market by 1913, with UK-manufactured boards becoming available from 1925 (Isaacs, 2015, p. 163).

Local manufacture commenced in 1925, with the establishment of the Builder's Composite Materials Ltd., Auckland. Their plasterboard, along with a UK product were awarded the Gold Medal at the NZ and South Seas Exhibition, Dunedin, 1926 (New Zealand Herald, p. 17, 28 April 1926) (Thompson, 1927, pp. 163–164). Initially the manufacturing process was manual but automated machinery was introduced in 1931.

12.3.7 Slates

Slate, a low-grade metamorphosed mudstone with good cleavage, was used in the UK sourced from mines in Scotland, Wales, and Cornwall. W.N. Blair's book on Otago and Southland building materials dedicates a section to roofing slates, concluding that while not every industry should be encouraged in a young colony, given the low price of slates and the high cost of freight "there is little wanted to turn the scale in favour of native production" (Blair, 1879, pp. 80–89). Although local attempts to mine slate were made in the 1870s, including at Red Rocks, Wellington and at Otepopo (30 km southwest of Oamaru), they were unsuccessful (Blair, 1879, pp. 87–88; Hayward, 1987, pp. 38, 56).

Until 1900, the majority by value of slate imports came from United Kingdom, although there was a sizable proportion from United States. From 1900 to 1920 imports from Belgium and France were noticeable, but not in later years. There were no recorded imports in 1945, but otherwise from 1940 to 1965 all imports were from the United Kingdom. In absolute values, slate imports fell from 0.9 slates per person in 1875 with total imports of 238,850 slates, dropping to 0.25 in 1900, to 0.15 in 1930, and to 0.01 in 1965 when just 21,935 slates were imported. From the mid1920s other roofing materials took over, although slate continued to be used for repair and replacement.

12.4 Discussion

The earliest days of European settlement saw the importation of manufactured products, notably cement, glass, corrugated iron, hand-forged nails, and slates. Where local industry could make a cost-effective contribution, local manufacturing began based on imports. Corrugated iron could be rolled from imported sheet and nails forged from imported rod. Cement manufacture, although possible with comparatively small investment, required larger plant to ensure consistency and quality, so was slower in commencing local production. Nails were critical for timber frame construction, but presumably the costs of imported product, whether cut or wire nails, coupled with a comparatively small market meant it was not until the late 1880s that local manufacture commenced, albeit with imported machinery.

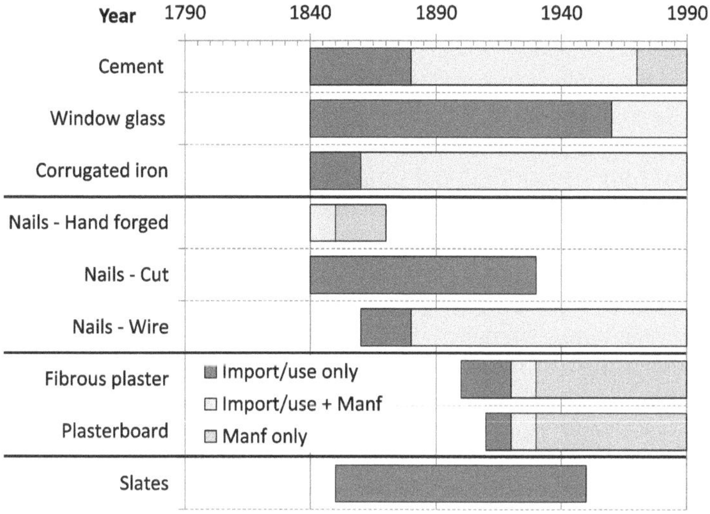

Figure 12.6 From import to local manufacture.

It was not until the 1920s that a sizable portion of the nail market was provided with local product.

Unlike the other technologies, local slate production never took off. Although primarily due to the lack of a suitable local material, alternative roofing products (including timber shingles and corrugated iron) provided lower-cost options. Slates, except for replacement purposes, largely stopped being used by the 1950s.

Fig. 12.6 plots the arrival of the selected building technologies, the period of importation, important and local manufacture, and just local manufacture. Although it is possible some of these, notably nails, arrived prior to 1840, they would have been in very small quantities. In the case of plasterboard and nails, the changes in the technology are shown separately.

A review of the use of 40 materials (including cement) and 24 technologies (including nails) found five different reasons for them to be no longer used: poor durability; poor structural or earthquake performance; poor fire performance; poor performance; or higher cost or speed of construction (Isaacs, 2015, p. 253). Of those considered here, Fig. 12.6 shows only handmade nails, cut nails, and slates stopped being used. This was most likely due to their higher cost or poorer performance compared to their replacements.

Fig. 12.6 closely follows the development periods set out in Table 12.1, with variations due to issues of scale or replacement. The cost of building a glass manufacturing facility meant this did not occur until the 1950s, while slate, as discussed, became an obsolete product, imported solely for replacement purposes. What is clear is that each product followed its own combination of import (often from a number of countries) and local manufacture, making detailed analysis of their environmental impacts difficult, if not impossible, based on existing documentation.

12.5 Conclusion

The data presented have illustrated even for these seven common building materials there is no simple way to attribute the source of the raw material or the finished product. The complexities of international trade coupled with changes in industrial development, both international and local, make the derivation of any performance metric a complex and difficult activity. In each case, the different impacts of raw material extraction, initial processing, bulk transport, manufacturing or processing to final product, transport to site, and final use would need to be quantified.

For products able to be closely tracked from raw materials to final use, modern research can find differences in orders of magnitude for apparently similar products. The complexities of tracing historic products make a comprehensive analysis of the environmental impact an even greater task. Furthermore, this chapter suggests that the value of such indicators is to help improve understanding through comparisons, instead of focusing on quantitatively fine (and potentially imprecise) differences.

References

Auckland Star, Electronic Version Accessed through: Paperspast.natlib.govt.nz (1870-1945). Auckland.

Blair, W.N., 1879. The Building Materials of Otago and South New Zealand Generally. J. Wilkie & Co., Dunedin.

Callan, L., 1989. Looking into glass. New Zealand Geographic, March, No. 1, pp. 68–84.

Carroll, B., 1987. A Very Good Business: One Hundred Years of James Hardie Industries Limited 1888–1988. James Hardie Industries, Sydney, NSW.

Census and Statistics Office, 1937. Statistical Report on Trade and Shipping in the Dominion of New Zealand for the Year 1936. Part I. Government Printer, Wellington.

Cottrell, W., 2006. Furniture of the New Zealand Colonial Era: An Illustrated History, 1830–1900. Reed, Auckland.

Cowan, H.J., 1998. From Wattle & Daub to Concrete & Steel: The Engineering Heritage of Australia's Buildings. Melbourne University Press, Carlton South, VIC.

CSR Viridian New World Glass, 2010. Manufacturing Float Glass. CSR Viridian – Glass Talks, 14 September. Available from: http://glasstalks.com/2010/09/14/manufacturing-float-glass/ (accessed 22 May 2014.).

Ernst, H., Cie, B., 1906. Berner Stuckfabrik (Bern Stucco Factory). Heimatschutz=Patrimoine. 1 (4). Available from: http://retro.seals.ch.

Evening Post, Electronic Version Accessed through: Paperspast.natlib.govt.nz (1865–1945). Wellington, New Zealand.

Hatschek, L., 1904. Process of Manufacturing Imitation Stone Plates, Slabs, or Tiles. US Patent 769,078. Reissued 12,594, 15 Jan 1907. Available from: www.uspto.gov (accessed 30 August.).

Hatschek, L., 1907. Manufacture of Imitation-Stone Plates, Slabs, or Tiles. Reissue 12,594 of US Patent 769,078. Available from: www.uspto.gov (accessed 15 January.).

Hayward, B.W., 1987. Granite and Marble: A Guide to Building Stones in New Zealand. Geological Society of New Zealand, Lower Hutt.

Isaacs, N.P., 2009. Nails in New Zealand 1770 to 1910. J. Constr. History. 24, 83–101.

Isaacs, N.P., 2015. Making the New Zealand House 1792 - 1982 (Doctoral Thesis), PhD, Victoria University of Wellington, Wellington. Available from: http://hdl.handle.net/10063/4804.

Leys, T.W., 1883. Brett's Colonists' Guide and Cyclopaedia of Useful Knowledge: Being a Compendium of Information by Practical Colonists. H. Brett, Auckland.

Lyttelton Times, Electronic Version Accessed through: Paperspast.natlib.govt.nz (1851–1869). Lyttelton.

McGrath, R., 1961. Glass in Architecture and Decoration. New ed. rev. Architectural Press, London.

Mornement, A., Holloway, S., 2007. Corrugated Iron: Building on the Frontier. Frances Lincoln, London.

National Asbestos Medical Panel, 2006. Asbestos Exposure in New Zealand 1992 to 2005 [Electronic Resource]. Department of Labour, Wellington. Available from: http://www.business.govt.nz/worksafe/information-guidance/all-guidance-items/asbestos-registers-national-annual-reports/asbestos-exposure-annual-report-2005.pdf (accessed 20.03.12.).

New Zealand Gazette and Wellington Spectator, Electronic Version Accessed Through: Paperspast.natlib.govt.nz (1839–1844). Wellington.

New Zealand Herald, Electronic Version Accessed through: Paperspast.natlib.govt.nz (1863–1945). Auckland.

New Zealand Industrial Exhibition, 1886. Official Record of the New Zealand Industrial Exhibition 1885. Government Printer, Wellington.

New Zealand Parliament, 1880. H-22 Colonial Industries Commission (Report of the), Appendix to the Journals of the House of Representatives — 1880 Session I, Vol. 2, Government Printer, Wellington, NZ. Available from: http://atojs.natlib.govt.nz/cgi-bin/atojs?a=d&d=AJHR1880-I.2.2.3.11&e=-------10--1----0 (accessed 20.06.13.).

New Zealand Window Glass Ltd., 1981. New Zealand Window Glass Limited - Report to Employees 1981. New Zealand Window Glass Ltd., Whangarei.

New Zealand Window Glass Ltd., 1982. Price List 10 May 1982. New Zealand Window Glass Ltd., 10 May.

Northern Advocate, Electronic Version Accessed Through: Paperspast.natlib.govt.nz (1887–1925). Whangarei.

Otago Daily Times, Electronic Version Accessed Through: Paperspast.natlib.govt.nz (1861–1920), Dunedin.

Progress, Electronic Version Accessed Through: Paperspast.natlib.govt.nz (1905–1910). Wellington.

Reed, A.H., 1964. The New Story of the Kauri, 3rd ed. rev. and reset, with new illustrations. Reed, Wellington.

Salmond, J., 1989. Old New Zealand Houses, 1800–1940. Heinemann Reed, Auckland.

Simpson, F.A., 1965. The First Century: A Centenary Review of Winstone Limited. Winstone Ltd., Auckland.

Statistics New Zealand, 2004. Statistical Publications, 1840 - 2000, Statistics New Zealand, Wellington. Available from: (accessed 10 April 2012).

The Monopolies Commission (Great Britain), 1973. Asbestos and Certain Asbestos Products: A Report on the Supply of Asbestos and Certain Asbestos Products, H.M.S.O., London. Available from: http://www.competition-commission.org.uk/rep_pub/reports/1970_1975/fulltext/066c01.pdf.

The Times, Accessed Through the Times Digital Archive (1785–1985). London, England.

Thompson, G.E., 1927. Official Record of the New Zealand and South Seas International Exhibition, Dunedin, 1925–1926. Coulls Somerville Wilkie for Otago Master Printers Association, Dunedin, available at: (accessed 11 March 2012).

Thomson, S., 2005. Wrinkly Tin: The Story of Corrugated Iron in New Zealand. Steele Roberts, Wellington.

Torres, F., Emeric, F., Guilhem, Y. de, 2006. A History of Lafarge Gypsum. Editions de Monza, Paris.

Wakefield, E.J., 1908. In: Stout, R. (Ed.), Adventure in New Zealand from 1839 to 1844: With Some Account of the Beginning of the British Colonization of the Islands. Whitcombe and Tombs, Christchurch.

Williams, G.J., McKee, T.J., 1974. Economic Geology of New Zealand: The T. J. McKee Memorial Volume. Australasian Institute of Mining and Metallurgy, Parkville.

Wilson, R., Snodgrass, K., 2008. Early 20th-Century Building Materials: Siding and Roofing. United States Dept. of Agriculture, Forest Service, Technology & Development Program. Available from: http://www.fs.fed.us/t-d/pubs/pdfpubs/pdf08732308/pdf08732308dpi72.pdf.

Yao, Y., Chang, Y., Masanet, E., 2014. A hybrid life-cycle inventory for multi-crystalline silicon PV module manufacturing in China. Environ. Res. Letters. 9 (11), 114001.

Thurgoona Campus: A living laboratory of healthy and sustainable materials

13

Marci Webster-Mannison
Melbourne Design Studios, Collingwood, VIC, Australia

13.1 Context

one of the first larger scale developments in Australia to comprehensively and rigorously pursue autonomy and sustainability.

Lindsay Johnston (2000)

This chapter reflects on the experiences of the builders and designers who, through their involvement with the Thurgoona Campus development, (re)discovered their craft and worked collectively to improve each project. It tells the story of the decision-making process for materials selection over the first 10 years of the design and construction of the Thurgoona Campus of Charles Sturt University from its conception in 1993.

Buildings articulated in rammed earth and recycled timber, glass, bricks, and steel invite questions about materials and construction. Buildings orientated for solar access, kinky angled roofs collecting daylight and solar energy, earth thermal exchange, night cooling, natural ventilation, and a windmill speak of renewable resources. The careful use of space and over 40 composting toilets at Thurgoona teach about waste. The integrated rainwater tanks, the meandering constructed creeks and wetlands animate the criticality of water in this dry arid region. The formation of vistas to the few stands of remnant trees and the "living museum" of endemic planting along the waterways gives a voice to biodiversity and ecosystems. The use of natural ventilation, daylight, and nontoxic materials breathe comfort and health.

The outcomes are observable (Fig. 13.1). One can see how the built environment influences how people live in this learning laboratory, breathe fresh air, bask in natural light, socialize with one another, plant permaculture gardens, and maintain free bicycles for use around the campus.

The challenges for a small team of 4–5 people covering the site planning, building design, and documentation for multiple campuses scattered around New South Wales, included the availability of information, constraints on time for thoroughly researching or recording the basis of decisions, and limited access to consultants. The projects relied on the designers' deep green design commitment and seemingly boundless energy to extend their knowledge and understanding of the impacts, risks, and consequences of decisions in a precautionary approach to design decision-making, rather than systematic analysis.

Materials for a Healthy, Ecological and Sustainable Built Environment: Principles for Evaluation.
DOI: http://dx.doi.org/10.1016/B978-0-08-100707-5.00013-7

Figure 13.1 CD Blake Lecture Theatre Complex (photographer: Dirk Spennemann).

13.2 The site

its design has been driven by creative responses to a local context.
Anne-Marie Willis (2000)

The 87-ha site on the northern outskirts of Albury-Wodonga, New South Wales, Australia, is comprised of former pasture land with very little remnant endemic vegetation and bad weed infestation. The desert-like climate is characteristically dry and hot in the summer with dry hot winds in the daytime and cool nights and a cold winter with icy winds and some frosts.

The tale begins with early adoption of an environmentally sensitive design approach during the process of negotiation and progressive purchase of the site, which ultimately dramatically changed the site boundaries, overturned the approved development plan, critically influenced the retention of a creek (originally destined to be piped), and provided frontage to major roads.

13.2.1 Early decisions

The designers took the staff, the University's Board of Governors, the local Murray Advisory Council, and local authorities on a journey of an ongoing series of meetings which walked people through the design process. A sequence of plans illustrating "layers" of site characteristics was used for analysis of the various layers, their overlaps, and interrelatedness, in order to explain the site planning. There was a deliberate attempt to "grow" the project from the adoption of a set of shared principles and values, into a site plan and on to further detailed decisions.

Simple agreed-upon and well-accepted principles and strategies were useful to explain the rationale for the site planning and architectural decisions. Elaboration of the agreed principles for minimization of environmental impact, responsible resource management, and low-energy design early in the design process fleshed out a more specific Environmental Design Strategy (Table 13.1).

Table 13.1 Environmental design strategy

Principles	Summary design criteria
Minimize environmental impact	
Minimize level of interference with "natural" systems: water, earth, and air	• Evolve plan from some sound thinking about the environmental framework and the organizational requirements of the brief, in relation to an analysis of overlays of the existing conditions vegetation, drainage, topography, climate, geology and soil conditions, services
Minimize pollution of soil, air, and water	• Make early decisions favoring passive techniques, water and soil conservation, and atmospheric hygiene
Maintain and where it has been disturbed restore biodiversity	• Enhance opportunities for conservation of ecosystems, habitats, and species
Increase awareness of environmental issues	• Make solutions visible to challenge building users to examine the underlying social, scientific, and cultural connections that operate, and how they cause environmental degradation • Engage in research • Seek publicity
Responsible resource management	
Minimize the depletion of "natural" resources, especially nonrenewable resources	• Reduce pollution and greenhouse gases by passive design and materials selection which consider embodied energy, sources, and production • Consider that timber construction stores carbon, whilst the use of some materials, such as steel, release carbon to the atmosphere • Reduce operational energy consumption such as heating, cooling, ventilation, lighting, and equipment • Use recycled timber to limit role in deforestation • Implement soil and water management plans to counter degradation, depletion of water supply, and the declining quality of water • Minimize pollution and waste disposal
Ensure social equity in the distribution of the costs and	• Follow economic theory founded on environmental principles

(Continued)

Table 13.1 (Continued)

Principles	Summary design criteria
benefits associated with the use of resources	• Recognize natural objects other than humans as having independent value • Ensure equity for people and nonhumans without market power • Consider social impact
Maximize the health, safety, and comfort of building users	• Design, use, and maintenance to promote a healthy building • Make places that people are happy to be in • Use nontoxic materials such as natural fiber wool and linoleum floor coverings, wool insulation, nontoxic paints and timber treatments, mesh protection from termites. Minimization of the use of PVC
Low-energy design	
Maximize the passive use of the building's form and fabric to collect, store, and distribute energy, considering gross and operational energy	• Thermal mass to act as a heat sink to stabilize temperature or lack of, to ensure fast response of structure to ambient conditions • Insulation to reduce the heat loss in the winter and heat penetration in the summer • Shading of direct sun to reduce heat gain in summer and adjust daylighting for comfortable use • Air circulation to assist homogeneous distribution of the air temperature in the room and for cooling effect • Cross-ventilation for the emission of fresh air and to exhaust heat. Air movement will also provide a cooling effect • Night cooling to flush hot air from spaces in the summer • Heating and cooling design based on an approach to thermal performance that utilizes passive techniques. Appropriate to the climate and conditions; take advantage of ambient conditions, thermal mass, insulation, high levels of daylighting, and natural ventilation • Lighting design based on energy conservation and flexibility of individual control suits variable user requirements by the integration of daylight and supplementary, direct and indirect task lighting • Recycled materials used wherever possible

Source: Webster-Mannison, M., 2006. Ecodesign for Large Campus Style Buildings. PhD Thesis. The University of Queensland.

13.2.2 Ecosystem decision-making

Site planning inherently impacts on building design, as well as decisions about materials for the site works. Consideration of the existing site conditions, including services, vegetation, drainage, topography, climate, geology, and soil conditions, shaped the site planning. Basically, the plan evolved from considering the environmental framework and the organizational requirements of a campus in relation to an analysis of the existing conditions.

The use of augmented natural drainage including existing dams and creek beds mitigated the need for piping of stormwater, reducing materials usage, and "building-in" flexibility for variable flows and future alteration. Narrow roads follow the contours of the site, punctuated by existing trees and crossed by rock-lined fords to carry stormwater, reducing the need for cut and fill and minimizing materials usage.

Ecosystem services obviate the need to connect to mains stormwater or sewerage, or associated materials-intensive centralized infrastructure. Water is separated into levels of purity, collected, treated, and reused. No water is wasted. Corrugated steel tanks store rainwater collected from the roofs of all the buildings for use in evaporative mist cooling systems and laundries. Wind and solarpowered recirculating stormwater wetlands keep the water moving, thereby preventing mosquito breeding and supplying water for irrigation. Graywater cleansing uses artificial wetlands for treatment and aeration. Dry composting toilets prevent the generation of black water.

Service runs are minimized by utilizing ring mains with secondary spokes distributing services to building clusters, thereby reducing services runs and distribution losses. No polyvinyl chloride (PVC) is used, being replaced by materials such as HDPE pipes and conduit and clay pipes.

13.3 Buildings

> *the greatest achievement of Thurgoona is in its absolute dedication to the creation of a healthy, stimulating, low impact environment and the progress that is evident from one building to the next.*
>
> *Jan Howllin (2002)*

The design starting point was to make the building form, fabric, and use of spaces responsive to daily and seasonal climate variations to optimize ambient conditions, and follow on with the incremental introduction of supplementary low-energy, active systems.

13.3.1 A decade of construction

Although the notion of a "green campus" had been accepted, the problem remained of what this meant, not only to the designers, but to others with differing perceptions of the appropriate interpretation of the stated green agenda, e.g., the following concerns were raised about the use of rammed earth:

- Maintenance of rammed earth and vulnerability to vandalism;
- Maintenance of timber-framed windows;
- Flexibility of rammed earth, e.g., future power points;
- Load-bearing capacity of rammed earth to support second and third storeys of future buildings;
- Longevity of solar hydronic heating systems, e.g., replacement of solar panels.

This Student Pavilion served as a prototype for the design of the office, teaching, and herbarium buildings and, in many ways, laid the foundation for the demonstration of the environmental design approach. For example, the use of rammed earth was initially rejected for use in the larger main buildings, however, following the opening of the Student Pavilion, those vehemently opposed to the use of such a "nuts and berries" material became some of its strongest proponents. The natural materials used in the Student Pavilion, feel, smell, look, and some even taste, good.

The decade of construction that followed may be roughly broken chronologically into the following parts:

- Site acquisition, mains servicing, inner ring road and parking, stormwater management systems and first building, Student Pavilion (1996);
- Stage 1: Graywater treatment wetlands, composting toilets, "living" Murray Valley herbarium planted along the constructed creeks, "phytogeographic" arboretum, main entry, outer ring roads and parking, School of Environmental and Information Sciences, GIS Mapping Building, Murray Valley Region Herbarium (1998);
- Stage 2: Endemic landscaping to public entry, CD Blake Lecture Theatre Complex, Information Technology Hub (2000);
- Stage 3: Buildings: Student Union; Student Housing (2001);
- Stage 4: School of Business Studies (completed 2002).

There is no one path to take to understand the decision stream for materials which is illustrative of the complexity of a design process. At each new stage of works, the designers, builders, University decision-makers, and occupants learnt from and responded to the experiences of past projects. The following discussion by and large relates to the Stage 1 office buildings.

13.3.2 Passive design strategies

Decisions about the building form and materials were driven by the primarily passive approach utilizing ambient conditions, thermal mass, insulation, high levels of daylighting and natural ventilation, thermal performance, and climate control. Climate-appropriate assumptions about the acceptable thermal comfort range, occupant satisfaction, and ultimately materials, took into account factors other than temperature; e.g., air movement, radiant effect of materials, humidity, the individuals' control over their space, 100% fresh air, daylight, variability of thermal perceptions of comfort with the climate and season, dress habits, nature of work, and connectedness with the outside.

These key initiatives were thrashed out during design workshops in which rough analyses were made to compare the cost and benefits of various elements in an attempt to rationalize the inclusion of specific desirable features in terms of the budget. For example, the cost of double glazing was about the same as a concrete ceiling to the upper level of the office buildings. The decision was made to integrate a concrete ceiling considering that the modeling indicated a more significant thermal contribution from the ceiling than the double glazing, the likelihood that windows might be opened by occupants at less than optimal times in terms of the thermal performance and other leakage issues, and cost savings gained by the deletion of the alternative ceiling.

13.3.2.1 Orientation and building forms

The development of the external and internal building forms responds to climate-specific passive factors such as solar shading, thermally massive construction to dampen temperature fluctuations, exposed surface area for heat transfer, adequate volume for ventilation flow, height of rooms for stratification, the means of controlling air flow during the day and at night for precooling of the exposed structure, and daylight penetration. On the other hand, the detailed development of the building forms generally related to the efficiency and functionality of spaces, site conditions, and characteristics of the materials including environmental impact, toxicity of production, construction and occupied space, multiplicity of uses as an integrated element, and available trades and skills.

The efficient use of space and form was a critical strategy for reducing materials use. For example, the roof shape was directly generated by the optimum pitch for direct mounting of solar collectors, which avoids problems of wind uplift, cleaning, and bird nesting. The pitching point of 2.7 m, the minimum by regulation, is set from the inner edge of the external wall generally, and lowered for portions of the building that extend beyond this point. The protruding thermal chimneys serve a dual purpose as outlets for the hot air and enhanced daylight penetration (Fig. 13.2).

13.3.2.2 Thermal mass

Albury's climate has a large average diurnal temperature range in the order of a minimum of 15°C and a maximum of 36°C, in the summer, and a minimum of 4°C and a maximum of 18°C in the winter. Internal mass was a logical materials choice to stabilize the internal temperatures by suppressing the swing in ambient temperature.

Neither externally insulated rammed earth walls, nor a lightweight, highly insulated external skin were practical or affordable. The introduction of another trade for the external walls would have complicated site access and responsibilities for coordination between the trades, and reduced the volume of rammed earth, thereby increasing the cost. Furthermore, the introduction of a framed structural system would have created problems with differential movement between materials and the

Figure 13.2 School of Environmental and Information Sciences, thermal chimney (photographer: Dirk Spennemann).

detailing of the rammed earth. The cost of construction called for a simple structure and led to the use of unreinforced, load-bearing rammed earth walls throughout the buildings.

The impact of increasing the 300 mm external and 200 mm internal wall widths as required structurally was modeled, resulting in the adoption of 400 mm external walls generally, and 300 mm for walls around the central corridor and every second wall between the offices. A 100 mm increase in thickness of all internal walls would have added 1600 mm to the overall building length and this was not affordable. The thicker walls carried the electrical and data/communications services to each pair of offices, and ensured adequate cover to the embedded conduits and sufficient strength where skirting ducts were recessed.

Concrete floors and ceilings for thermal mass worked with the hydronic radiant heating and cooling as a supplement to the passive strategies. The underside of the exposed concrete ceiling slabs is ribbed to increase the surface area of the thermal mass, and thus, improve convection efficiency of the ceiling, allowing the thermal mass of the ceiling to emit the stored heat to air quicker than a flat surface. The designers experimented with commercially available roof and wall cladding profiles as custom-made formwork was not affordable. The selected profile achieves an approximate 24% increase in surface area compared to a flat surface. The ribs were placed parallel to the direction of airflow from the external wall to the opposite internal wall which, although not possible to model, made sense to facilitate airflow and light penetration. Important for progressive reuse throughout the project, four ribbed formwork sections to suit the range of room configurations were incorporated into the standard flat slab formwork.

Ideally for the best use of the thermal mass, the concrete floor slabs would be left uncovered, perhaps polished or covered using a material with little or no insulation value. However, the University standard of a carpet finish to the office areas was followed. The carpet was 100% wool with jute backing, and was laid on a natural fabric underlay with a smooth edge, not glued.

13.3.2.3 Natural ventilation

The early decision to use both cross-ventilation and buoyancy-driven ventilation strategies to deliver fresh air without the use of mechanical systems or ducts, dictated decisions regarding the height and depth of the rooms and the detailed design of inlet and outlet openings. For example, the consequence of increasing the quantity of material due to the potential need to increase the height of the room to allow for stratification of hot air above head height was avoided by eliminating suspended ceilings in favor of the exposed underside of the concrete slab which was also essential for engaging the thermal mass.

The designers worked with potential window joiners, the rammed earth contractor, carpenters, and construction manager to design an affordable window unit. The recycled timber-framed window comprises three distinct components; low-level automatically opening vents for night ventilation, the lower pane of the double-hung window for view and occupant-controlled ventilation, while the upper pane channels daylight deep into the room and allows for further occupant control for high-level release of hot air and draught-free fresh air intake under windy or cold conditions. Similarly, manually operable glass louvres equal to the external openable area were integrated with the internal recycled timber door frame allowing a degree of acoustic attenuation between the offices and the corridors when required, and some borrowed light.

The original design of the thermal chimneys (thermally massive rammed earth) was altered to timber-framed and clad because of the high construction costs of working from a roof platform and the structural economy of limiting the concrete ceiling to the span of the offices where thermal performance was more critical.

Double glazing was recommended as part of the optimum modeled solution, however the high cost of double glazing was well beyond the budget. At the time, the range of standard double-glazed units available on the market was limited. Timber units with high insulation values were the most expensive and made from imported, ecologically unsound timbers; many of the cheaper aluminum units did not incorporate a thermal break, and PVC units, the cheapest option, were considered unacceptable due to toxic off-gassing, the significant environmental contamination of the manufacturing process, and inability (at the time) for recycling.

13.3.2.4 Rain and sun shades

The rammed earth contractor, steel fabricator, the construction supervisor, carpenters, architects, facility manager, thermal performance consultant, mechanical engineer, electrical engineer, and structural engineer were all involved in the construction of a series of prototypes.

Various options for each orientation were evaluated and the preference was for operable external timber louvres considering heat gain/loss and glare control, but these were not affordable, nor was a custom shade for each orientation and window size.

Analysis of the various shading requirements and window sizes, combined with discussion and prototyping with a local metal worker, the rammed earth contractor, and the construction supervisors, led to the design of a standard metal frame fixed to the rammed earth either side of the window infilled with horizontal and vertical shading elements to suit each particular condition.

The galvanized steel frames of the sun shades infilled with corrugated iron sheets are held off the wall, allowing heat to escape on all sides and preventing heat build up under the shade. This worked well for fixing, which had to be a minimum of 200 mm from the edge of the rammed earth wall. The initial design was for a timber frame infilled with timber slats, however this was altered during construction due to the University's concern about maintenance of timber.

13.3.2.5 Insulation

Insulation was specified to be placed around and under the footings and to extend about 1200 mm under slab, however, this proved quite impractical on-site due to the 1 m depth and the proportion of the footings. Insulation was placed to the edges of the slab and footing and under the slab for 1200 mm, which related to the standard sheet size, the logic being that the greatest heat loss would be from the edges and that the area under the slab would stabilize over time.

The stud walls and roof spaces were insulated with wool to reduce the heat loss in the winter and heat penetration in the summer. The roof space is ventilated through slotted plywood eaves and an open ridge for summer cooling of roof space.

Originally, the underside of the roof of the office buildings was insulated, but not the top of the ceiling slab. Consistently, the BMS indicated that the first floor was calling for greater heating than the ground floor, which is the opposite of what one might expect given that heat rises. The cooling demand was the most critical, and the top of the ceiling slab was retrofitted with wool batt insulation.

13.4 Materials: evolving the detail

> *skilled construction techniques and a certain organic, rhythmical legibility. . . . shows all the characteristics of technical innovation, material richness and special generosity, which distinguishes all the work of the CSU team.*
>
> *Andrew Mackenzie (2005)*

Decisions about the main materials of rammed earth walls, concrete floors and ceilings, recycled materials, and windows and infrastructure were made with relatively little information to aid in understanding the lifecycle of materials. Neither time nor resources were available for systematic evaluation of the impacts of embodied energy in production and transportation, pollution in production, use and demolition, use of the raw materials, production and construction methods, and health implications of the materials and the design implications for the form, materials, detail, use, operation, and maintenance of the buildings.

Some initial simple judgments were assumed in the design process:

- Reduce resource use through efficient functional space use and reduced quantities of materials;
- Do not use nonrenewable materials with high embodied energy and water;
- Reduce materials loss through improved design, handling, and construction practices;
- Use less energy-intensive building operation through elimination of active systems such as lifts and mechanical air handling and conditioning;
- Durable buildings require less maintenance;
- Flexible, fit-for-purpose space needs less frequent refurbishment;
- Use labor-intensive rather than production-intensive construction techniques;
- Source materials locally to avoid energy and pollution impacts of transport.

13.4.1 Rammed earth

The choice of rammed earth for wall construction was simple, as it is an economic, environmentally sound, and locally sourced solution involving the compaction of soil in formwork to produce walls that are thermally massive, very durable, non-toxic, give acoustic attenuation, require little maintenance, breathe, and are resistant to fire, termites, and the weather. They also look good.

13.4.1.1 Soil selection and mix

The use of the variable on-site soil would have required assessment of mix requirements for each batch, which was considered impractical and too expensive. Reliance on laboratory analysis was not feasible given the number of days needed for testing and the construction process of mixing numerous small batches. The option of on-site assessment of the soil and mix would have required an experienced rammed earth contractor to assess each batch for the particular soil characteristics to determine the mix. There simply were not enough people with the necessary experience to manage such a large job. Sieve screening for easier mixing and a more even finish was considered, but deemed impractical by the contractors. The use of quarry overburden, which had relatively consistent characteristics, significantly simplified the mix process.

The rammed earth contractors relied on the on-site assessment of the mix requirements which was supported by the structural evaluation of sample walls. In assessing the mix, the rammed earth contractor looked at the characteristics of texture, compressibility, plasticity, and cohesion through simple tests such as looking at the grain size and percentage, the ability to press the mix into a ball and for the ball to just hold together in the hand, and water absorption of the mix.

The quarry overburden was sourced from a nearby quarry. The raw material was mixed with water and sand and stabilized with approximately 2% off-white cement to achieve the desired consistency. The addition of cement was based on simple water resistance tests of sample walls which indicated that a small addition of cement improved resistance to erosion as well as improving the working consistency of the mix. The soft earthy pink color of the finished wall was a result of the color of the raw material. A light color material was selected to reflect heat, cooling being a priority.

The rammed earth contractor took responsibility for soil analysis which included lineal shrinkage, sieve analysis, and compressive strength compliance to the recommendations of the CSIRO (1987). This worked well for the Student Pavillion and Stage 1 buildings which were delivered as a trade contract directly managed by the University. However, the Teaching Complex, which was lump sum delivery process, produced problems. Neither the builder nor their rammed earth subcontractor, a local concrete company, had any direct experience with rammed earth. The raw material was sourced from a different quarry and contained larger grains at a higher percentage, reducing robustness of the finished walls. A mechanical mixing and handling process was used and moved the mixed material around with chutes similar to on-site concrete mixing requiring a wetter mix. The hydraulic ramming was less consistent, resulting in the wavy appearance and crumbly finish of the walls. The walls were then spray-sealed rather than drenched as specified. The designers were not involved in the formal administration of the contract and the Superintendent did not require the contractor to complete the building as specified, significantly, regarding the sealing of the rammed earth walls and the belowground drainage around the walls.

13.4.1.2 Wall height and form

Key to affordability was the decision to use the rammed earth walls structurally, set out to the standard formwork dimensions, and to eliminate reinforcement, saw cutting to the top of walls and lintels and overfills. The unreinforced, load-bearing, rammed earth walls support the exposed concrete, first floor and ceiling level slabs and take the roof loads of the two-storey buildings. The use of load-bearing rammed earth influenced the building form; walls were always returned at the ends forming L or Π plan shape for structural stability.

Although the Student Pavilion was a rectilinear form, the use of the curves suited the subsequent larger buildings which wrapped around the natural site contours. The advice from rammed earth contractors was that walls with simple standard curves were not more expensive than straight walls, partly due to the scale of the job and largely dependent on the formwork owned by the particular contractor. Curving of walls also contributed to structural stability.

The rammed earth walls finish level at a consistent height, generally of 2700 mm, and wool-insulated, plywood stud walls are built up on the outer edge to infill between the top of the wall and the roof where required to end walls. The tops and the sills of the rammed earth walls are splayed, as recommended by the contractor for the Student Pavilion, to ensure quick run off of water. The alternative of capping with tiles or mortar would have increased construction cost and materials usage.

The location of the opening in the walls reflected a balance between the practicalities of construction, and thereby the cost, and the idealized form for passive ventilation and daylighting. Angled night cooling vents were originally envisioned in each corner of the external wall to the Stage 1 buildings to enable even distribution of air throughout the room and limit short circuiting, and included a centrally located single point of control. A number of problems with this approach were identified. The rammed earth contractor was concerned about the labor costs involved to press the molds in while the earth was still fresh to make the angled opening, and to ram two additional holes, rather than leaving one window opening. The structural stability of the wall would be weakened as part of the underlying design logic was to return rammed earth walls at the ends. From a user perspective the need to maintain the free flow of air through the vents would unduly restrict furniture placement. The three key options were a pair of vents in corners with a central window for daylight, a pair of vents in corners with window over each, or a single central vent with a window over.

The resolution was a single opening with a combined window and vent unit (Fig. 13.3). The opening and frame extend to the underside of the ceiling slab eliminating the cost of a lintel and wall infill over the window. A 200 mm high

Figure 13.3 School of Environmental and Information Sciences, sunshades (photographer: Dirk Spennemann).

hob of rammed earth under the window reduces dust by keeping the vents off the ground and allows for a continuous skirting recess. The joinery and fit-off costs of a single frame were significantly lower than the other options, partly related to the relative degree of difficulty in meeting up with rough earth walls. The detailed dimensioning of the walls and openings relates directly to the location of vertical and horizontal construction joints. Anchors were required in a number of details, e.g., connection of sun shades, and an "environmentally friendly" anchor was sourced.

The rammed earth contractor prepared 3D drawings of every wall in the Stage 1 buildings, which was an invaluable tool for services coordination and used throughout construction to coordinate with the other contractors on-site. Careful programming was critical, for example in allowing for the specified curing time before the first floor slab could commence, and likewise allowing for the curing time for slabs before the ramming could commence. The specified rate for construction for 50 m^2 rammed earth wall per day was achieved with four rammed earth teams working simultaneously on different walls of simple design, however, this was difficult to maintain in more complex areas such as columns and underfills.

13.4.1.3 Formwork

The rammed earth contractor's prefabricated formwork was used repeatedly with marine ply siding replaced on-site as necessary. The wastage involved was not quantified but was that of a nontoxic, biodegradable, renewable resource, albeit a processed product.

Wall junctions were generally designed with standard right angles to minimize formwork cost and materials wastage, with the addition of a few "special" angles which, although not standard, were limited in number and designed for repeated use.

Adaptability and flexibility of servicing and redundant capacity were built-in quite literally. Data and electrical cabling conduits were cast into the slabs and walls to feed into the skirting ducts recessed into the base of the walls using blockouts in the formwork. An environmentally sound suitable off-the-shelf skirting duct could not be found. The commercially available products were also expensive compared to the solution of a formed skirting duct recess covered with a removable screw-fixed recycled hardwood timber cover plate. The foam blockouts were inexpensive and easy to glue in place and added little to the cost; the recycled timber cover plates were very cheap compared to the extruded aluminum and PVC commercially available products. Additionally, the main cost of the solution of recessed duct with a recycled timber cover plate was the labor component.

Wall strip lighting recesses followed the same approach of using blockouts in the rammed earth walls to provide flexibility and keep costs of fittings low. Picture rail recesses provide flexibility to hang notice boards and artworks without the associated costs of installation or maintenance. Various wall niches were blocked out to provide for installation of future artworks.

13.4.1.4 Finish

The external walls and foundations of the Student Pavilion were completely protected by the verandah form. A verandah building form with circulation externally and no internal corridor was considered an ideal affordable type for the future office buildings, however the future building occupants felt that it was not acceptable to be "forced" to use external circulation to access their offices. It was not affordable to provide sufficient overhang to protect the rammed earth walls of the resultant office buildings designed with internal corridors.

The off-form smooth finish of the exposed external walls was the cheapest, requiring the least labor, and results in the expression of the texture of the material and ramming layers. The broomed or brushed alternatives more commonly used give an even appearance, not dissimilar to concrete.

Given the high level of exposure of the external walls, generally, a sealant was needed for weather protection; a number of sealants were trialed on site in the effort to find an environmentally friendly product with a satisfactory performance. The simple test comprised the drenching application of various sealants to salvaged parts of a rammed earth wall which had been knocked down due to defects. Wall panels were hosed with water for a number of days and observations were made on the extent of erosion over time. The conclusion was drenching with a solvent-based sealant for the exposed walls and a nontoxic water-based sealant internally.

The climate extremes of hot dry summers and cold wet winters, combined with the highly reactive ground conditions led to a simple solution to the effects of varying moisture levels. A brick paved skirt laid around the perimeter base of all buildings was sloped to direct water runoff away from the walls and protect the foundations from the extremes of soil shrinkage and expansion.

Architectural detailing grappled with the junction of the rammed earth and the exposed concrete slab by measuring the deviation using a cornice detail that allowed a ± 15 mm tolerance.

Wet areas use tiles to protect the rammed earth walls where they are directly exposed to water. Unfortunately, tile seconds and broken tiles were not a possible source, partly as unsaleable tiles are crushed at the factory, but also because of the reluctance of manufacturers to sell damaged tiles.

13.4.1.5 Insulation

The resolution of detail was often a highly interactive process throughout construction, particularly given the trade package delivery process of most of the buildings.

For example, the extent and type of insulation to the slab edge were not resolved before construction commenced. The designers sought input from the rammed earth contractor, concreter, brick layer, construction supervisor, and supervising engineer. The concrete slab required a minimum area for bearing on the rammed earth walls, leaving limited space available to insulate and infill the slab edge. The resolution of the detail involved weather protection using brick infill with wool insulation to the slab edge. The anticipated differential movement between the rammed earth, brickwork, and concrete is accommodated by slip joints between the rammed earth and the slab and the absence of mortar between every sixth brick.

A brickwork base is built up from the footing incorporating a damp-proof layer and termite protection mesh. Lime mortar to assist with future recycling was considered, however it was not used due to structural engineering concern about the mortar strength.

13.4.1.6 Embodied energy and water use

At the time of the decision to use rammed earth, there was very little information quantifying embodied energy and water, however it seemed a sensible assumption that rammed earth was the lowest of all the thermally massive choices such as bricks, concrete blocks, and in situ or precast concrete. Earth bricks, in comparison, involved more skilled labor, and critically time and site organization, as well as being relatively more complex to detail, particularly in terms of services. Stone construction was considered too expensive and of high environmental impact due to quarrying.

Estimates of embodied energy for rammed earth vary dramatically. Project-specific factors, such as the availability of on-site material, transport of raw materials, cement content, ramming and mixing method and design, are critical. The primary energy component of rammed earth is relatively low as there is no manufactured product, however it includes energy consumption in the extraction of the raw materials. Secondary energy consumption is also low as there is no manufacturing process with machinery, lighting, conditioning, plant maintenance, etc., or energy for transport of processed materials, with the exception of the transport of the raw materials to site, on-site mixing and materials movement, formwork, and hydraulic ramming.

The energy used to transport the earth was only that of moving the raw material to the site. The cement content is the highest energy input of rammed earth construction, and can be as high as 8%–10% (Lawson, 1996, p. 36; Treloar et al., 2001, p. 102), compared to 2% typically used at Thurgoona. The labor involved is substantially on-site, and varies dramatically with the design ranging between 0.3 m^2/person/hour (Berge, 2001, p. 221)—8.7 m^2/person/hour (Easton, 1996, p. 108), whereas, materials such as concrete and bricks require a substantial additional off-site production component. The rate of rammed earth construction at Thurgoona was about 3 m^2/person/hour (Webster-Mannison, 2006, p. 224).

Water use in rammed earth is very low compared to concrete and brick production (Berge, 2001, p. 21), largely involving the addition of water to the mix and some cleaning of mixing areas and tools.

13.4.1.7 Pollution and environmental damage

The production of rammed earth creates little environmentally damaging wastes when locally sourced quarry overburden (usually treated as a waste product) is used (Lawson, 1996, p. 35). In contrast, the firing of clay bricks emits polluting sulfur and fluorine compounds (Berge, 2001, p. 119), and uses large amounts of water. The quarrying of materials may damage groundwater and local biotopes (Berge, 2001, p. 119).

13.4.1.8 Toxicity

The potential presence of radon in concrete, bricks, or rammed earth was a concern and varies with the source of the material. No specific local information was available and it was not clear that the methods for testing were adequately understood in terms of determining the acceptable levels of radon for building materials, and, regardless, commercially available tests were far too expensive.

13.4.1.9 Reuse

Unreinforced rammed earth is fully and easily recyclable as a building product (Berge, 2001, p. 120), or it may be returned to the earth through natural decomposition. The use of crushed recycled bricks as the base material was explored and prototyped by Rick Lindsay of the Earth Structures Group (contractor for most of the projects), with excellent results in terms of the finished product, however, it was quite expensive.

13.4.1.10 Indoor health

The thermal mass of the rammed earth helps to stabilize the indoor temperatures, keeping the indoor temperature closer to the outdoor average. Additionally, rammed earth helps to stabilize indoor humidity by absorbing and releasing moisture to the interior spaces.

There is no toxic off-gassing due to the use of a nontoxic, water-based permeable sealant internally to reduce dust. A solvent-based sealant was used externally, where off-gassing was not considered to be a problem.

13.4.2 Recycled materials (timber, glass, steel, bricks)

Salvaged prefabricated library stacks, timber, rammed earth walling, and bricks are amongst the variety of recycled materials used extensively throughout the project. The environmental benefits of using recycled materials were not quantified, however a quick comparison with the impacts of new materials reveals the "savings" to be made by the substitution of recycled material.

13.4.2.1 Prefabricated library stacks

Originally installed in the State Library of New South Wales in 1930, the Dixson stacks were a three-storey integrated structural system of glass flooring and approximately seven kilometers of library shelving. In October 1995, the stacks went to tender but were unable to attract a buyer. The project manager for the State Library project faced not only the lost credit for the sale of the material in the cost plan, but the additional costs for removal and dumping. The University agreed to take the material for the cost of transport which was $10,000.

The 24 mm thick cast glass flooring panels typically 900 × 600 mm, have a sand-blasted finish to one side. The structural flooring and shelving components include steel rhs posts (3 + ″ × 2″), steel I beams (3″ × 3″ × 3/8″), head base plates, shelf supports, steel angles for support of glass panels, bracing, shelf end panels, steel shelves, and trim (2″ × 2″).

The 24 mm thick cast glass flooring was underwired to meet the standard for trafficable floors and installed using the original steel beams and frames for support around the voids and in the corridors of the two-storey buildings to enhance daylight penetration, as well as in the foyer of the CD Blake Lecture Theatre. The reassembled shelving components were mounted on recycled timber plates fixed to the rammed earth walls, and the structural steel was reused for floor supports and pergola beams.

The material was stored on-site, and hardly a dent was made in the overall quantity, however it is unknown what remains of the material today, as the commercial architects that took over the design a decade later did not continue to use recycled materials.

13.4.2.2 Windows and doors

The recycled oregon timber window and door frames were a cost-effective solution when considered as part of the whole wall assembly (Fig. 13.4). The cost-effectiveness of the window units was demonstrated by comparison with a standard aluminum window installation, e.g., taking into account the elimination of lintels and overfill. Less obvious indirect cost savings, such as the elimination of air conditioning and ducts, provision of a thermal break (not included in cheaper aluminum windows) were not explicitly considered. Aluminum and PVC were eliminated by the selection of the "Anetta" frameless double-hung window. Although relatively expensive compared to the cheap standard windows that typified University projects, the economy of combining the framing for the vents and windows in recycled timber went some way to justifying the selection. Prototyping by the joiner and construction supervisor led to a very neat and affordable framing solution detailed with a horizontal recess under the sill to form a shadow line to fit the variable dimensions of the rammed earth, allowing for ± 15 mm tolerance to the top of the wall and ± 10 mm tolerance to width of opening.

The Student Pavilion used recycled timber exclusively with the exception of marine ply cladding. Future projects mainly used recycled timber for joinery

Figure 13.4 School of Environmental and Information Sciences, window unit (photographer: Dirk Spennemann).

including windows, doors, skirtings, picture rails, etc., and large structural beams, with roof and wall framing in sustainably sourced timber. The team progressively redocumented areas and details as necessary to accommodate the quantities, sizes, and stress grades of the recycled timbers that were available at the time of construction; in some cases material identified during design was able to be "reserved" in advance of construction because of the goodwill and enthusiasm of local suppliers, tradespeople, and the construction supervisor.

13.4.2.3 Embodied energy and water

The reuse of materials significantly reduces embodied energy compared to using new materials (Treloar et al., 2001, p. 140), especially the embodied energy "discounted" in consideration of the extended life of the reused material. The largest energy input was likely to have been the transport of the material, sometimes including interim storage after demolition and before delivery. Recycled timber was sourced locally and energy costs were not significant compared to sourcing new timber which would have come from further away. The recycled glass was delivered direct to the site and transport was less than the total transport costs of a new product.

Reprocessing impacts were minimized by reuse of steel and glass, mostly in their original forms. Steel was hot dip galvanized and some of the steel members were bent into curves. Glass flooring was cut to size, underwired, and rehoused in its original steel frame. The library shelving was cadmium-plated to provide nontoxic corrosion protection, which is an energy-intensive process using a nonrenewable and limited resource (Woolley and Kimmins, 2000, p. 142). In the case of window selection, an alternative to aluminum and the associated high embodied energy and water content and the highly toxic production process was sought, and eliminated with exception of the night purge vents. Off-the-shelf, anodized aluminum vents were cost-effective and low-maintenance, as well as recyclable. There is a need for alternative products and a change in perception about maintenance "costs" which may be largely a labor cost with minimum associated environmental impact. Timber windows can be repaired and renewed, whilst cheaper and "maintenance-free" products, such as aluminum windows need to be replaced. Appropriately designed timber-framed windows, even allowing for maintenance, have the least environmental impact of other window choices (Woolley et al., 1998, p. 123). The embodied energy and water use of timber is very small compared to aluminum or steel (Berge, 2001, pp. 20, 22; Sturges and Lowe, 2000, p. 186; Lawson, 1996, p. 13; Treloar et al., 2001, p. 140), and the use of recycled timber for the window frames would reduce this significantly.

13.4.2.4 Toxicity

The environmental and health effects, on both the applicators and occupants, of commercially available finishes suitable for the timber windows and doors were considered. Timber was prefinished with a full coating all round with "Organoil," made in New South Wales, Australia. "Organoil" comprises gum turpentine, tung oil, citrus oils, eucalyptus oil, and bees wax (ingredients less damaging in their manufacture than petrochemical products), and being plant-based, has a less energy-intensive manufacturing process than synthetics, waste products that are biodegradable and a very low toxicity rating (Woolley et al., 1998, p. 136).

13.4.2.5 Environmental impact

Recycled timber is fully seasoned and therefore very durable and stable. A local joinery made the window frames and doors from locally sourced recycled timber and these were prefinished with a nontoxic, plant-based oil. Recycled materials inherently have a very low environmental impact, and where appropriately sourced and detailed are cost competitive, the higher labor cost being outweighed by low material cost.

Reused metal was galvanized or plated for corrosion protection, which was considered to have less environmental impact than protective paints mostly because the galvanized or plated finish will last longer and the base material will be easier to recycle.

13.5 Conclusion: the issues are complex

It is not possible, in a practical sense, to elaborate on every materials selection decision stream, rather, some particular decisions about materials that were consistently applied, and their interaction with the design of other elements have been explored in this chapter.

Informing this story is the interrelatedness of decisions about ecoservices and materials for the campus and its buildings, and the archetypical determinants of passive design strategies, the local availability of materials and skills, ecological impact, and toxicity. Critical to "protecting" the design from cost cutting was to ensure a clear multiplicity of purposes for every facet of the design.

There is no easily applied prescription for green materials selection and it is inseparable from early design decisions, however, prioritizing a passive approach leads to radically different decisions about building form and materials, compared to the application of traditional mechanical heating, ventilation, and air conditioning.

References

Berge, B., 2001. The Ecology of Building Materials. Architectural Press, Oxford.

Easton, D., 1996. The Rammed Earth House. Chelsea Green Publishing Company, Vermont.

Howllin, J., 2002. Truly, Wholly, Deeply. Steel Profile No 78, March. BHP, Sydney.

Johnston, L., 2000. Deep Green, Architectural Review Australia No 073, Spring.

Lawson, B., 1996. Building Materials, Energy and the Environment. Royal Australian, Canberra.

Mackenzie, A., 2005. Learning through example, Arch. Rev. Aust., 79. pp. 84–89.
Middleton, G., 1987. Earth-wall Construction, CSIRO Bulletin 5. Australian Government Publishing Service, Canberra.
Sturges, J., Lowe, R., 2000. Construction Materials and Sustainability, the Limitations on Present Methods of Selection. International Conference, Sustainable Building 2000. Netherlands: Æneas.
Treloar, G., Owen, C., Fay, R., 2001. Environmental assessment of rammed earth construction systems. Struct. Survey. 19 (2), 99–105.
Webster-Mannison, M., 2006. Ecodesign for Large Campus Style Buildings. PhD Thesis. The University of Queensland.
Willis, A.-M., 2000. Australian Building Energy Council Building Case Study: Charles Sturt University, Thurgoona Campus. Available from: www.abec.com.au/studies/studies.
Woolley, T., Kimmins, S., 2000. Green Building Handbook Volume 2: A Guide to Building Products and Their Impact on the Environment. Spon Press, London.
Woolley, T and ACTAC (Association) (1998) Green Building Handbook: a guide to building products and their impact on the environment, London: E and FN Spon.

The Hockerton Housing Project: A case study of the use of concrete

14

Robert Vale
Victoria University of Wellington, Wellington, New Zealand

14.1 Thermal mass in zero-heating houses

Concrete is a material with high embodied energy and high carbon emissions (see Section 3.3.4), so should its use be avoided in a building which attempts to be "green"? This chapter explores this by comparing three examples of "green" housing, two in the United Kingdom and one in Australia, to find out the effect of using concrete on their overall environmental impact. Two of these houses use concrete as a way of storing heat, because in a 1993 paper Lund proposed that so-called "superinsulation," the use of very thick insulation, multiple glazing, and controlled ventilation, to reduce the escape of heat, combined with thermal mass to store heat, could produce adequate indoor temperatures for a zero-heating house in a cold climate (Lund, 1993). Lund's paper suggested that a temperature range in winter of 22°C ±5 degrees could be achieved with no additional heating. Lund's theoretical study appears to suggest that the extreme situation of a zero-heating house may require extreme levels of both insulation and thermal mass, but if that mass were in the form of concrete, would that invalidate the environmental benefits of not needing to use energy for heating the house?

14.1.1 The Hockerton Housing Project

The first attempt in the United Kingdom to build a zero-emission community, the Hockerton Housing Project (HHP), was completed in 1998. In 1993 a group of people, led by builder Nick Martin, set out a proposal to build houses on a 10-ha site outside the village of Hockerton in Nottinghamshire, which is a few miles from the market town of Southwell. The houses were to be the center of a sustainable development using renewable energy and which would attempt to have a low overall environmental impact. The houses, while being "green" in their own right, were also seen as the focus of a whole way of living that would be more sustainable, with the planting of trees, the production of organic food, and the creation of a lake which would encourage wildlife and where fish could be farmed. At the same time employment would be generated in ecologically sound businesses. The Hockerton Project comprises a row of five single-storey earth-sheltered houses which are autonomously serviced; in other words, they are self-sufficient for all their servicing needs. The houses use a repetitive modular structure to simplify the construction

Materials for a Healthy, Ecological and Sustainable Built Environment: Principles for Evaluation.
DOI: http://dx.doi.org/10.1016/B978-0-08-100707-5.00014-9

and they make use of off-the-shelf components to keep costs down. Each house has a single-aspect south-facing plan to maximize the solar gain for space heating. A very large insulated tank in each house is used for domestic hot water. Sewage is treated in a septic tank, a series of reed beds, and a landscaped lake. Rainwater provides all water for the households, it is collected from the whole site and stored in a reservoir formed from the spoil dug out to make the excavation for the houses, which are built on a gentle, sunny south-facing slope. Electricity for the entire project is supplied by two 5 kW wind turbines and by a row of photovoltaic panels totaling 7 kW along the edge of the roof. The solar and wind generators are linked with the local electricity grid, allowing surplus power to be supplied to the local area. Construction commenced in 1995 and the houses began to be occupied in December 1997. For more details see HHP (1998).

14.1.2 The Autonomous House

The Hockerton development was inspired by an earlier project for which Nick Martin had been the builder. This was the Autonomous House in the small town of Southwell, Nottinghamshire. For a very full account of the house, see Vale and Vale (2000). The house was first occupied in December 1993, the same year that Lund published his paper on thermal mass (Lund, 1993). The Autonomous House, the first of its kind in the United Kingdom, uses superinsulation to eliminate the need for fossil fuel heating, makes its electricity from photovoltaic panels, with surplus being exported to the national grid, collects its total water supply from the rainwater falling on the roof, processes its own wastewater in the garden, and turns its sewage into garden compost. It is located in the center of the town in a heritage conservation area. The house is relatively massive in construction compared to a conventional English house, but its construction is basically similar to most UK housing, using masonry cavity walls, and a tiled roof on a timber roof structure. However, in an attempt to provide additional mass to store heat, it uses precast concrete flooring and dense concrete blockwork for the internal loadbearing walls. The four-storey house has a cellar containing the rainwater tanks (to prevent freezing) and the composting chamber for the waste treatment system. In appearance the Autonomous House is extremely traditional, partly in respect of its location in a designated Conservation Area in the center of an historic town but also because the durability of traditional materials is well known, so it is possible to use such materials and confidently design for a life measured in hundreds rather than tens of years. The house has a roof of clay pantiles, walls of brick, and wooden windows with the flat arches over the window openings made of clay tiles in the same way that the ancient Romans used to build.

14.1.3 The two projects compared

The rural, grouped, attached, single-storey and underground Hockerton houses are in many ways the complete antithesis of the town center, one-off, detached, four-storey and above-ground Autonomous House. The greatest visible similarity between the two projects lies in their internal planning and their construction.

The construction designed for the cellar of the Autonomous House formed the basis of the construction for the external retaining walls of the Hockerton houses, using two skins of concrete blockwork as permanent shuttering for a reinforced concrete structural core, waterproofed by an externally applied membrane. The above-ground walls of both projects use an insulated masonry cavity construction with the same facing brick (locally produced and fired with methane from decomposing garbage) used on both. However, at Hockerton, only the south-facing wall and the conservatory are above ground. Internal planning of both projects is based on a simple repeated bay structure. This repetitive planning keeps all structural components to the same dimensions and makes for low-cost construction both in terms of the simplified ordering of materials and the simplicity of construction, which saves time on site. The structural bays are spanned by precast concrete beams in an inverted T shape with concrete block infill between them, then the bays are subdivided by nonloadbearing partitions to create the required internal spaces. The bay width (wall-to-wall) for the Autonomous House is 3.5 m, but at Hockerton this was reduced to 3.0 m to cope with the considerable loading on the roof of 400 mm of topsoil on top of 300 mm thick insulation. This earth-covered design is probably responsible for the fact that the scheme received Planning Permission in spite of being in the Green Belt and outside the "village envelope" of Hockerton. The earth-covered design, planted with fruit trees, makes the houses invisible from the adjacent main road, they look like an orchard.

14.2 Thermal mass in the two projects

14.2.1 Thermal mass in the Autonomous House

As well as having many design similarities, it is in their thermal strategies that the two projects are most connected. The principal factors in the thermal design of the original Autonomous House were the use of high levels of insulation combined with thermal mass. Theoretical studies suggested that a house in the United Kingdom, with low levels of solar radiation available in the winter, would need to be thermally massive to allow it to operate without conventional space heating (for details see Vale, 1995, pp. 106–124; Vale and Vale, 2000, Chapter 4). The need for massive construction, particularly in relation to a house which attempts to avoid the use of any conventional heating system, is supported by Lund (1993) who said that a superinsulated house with no solar components in northern Scandinavia would need 100 m^3 of concrete (with a deduced mass of roughly 210 tonnes) to provide sufficient thermal mass to meet a zero space heating demand. Lund's paper assumes a house with an internal temperature varying on an annual basis between 17°C and 27°C.

The Autonomous House gains its thermal mass by having all structural elements made of materials with a high thermal capacity. In practice, this meant using concrete and masonry construction, which is the conventional construction technique for houses in the United Kingdom. Normal construction practice was varied to the extent that the external brick cavity walls were built with the inner skin of 100 mm thick dense concrete blockwork; the internal loadbearing walls were of 150 mm thick dense concrete blockwork, and all floors used precast concrete beam-and-block construction. In conventional UK construction, the inner skin of external walls would be lightweight aerated concrete blockwork, the internal loadbearing walls would be only 100 mm thick blockwork, or probably timber studwork and plasterboard, and the floors would be of timber construction. However, no elements in the Autonomous House were made larger than was needed for structural purposes in order to provide increased mass, and all components were off-the-shelf to save cost and to demonstrate that a zero-heating house could be built with the materials and techniques that were already available to the building industry in the early 1990s. The total thermal mass of the house (i.e., that part of the building fabric that is within the insulated envelope and able to take part in thermal exchange with the interior space of the building) is 127.2 tonnes, which represents a volume equivalent to about 69 m^3 of concrete, about two-thirds of what Lund recommended. It is probably more useful to consider the thermal storage in terms of a square meter of floor area of the heated space. In this case, the Autonomous House provides thermal storage of 0.22 kWh or 0.78 $MJ/K/m^2$. The distribution of the thermal mass is shown in more detail in Table 14.1.

The mass is nearly all concrete, except for the roof decking, which is 70 mm thick double-tongue-and-groove softwood boarding, forming a loadbearing structural plate that spans between the internal crosswalls, and forms both structure and ceiling in one element, and the nonloadbearing internal walls, which are made of the same material. The floors are covered in clay quarry tiles so that the thermal mass is not decoupled from the space, as would be the case if carpet were used. The total thermal mass of about 127 tonnes is low compared with Lund's suggested figure of 210 tonnes. As a comparison, the thermal mass of a house of conventional UK construction is shown in Table 14.2 in comparison with the thermal mass of the Autonomous House. It can be seen from Table 14.2 that the Autonomous House has thermal mass more than five times that of a conventional house, although none of the elements of the house is larger than they need to be for structural purposes.

14.2.2 *Thermal mass in the Hockerton Housing Project*

At Hockerton the thermal mass is increased considerably compared to the Autonomous House. The ground floor concrete slab is 300 mm thick, twice the

Table 14.1 Thermal mass in the Autonomous House, heated floor area 176 m^2

Element	Volume (m^3)	Mass (tonnes)	Thermal storage		% of total
Floors	28.1	51.5	51.1 MJ/K	14.2 kWh/K	37
Roof decking	8.5	5.5	15.1 MJ/K	4.2 kWh/K	11
External walls	16.2	35.8	36.0 MJ/K	10.0 kWh/K	26
Internal walls	16.4	34.5	35.6 MJ/K	9.9 kWh/K	26
Total	69.2	127.3	137.8 MJ/K	38.3 kWh/K	

Table 14.2 Thermal mass of a conventional UK house compared with thermal mass of the Autonomous House

Conventional House	
Floor Ceiling External walls	Carpet on underlay Plasterboard on timber joists Gypsum plaster on aerated concrete block
Thermal mass: 0.15 MJ/K/m^2	
Autonomous House	
Floor Ceiling External walls	Clay tiles on concrete Plaster on concrete Plaster on concrete
Thermal mass: 0.78 MJ/K/m^2	

Table 14.3 Thermal mass at Hockerton, heated floor area 114 m^2

Element	Volume (m^3)	Mass (tonnes)	Thermal storage		% of total
Floor	36.4	76.0	63.6 MJ/K	17.7 kWh/K	31
Roof decking	29.9	62.1	52.5 MJ/K	14.6 kWh/K	26
External walls	26.4	55.1	46.5 MJ/K	12.9 kWh/K	23
Internal walls	22.6	46.2	39.4 MJ/K	11.0 kWh/K	20
Total	115.3	239.4	202.0 MJ/K	56.2 kWh/K	

norm for conventional construction, but the slab is also the foundation, and is of similar design to the slab that forms the cellar floor in the Autonomous House. The internal walls are 200 mm thick, which is the thickness needed to provide the necessary 100 mm bearing on top of the wall for each end of the concrete roof beams. The roof comprises prestressed precast inverted T concrete beams and concrete infill blocks, with an in situ concrete slab laid on top to bond the assembly together. The rear (northern) wall and the end walls of the terrace of five houses are made in the same way as conventional masonry cavity walls, but the cavity has been made 300 mm wide, and instead of being filled with insulation, as in the Autonomous House, it is filled with reinforced concrete 300 mm thick. This provides the retaining wall function for the buried side of the houses, and results in a mass concrete wall 500 mm thick overall. The whole terrace of five houses is surrounded with 300 mm thick expanded polystyrene insulation and a polyethylene tanking membrane, so all the concrete is in contact with the interior so that it can all serve as thermal mass. The net result is shown in Table 14.3, set out in the same format as the data in Table 14.1. Total thermal storage capacity is 1.77 MJ/K/m^2, which is more than double that of the Autonomous House at 0.78 MJ/K/m^2. It can be seen that the mass is distributed more evenly between the structural elements of the

Hockerton houses compared with the Autonomous House. The elements of construction at Hockerton are still appropriately sized for their constructional purposes, it is just that the construction is considerably different from the norm because of the client's wish to have the houses partly buried. Total thermal mass at 239 tonnes per house exceeds Lund's figure of 210 tonnes.

14.3 Thermal performance of the projects compared

Table 14.4 shows the temperatures achieved in the living room of the Autonomous House in the winter of 1995–96. These temperatures were achieved with four occupants and a consumption of about 350 kg of wood fuel in the 4.5 kW woodburning stove in the ground floor entrance hall. This represents a useful heat input from wood fuel of about 900 kWh over the heating season from a delivered energy input of roughly 1400 kWh. The use of the woodburner means that the house fell short of its "zero-heating" target, although it is possible to argue that the use of biomass as occasional fuel over the winter is no different from the use of passive solar gains, at least in terms of greenhouse gas emissions, or the eating of food by the occupants of the house. As a comparison with the woodburner, during the 6-month period for which temperature measurements are shown in Table 14.4, the occupants of the house consumed food with a calorific value (i.e., delivered energy value) of 2400 kWh, whereas the woodburner consumed fuel with a delivered energy value of 1400 kWh.

It can be argued from the temperature data in Table 14.4 that the effect of the mass of the house has been to shift the fall in internal temperature until it lags about 3 months out of phase with the lowest external temperature. After March the indoor temperature rose again. If Lund's hypothesis is correct, the Hockerton houses should attain higher temperatures than the Autonomous House as a result of their greater mass. Table 14.5 shows this to be the case.

Table 14.4 Winter living room temperatures in the Autonomous House 1995–1996

Month	Average external temperature (°C)[a]	Average sunshine (hours)[b]	Average living room temperature
Oct	10.7	87	22.4°C (1995)
Nov	6.5	48	18.3°C (1995)
Dec	4.5	42	18.0°C (1995)
Jan	3.3	43	17.7°C (1996)
Feb	3.5	57	17.7°C (1996)
Mar	5.7	97	17.5°C (1996)

Specific heat loss for the house (including ventilation) is 0.63 W/m^2K.
The site has 3344-degree days to an 18°C base (Page, 1986, p. 245).
[a]Data from 1941–1970 (Page, 1986, p. 224).
[b]Data from 1941–1970 (Page, 1986, p. 179).

Table 14.5 Winter living room temperatures in a Hockerton house 1998–1999

Month	Average living room temperature (°C)	Autonomous House for comparison (°C)
Oct	21.9 (1998)	22.4
Nov	20.9 (1998)	18.3
Dec	19.2 (1998)	18.0
Jan	18.9 (1999)	17.7
Feb	19.3 (1999[a])	17.7
Mar	n/a	17.5

Specific heat loss for the house (including ventilation) is 0.99 W/m^2K (Michaelis et al., 1998, p. 57).
The site has 3344-degree days to an 18°C base (Page, 1986, p. 245).
Measurements from 2 Mystery Hill, Gables Drive, Hockerton.
[a]Temperature up to and including February 25.

These figures suggest that the greater mass is indeed performing as expected. This impression is reinforced by the fact that the lowest temperature recorded in the living room of the Autonomous House over the winter of 1995–96 was 15.5°C, whereas at Hockerton the lowest temperature for the winter of 1998–99 was 17.9°C on January 13. This is in spite of the considerably lower specific heat loss rate of the Autonomous House compared with Hockerton, 0.63 W/m^2K as opposed to 0.99 W/m^2K and in spite of the input of 900 kWh from the woodburner in the Autonomous House. The temperatures at Hockerton are also more stable, with little variation from month to month compared to the Autonomous House. It is interesting to note that the household who supplied the Hockerton temperature data reported that they were thinking of opening the bedroom windows at the end of February, as it was beginning to get too hot (in the English winter in a house with no space heating system) for comfortable sleep (White, 1999).

The comment can be made that the results from the two houses are not comparable, as they are from different years; but the figures are compared here to indicate tendencies. It can also be pointed out that the Hockerton houses are much better planned for passive solar gains than the Autonomous House, which is severely constrained as regards passive solar gains by urban design contextual considerations on its site which mean that its long axis runs north–south, while Hockerton runs east–west. It should also be noted that the Hockerton houses have no form of auxiliary heating, such as the woodburner at the Autonomous House, and that they are unlikely to have dried out following construction and reached final moisture equilibrium by the date that the measurements presented in Table 14.5 were made.

14.3.1 Embodied energy considerations

Increased thermal mass seems to have improved the thermal comfort of the Hockerton houses compared to the Autonomous House but in a study of materials the other question to be considered in the use of such large amounts of thermal mass is that of embodied energy. Does the inclusion of such a very high-mass make the energy balance untenable?

14.3.2 Embodied energy comparison of Hockerton with the Green Home

A study was made comparing the embodied energy and lifecycle energy (embodied energy, energy of maintenance, and operating energy) of a Hockerton house and the Green Home, a relatively conventional passive solar house in Melbourne, Australia, designed in the 1990s by David Oppenheim (Fay, Vale, and Vale, 2000). The climate data for the two locations are shown in Table 14.6.

The Melbourne house has a concrete slab floor, but little thermal mass in its timber-framed walls, which have brick-veneer outer cladding, so there is mass in the walls but it is outside the insulation and so not available to store useful heat. There is a modest amount of insulation in walls and roof, and windows are single-glazed. The study by Fay et al. found that the initial embodied energy of a Hockerton house was slightly, but not significantly, higher than that of the conventional house, being 16 GJ/m^2 as opposed to 15 GJ/m^2. This is a counterintuitive result, as the Australian house has a "green" timber frame (albeit with brick-veneer cladding) whereas the Hockerton house is made of thick reinforced concrete which is considered to be very "un-green." Table 14.7 shows where the differences occur.

Table 14.7 demonstrates that the initial embodied energy of a Hockerton house is greater than that of the Melbourne house, but surprisingly, not that much greater, particularly when expressed in terms of the floor area of the house. It can also be seen that in all categories other than the basic building shell, the Hockerton house has a similar or lower embodied energy figure. The conclusion is that the incorporation of sufficient concrete thermal mass and insulation to allow zero-heating operation of a house in a climate considerably colder than that of Melbourne does not cause a serious increase in its embodied energy.

If the lifecycle energy consumption is considered, the picture looks even better. Again taking the comparison between the Melbourne house and a Hockerton house, over a 100-year life the figures are as shown in Table 14.8. In Table 14.8 the embodied energy of maintenance is taken into account as well as the energy used to run the house (heating, hot water, lights and appliances, cooking). This demonstrates that the strategy of designing for zero heating in a climate where heating is typically a large part of annual energy demand has proven to be effective in minimizing the lifecycle energy, and therefore the environmental impact, of the Hockerton houses. The use of a large amount of concrete to provide thermal mass has been an important part of this strategy. It can be argued that the

Table 14.6 Climate data for the two locations

Hockerton	53°0′ N	3344-degree days	1296 hours of sunshine
Melbourne	37°49′ S	1377-degree days	2080 hours of sunshine

Table 14.7 Initial embodied energy of the Melbourne and Hockerton houses

Element	Melbourne	Hockerton
Substructure	280 GJ	780 GJ (includes insulation)
Upper floors	50	nil
Staircase	2	nil
Roof	250	510
External walls	660	490 (includes insulation)
Windows	50	260 (includes conservatory)
External doors	5	2
Internal walls	70	140
Internal doors	20	2
Shell total	1387 GJ 10.8 GJ/m^2	2184 GJ 12.9 GJ/m^2
Wall finishes	20 GJ	10 GJ
Ceiling finishes	2	2
Finishes total	22 GJ 0.2 GJ/m^2	12 GJ 0.1 GJ/m^2
Fitments	30 GJ	20 GJ
Sanitary fixtures	50	30
Plumbing	70	70
Electrical	60	40
Other structures	60	nil
Fittings total	270 GJ 2.1 GJ/m^2	200 GJ 1.2 GJ/m^2
Appliances	70 GJ	40 GJ
Furniture	70	70
Chattels total	140 GJ 1.1 GJ/m^2	110 GJ 0.7 GJ/m^2
Direct on-site energy	120 GJ 0.9 GJ/m^2	150 GJ 0.9 GJ/m^2
Total	1939 GJ 15.1 GJ/m^2	2656 GJ 15.8 GJ/m^2

Table 14.8 Lifecycle energy consumption of the Melbourne and Hockerton houses

	Initial	25 years	50 years	75 years	100 years
Melbourne	15 GJ/m^2	30 GJ/m^2	48 GJ/m^2	64 GJ/m^2	80 GJ/m^2
Hockerton	16 GJ/m^2	20 GJ/m^2	25 GJ/m^2	33 GJ/m^2	38 GJ/m^2

comparison between a project in the United Kingdom and one in Australia is not valid, as like is not being compared with like. However, the embodied energy data used in this analysis, based on the work of the late Dr. Treloar of Deakin University (see, e.g., Treloar, 1998), are probably considerably more inclusive and detailed than data available at the time for the United Kingdom. This means that the embodied energy figures in this study are more stringent than they would be if other data were used. An additional consideration is that the Hockerton houses are operating in a climate that has 240% of the heating degree days but only 62% of the sunshine of the Melbourne site. In spite of these handicaps, the Hockerton houses still perform better than the house in Melbourne, with their lifecycle energy (embodied, operating, and maintenance) being less than half that of the Melbourne house after 100 years.

14.4 Other considerations

14.4.1 Cost

The Hockerton houses when completed in 1998 cost just over £91,000 for a house of 114 m^2 plus a conservatory of 57 m^2. It is difficult to compare these costs with those of a conventional house, as a conventional house in the United Kingdom is smaller and provides no conservatory. If the autonomous services and the conservatory at Hockerton are taken out of consideration, the houses cost about £60,000. A conventional house of the same floor area as Hockerton and built at the same time would have cost just over £53,000 (data from Michaelis et al., 1998, p. 49). The difference in cost is only £7000, or an addition of 13% on the cost of a normal house of the same size. Given that the Hockerton houses are the first of their kind, it is likely that future versions might be built for less, once the techniques for building them become part of mainstream construction practice.

14.4.1.1 Acceptability

The figures show that the Hockerton houses work well, but if people do not like to live in them this type of building is unlikely to be taken up more widely. Evidence suggests that the Hockerton residents still seem satisfied with their houses. One who moved after 16 years said it was "a wonderful place to live" because of its beautiful location, very low bills, and the sense of being able to make a contribution to reducing carbon emissions (Rajan, 2014). Back in 2001 the Hockerton houses were cited as an example of how sustainable products could be beautiful (Datschefski, 2001). In a more recent book, the HHP, now nearly 20 years old and still working as intended, is used as an exemplar of "one planet living" (Thorpe, 2014, pp. 318–321).

14.5 Conclusions

It would seem from these preliminary results that very high-mass superinsulated construction is an effective way to provide relatively low-cost "zero-heating" houses. The high-mass makes the most of low solar gains in the winter. The Hockerton houses appear to support Lund's theoretical premise of the value of extreme amounts of thermal mass in the achieving of comfortable zero-heating performance. At the same time it seems that the inclusion of high thermal mass, which implies the use of large quantities of concrete, need not be a cause of excessive initial embodied energy. Indeed it can be demonstrated that very high-mass houses can show substantial savings in lifecycle energy compared with conventional forms of construction. These findings suggest that the choice of materials may need to be a lot more subtle than simply having a list of materials that are "unsustainable." Sustainability depends much less on the choice of materials and much more on how they are used.

References

Datschefski, E., 2001. The Total Beauty of Sustainable Products. RotoVision, Hove.

Fay, R., Vale, R., Vale, B., 2000. Assessing the importance of design decisions on life cycle energy and environmental impact. In: Steemers, K., Yannas, S. (eds.) Architecture, City, Environment: Proceedings of PLEA 2000, July 2000, Cambridge, United Kingdom. London, James and James, pp. 164–169.

HHP, 1998. Hockerton Housing Project Launch Brochure. Hockerton, Hockerton Housing Project Ltd.

Lund, P., 1993. Optimum solar house: interplay between solar aperture and energy storage. In: Proceedings of the ISES Solar World Conference, Budapest, August 23–27. pp. 105–111.

Michaelis, C., Hardwick, G., Robson, D., Stankov, V., Burford, N., 1998. Hockerton Housing Project Construction Report. Birmingham, Data Build Ltd.

Page, J., 1986. Climate in the United Kingdom. HMSO, London.

Rajan, P., 2014. Quoted in EcoHouse: Moving from LONDON to Luxury Environmentally Friendly Houses Sep 15. London, Barcroft TV. Available from: https://www.youtube.com/watch?v = doV3uWuFESA (accessed 21.07.16.).

Thorpe, D., 2014. The 'One Planet' Life: a blueprint for low-impact development. Routledge, London.

Treloar, G., 1998. A Comprehensive Embodied Energy Analysis Framework. PhD Thesis, Geelong, Deakin University.

Vale, R., 1995. The Autonomous House in Theory and Practice. PhD Thesis, Nottingham, University of Nottingham.

Vale, B., Vale, R., 2000. The New Autonomous House. Thames and Hudson, London.

White, N., 1999. Private Communication, 25 February. Hockerton, Hockerton Housing Project.

Lambie House: deconstruction and eco-refurbishment

15

John Storey
Symbiosis Architects and Consultants, Nelson, New Zealand

15.1 Introduction: a domestic eco-refurbishment

In 1998 the author was commissioned to carry out a major domestic eco-refurbishment and upgrade of a 1950s house in Wellington, New Zealand. The clients wanted an architectural solution which was at once ecologically and environmentally sound, aesthetically pleasing, comfortable, healthy, and life-enhancing, but also built within the standard budgetary constraints of a normal home. The house also had to be adaptable to the changing needs of the owners and be a sound financial investment. A benefit of refurbishing this existing house was that it was built with good-quality, durable materials, mainly heartwood from first growth native New Zealand timber. Using this kind of timber is of course not a sustainable, or indeed possible option for new buildings (see Chapter 2), so this was an added reason to ensure its careful reuse. Despite being built well, the house was poorly planned, spatially mean, crudely built, and not compatible with the user's existing or future lifestyle intentions. It did however occupy a wonderful site, an easterly facing steeply sloping section, sheltered from the predominant northerly and southerly winds, which had delightful and expansive views across Wellington harbor and had good exposure to the northerly sun. The redesign created an expansive, highly crafted, comfortable, and life-enhancing home. Scenario planning was used to ensure that the home could be readily adapted to changing client lifestyle demands. The design successfully incorporated passive solar design principles, the use of sustainable and healthy new materials, and a series of crafted elements to give continuing delight and pleasure to the owners.

One of the major design strategies involved was to reuse as many of the existing building components and materials as possible, consistent with satisfying the stringent design intentions and the New Zealand Building Code. This involved a significant amount of deconstruction of the existing building.

The New Zealand Building Code requires that all building materials used structurally, whether new or reused, must be certified to last for at least 50 years in the specific context of use by a structural engineer. New or reused cladding materials must have a certified lifespan of at least 15 years and all other materials 5 years. If a designer chooses to reuse components and materials they become the de facto certifier and if failures occur within the designated durability provisions, the "certifier" can be sued. In this case reused structural members were examined by and certified by the structural engineer and the architect examined and certified the other reused components.

Materials for a Healthy, Ecological and Sustainable Built Environment: Principles for Evaluation.
DOI: http://dx.doi.org/10.1016/B978-0-08-100707-5.00015-0

15.2 The builder and the clients: commitment to resource savings

While both the architects and the clients were committed to creating an ecologically and environmentally friendly outcome it proved to be extremely difficult to obtain the services of a building contractor who shared these intentions and was still able to produce good-quality work at an affordable price. In the end, a builder was employed who had a good track record in conventional building work but little knowledge of sustainable construction.

A good deal of extra time and effort were spent on preparing documentation and in explaining the ecological design intentions to the main contractor and his workers as well as his subcontractors, but this approach met with only limited success. Fundamentally the principal contractor sought to avoid any extra work or costs that the ecological and environmental intentions of the project imposed and did not instruct his own workers or his subcontractors concerning associated contractual requirements. It was left to the architect and the clients to catch deviations from the specifications and drawings.

The clients remained on site for the whole of the duration of the contract and were extremely conscientious about informing the architect of possible deviations from the terms and conditions of the contract. The architect spent considerable extra time in following these reports through and insisting on compliance with the contract. In the end the contractor realized that it was going to cost him less time and expense to comply with the contract documents than to attempt to circumvent their requirements. Things started to improve from then onwards but with a constantly changing workforce, numerous subcontractors, and an essentially uncommitted main contractor, compliance remained an uphill battle throughout the duration of the contract.

15.3 Reusing windows and doors

The reuse of doors and windows was seen from the very beginning of the design as both an economic and a resource-efficient approach to the redesign of the house and a worthwhile objective in itself.

Windows and doors have a major impact on the visual coherence of any building. Great care was therefore taken to integrate existing and new windows and doors and to create a unified architectural expression for the building as a whole and for each of the spaces, while still seeking to create a modern architectural statement. Externally, painted timber componentry is the obvious unifying factor but fitting the new windows into the existing overall architectural form of the building and retaining a similar scale and proportion in the new work are in fact the primary contributing factors to the deliberate, visually cohesive integration of old and new. Modern timber sections were deliberately used in new work rather than seeking to replicate the original sections, as the new sections work better and were less expensive and resource-intensive than trying to match the existing sections in the new work.

Internally there were three conditions, one where the original component was retained in its original location, one where existing components were relocated and reused, and one where new components were used. New windows and their linings and architraves were clear finished internally; reused windows and their new linings were painted internally but their architraves were clear finished; and existing windows with their linings and architraves were simply repainted. Only new windows or reused windows or existing windows were used in any single space. They were never mixed. The notion here was to provide a resource conservation and historical code that was understood by the clients but was not readily apparent to the casual observer.

All the existing windows except one small unit and all the existing external doors except one were reused in the upgraded building. All interior doorsets except three were also used. All internal window and door sills, linings, and architraves had to be replaced in the reused windows and exterior door and door screens because the original sills and linings were inevitably damaged during extraction of the windows and doors from their original locations.

Some limited adaptation of three of the existing windows occurred by way of removal of one of the sashes in these multisash windows, but in all other instances the whole window was reused. All reused windows and exterior doors had their hinge actions eased and were weather-stripped to control draughts and heat loss. New hardware was fitted to provide additional security and design consistency and because, after nearly 50 years of constant use, much of the existing hardware was in poor condition.

Interior doors were paneled to unify them visually with new doorsets and consequently the rebates in existing frames had to be deepened. Work on all relocated windows and doors was undertaken in the contactor's own joinery shop while work on windows and doors retained in their original positions was carried out on-site by joinery shop staff. New architraves and hardware were provided for the reused interior doorsets. Three interior doorsets were riddled with borer and could not be reused. Financially the reuse of the existing windows and exterior door components generated significant cost savings, but the reuse of interior doors was less financially beneficial and resulted in only marginal savings. Both resulted in considerable resource savings.

15.4 Reusing the roof

The clients were very concerned at the lack of head room in the existing house and the architects were required to address this deficiency in the redesign. The existing roof was a minimally angled monopitch, with a multilayered bituminous felt finish laid on timber boarding and beams, with an uninsulated flat ceiling below. The felt was at the end of its useful life; it leaked and was very unsightly. External appearance was important as the building is approached from above and the roof is in effect the fifth elevation of the house.

The architects wished to retain the existing sound roof structure and decided to add a new weather skin above the existing roof supported by this existing roof structure. Initially the architects designed a solution which involved cutting the existing roof boarding and weather skin along the line of an existing, longitudinal structural wall and lifting the roof above this to create a double-pitched roof with a skylight ridge. This involved pivoting the roof structure on the external walls and building up the central structural wall. A specially designed, structural, gusset plate connector was to be utilized to close the gap created at the ridge.

Client, architects, structural engineer, and quality surveyor were all happy with this solution but none of the tendering building contractors would quote for carrying out this part of the work. All the contractors were concerned about the roof being lifted off the wall and damaged by the wind during the pivoting operation. Wellington can be quite windy but in fact this site is very sheltered.

Reluctantly the architects redesigned the roof. In the final design the existing roof was left in place and a new monopitch corrugated metal weather skin fixed on top of this on battens, with a ventilation gap between the two roof finishes as in the previous design. The flat ceiling was removed and a sloping ceiling was incorporated which followed the underside of the existing beams to improve head height in the main living spaces. The intention was to reuse the ceiling battens to construct the new ceiling. However, some of the existing beams and battens had warped and for the eastern two-thirds of the ceiling it was necessary to remove the existing ceiling battens and replace them with low-profile metal levelling laths. This lost a very precious 25 mm of the floor to ceiling height but resulted in a very good-quality ceiling plane in the large living area spaces. Existing ceiling battens were reused in the westerly third of the ceiling and within the utility and kitchen areas. Wool insulation made from recycled carpet offcuts was incorporated throughout between the existing beams.

Initially the changed design was regarded as something of a compromise but it worked out very well. The final design was less costly than the original and the increased height generated within the living spaces by taking the existing ceiling void into the rooms proved to be perfectly satisfactory to the clients. Indeed, the new living areas now have a very generous, spacious feel. The main living spaces open to the sun rather more in the final design than in the original redesign. This has the effect of making views more expansive, but increasing solar penetration in summer causes some overheating on summer mornings. This is easily dealt with by opening doors and windows at either side of the house for a few minutes.

There were a number of additional bonuses associated with reusing the existing roof. Metal roofs are notoriously noisy. The noise of raindrops hitting the roof and creaking associated with thermal and wind stressing is often transmitted to interiors, and can be quite irritating. In this building the noise transmission is well muffled so interiors are significantly quieter in all conditions. Overheating of the interior through the roof in summer is mitigated by the double roof and the ventilated roof cavity. The internal insulation value of the roof is also increased.

The weather protection created by retaining the existing roof enabled work to proceed on-site even on very rainy days and protected the existing hardwood floor and the lower storey of the house.

15.5 Recovering existing fittings and fixtures

The specification required all existing fittings and fixtures which were not required for the renovated house to be carefully removed and handed over to the client. The client in turn made strenuous efforts to reuse or sell the recovered fixtures and fittings. One of the kitchen worktops was adapted and now serves as a desktop in one of the studies. The other kitchen cupboards and worktops were adapted by the client's son and now serve as workshop benches, cupboards, and shelving in the garage. An existing bookcase was moved and adapted for use in a study area. A number of other fittings were disassembled and the timber stored for future use.

Recycled native hardwood was used to make a purpose-designed vanity unit for the main bathroom. This was costed and proved to be no more expensive than an off-the-shelf unit using new materials. The purpose-built recycled hardwood vanity unit is a very pleasing and successful feature of the bathroom area.

15.6 Materials: minimizing waste

The general policy was to retain as much of the existing materials on-site and minimize waste sent to landfill. Considerable efforts were made to devise ways to incorporate as much of the existing materials as possible into the renovated building.

15.6.1 Weatherboarding recovery and reuse

The whole of the cedar weatherboard siding on the upper floor of the south end of the existing house had to be removed to make way for the extension. This siding was in very good condition and the extension was designed to incorporate all of the recovered siding and avoid the need to purchase new weatherboards.

The builder claimed that the cedar was very brittle, and split very easily. There was an accretion of nearly 50 years of paint on the boards, representing at least five layers of paint and the builder had not thought to run a knife along the interface between the boards to cut though this film. It also eventuated that the people assigned this task had little experience in this operation and had not been shown how to maximize recovery. A seemingly conservative recovery rate of 70% was presumed in the design but in the end a recovery rate of just over 40% was achieved. The suspicion is that insufficient care or skill was exercised by the builder in springing the boards.

It did however prove possible to reuse recovered hardwood flooring from the house to board over the old stairwell and repair areas of flooring damaged during deconstruction and so avoid the need to bring in new materials.

15.6.2 Relocating excavated material

The intention was to distribute all excavated material around the site. Between 100 and 120 m^3 was dispersed on site and only 20 m^3 had to be sent to the local cleanfill. Although disposal locations were agreed in advance the builder sought to renege on this agreement. This resulted in acrimonious arguments between the builder and the architect/clients. Both parties learnt from the encounter and in the end most of the excavated materials were located sensibly.

15.6.3 Gypsum wallboard

The intention was to retain as much of the existing gypsum wallboard linings as possible. However, a combination of circumstances militated against this. Almost all of the plasterboard had to be removed to allow for the incorporation of the wool insulation in external walls and ceilings or due to the reconfiguration of the existing internal spaces or match board thicknesses. This meant that there was very little sound gypsum board left. It would have taken more effort and resources to retain it and integrate it with the new wallboards than to remove it. To this day there is no gypsum recovery program in Wellington and so once the gypsum board was removed there was little option other than sending it to landfill, despite the aversion of both clients and architects to this practice.

In retrospect it might have been possible to use the gypsum onsite as a soil conditioner. However, this option was not considered at the time.

15.6.4 Waste timber: energy recovery

Waste timber was handed over to the clients who stored it and cut it up for use in their chip heater. The chip heater is the primary method of generating hot water in the house. This timber was untreated and not contaminated with lead-based paint, either of which would have rendered the timber unsuitable for use in this way.

15.6.5 Other waste materials

The only other items that were sent to the landfill were small quantities of concrete, asbestos-based tiles, and preservative-treated timber and general rubbish. Following the regulations at the time, which have since become stricter in New Zealand (see Chapter 6), the asbestos tiles were carefully removed by the contractor without creating dust and were double-bagged before being sent to the landfill where they were placed in special hazardous waste areas along with small amounts of preservative-treated timber. The contractors felt that it was unnecessary to use specialist asbestos removal experts in this instance as the asbestos tiles were on an external concrete deck, were embedded in bitumen, and were therefore easy to remove with their bitumen backing without breaking them. Operatives were issued with protective clothing and masks. The small amount of general rubbish was disposed of at the local landfill, while concrete waste was sent to the local cleanfill.

15.7 Conclusions: intentions, attitudes, and realities

The attitudes of both clients and contractors are key factors in achieving successful outcomes in the area of deconstruction and resource reuse and recovery. In this instance the architects and clients were in total accord on this issue and sought to practice resource recovery in all aspects of this project. These intentions were backed up by carefully written specifications and contract documentation and by close observation of the works by client and architect.

Despite this, not everything went smoothly. Some of these difficulties can be attributed to the unsympathetic attitude of the contractors to resource conservation issues and sustainability in general. This attitude can be observed in the very low recovery rate achieved for the cedar siding. Being more careful would have probably resulted in a significantly higher recovery rate but the contractor believed that it would have cost him more time while using new materials would "only" cost the client and the planet.

Interestingly, it proved quite difficult for the contractor to source materials of adequate quality to match the existing weatherboards. This caused delays in achieving weathertight enclosure which had quite significant "knock-on" effects. In the end these delays probably cost the contractor more than the extra over cost paid by the client for the new materials, so everyone was the loser in this instance.

The attitude of the contractor made achieving resource recovery measures and getting sustainable new materials delivered something of a challenge, but his work was of a reasonably high standard and generally once he had committed to a course of action the result was good. This was the case in relation to the roof, the reuse of windows and doors, and the removal of fixtures and fittings. Only in the areas of disposal of excavation materials and recovery of weatherboard sidings was his performance disappointing.

The diligent observation of the works by the clients and the architect, together with the insistence on compliance with the contract specification by the architects throughout the project encouraged the builders to stick more closely to the contract requirements for resource conservation than they might otherwise have done.

The obvious lesson here is to only employ contractors who are sympathetic to the objectives and intentions of the clients and architects in relation to deconstruction and reuse of components and materials. This is quite easy to say but not quite so easy to do.

Wellington would benefit from a "Green Builder" program like that established in Austin, Texas in 1992. In this program, builders are encouraged to learn about and adopt sustainable construction practices by participating in an active education and support program provided free of charge by the City of Austin. Registered Green Builders are able to rate the houses they build under an environmental rating system controlled by the City and thereby obtain a marketing edge over their rivals. Providing both the educational support and business incentives to encourage the involvement of builders is a vital aspect of achieving sustainable architecture in practice.

On the positive side there was very little waste. Expensive, resource-intensive elements like windows, doors, and kitchen cabinets were virtually all reused or were handed over to the client for future use. The roof was also reused. Sanitaryware and hardware was, after 50 years of constant use, quite worn out and beyond salvage. Only small amounts of concrete, timber, and assorted rubbish and modest amounts of excavated material were landfilled. Despite the small quantities involved, it would have been preferable to dispose of the asbestos-based tiles and treated timbers in a safer manner. Instead they were disposed of with other material in the hazardous materials section of the local landfill. There was at the time no alternative disposal method available in Wellington. Table 15.1 summarizes the recycling, reuse, and disposal methodologies employed for each material.

In the end this renovation was an exercise in the art of the possible. Nothing was done in this project that could not be done easily in any other project. The deconstruction and resource conservation measures taken resulted in both financial and resource savings. It could be argued that both the clients and architects had to put in considerable extra time and effort to achieve the results obtained but both were committed to this course of action from the beginning and both knew that the deconstruction, reconstruction process embarked on would require them both to put in this extra effort and commitment to obtain the desired results. However, in an age of competitive fee bargaining not many architects feel able to devote the considerable extra time to save their clients money or the world resources, and even fewer clients are prepared to pay extra fees to make such resource savings. This is a considerable disincentive for architects to undertake resource conservation measures in the normal course of events.

Table 15.1 **Recycling, reuse, and disposal methodologies employed**

Material/component	Disposal method	Reason for treatment
Gypsum wallboard	Landfill	Badly damaged by disassembly No gypsum recovery system available
Excavated earth	85% redistributed on site 15% landfilled	Ran out of space onsite. Lowest quality excavated material landfilled
Timber	100% exterior wall framing reused 50% interior wall framing recovered and reused 40% recovery and reuse of ceiling battens All timber that was not reused on site or stored for future use was retained and used to fuel the water-heating wood burner stove.	High-grade timber with little sign of deterioration after 50 years of use A significant percentage of the lower-grade timber used in interior walls and as ceiling battens had warped badly and could not be reused
External timber doors and screens	100% recovery 66% reuse in house 33% retained for future reuse (one item)	It is likely that the unwanted door will eventually be sold to a building recycler. There is a good market for such items
External timber windows	100% recovery 92% reuse 8% retained for future reuse (one item)	It is likely that the unwanted window will eventually be sold to a building recycler. There is a good market for such items
Asbestos roof soffit lining board, interior floor and exterior decking tiles	Bagged in two layers of polythene and landfilled	At the time this was the only approved method of treatment available (this would no longer meet regulations)
Kitchen cabinets and worktops	Recovered and adapted and reused as workbenches and cabinets in garage One section of worktop was adapted and reused as a desktop in home workspace	Existing cabinets and worktops of good-quality but were unsuitable for the kitchen configuration required by owner

Meridian: New Zealand's first Green Star-rated building

16

Guy Marriage
Victoria University of Wellington, Wellington, New Zealand

16.1 Introduction to New Zealand built environment sustainability

This chapter examines some of the challenges and practical realities of choosing materials on building projects, and uses as a case study the construction of New Zealand's first Green Star-rated building: the Meridian building in Wellington. It looks at the aims and aspirations of this project and notes that sustainability is easier to apply in theory than in practice, especially when harsh economic realities and practical constraints have to be accommodated in building projects.

16.1.1 New Zealand

New Zealand is a rural country with few large cities and an international tourism marketing line of "100% Pure." A strong worldwide image of green fields and lush, exuberant, untouched wildness may give the impression that it is one of the world's most eco-friendly destinations. Sadly, the reality is quite different. Much of the country is covered with fertilized pasture for sheep and dairy cattle, the latter contributing to a marked decline in the quality of water in formerly pristine rivers due to nitrate runoff (Ministry for the Environment, 2015).

Other parts of the country are covered in huge forest plantations of pine trees. The majority of native indigenous forests (at least 75%), which covered almost all of the country before human settlement have been destroyed in the last 175 years (MFE, 2013; see also Chapter 1). Situated on the edge of the Pacific "Ring of Fire" and in the middle of the extremely windy "roaring forties," New Zealand also has active volcanic features and is subject to strongly seismic activity over much of the country. These natural features and weather conditions have had a large effect on the way that the country's building materials have been used.

16.1.2 Sustainability in New Zealand

Being a small country, with the majority of housing being small-scale timber buildings, New Zealand has moved in recent years to start to embrace sustainable building practices, initially focused on commercial office buildings. The New Zealand Green Building Council (NZGBC) was established in 2005, and then produced a

Materials for a Healthy, Ecological and Sustainable Built Environment: Principles for Evaluation.
DOI: http://dx.doi.org/10.1016/B978-0-08-100707-5.00016-2

New Zealand version of the Green Start building rating scheme. The first Green Star-rated building was the Meridian building in Wellington and over the following years a growing number of commercial buildings have been designed and constructed to gain NZGBC's Green Star status (NZGBC,2015).

The Labour government (1999–2008) required that all new government buildings would need to be Green Star-rated, but this requirement was abandoned by the subsequent National governments (from 2009). Nevertheless, the NZ commercial building sector has since thoroughly embraced the concept of "green" buildings, and regardless of government diktat, most new quality office buildings are indeed designed for sustainability and seek Green Star accreditation by the NZGBC (Prins, 2013). For a certain sector of the commercial market, that includes many government departments, "green" credentials certainly sell.

Many of the leading NZ architectural practices offer Environmentally Sustainable Design (ESD) as part of their services, with NZ's biggest practices actively involved in this. One of the lead practitioners involved in sustainability is Warren and Mahoney, a leader in NZ architecture and rapidly also becoming a leader in sustainable building. Having secured a large slice of the official construction work in the rebuilding of Christchurch (the city of Christchurch, on the South Island's Canterbury plains, was badly damaged in a series of earthquakes in 2010–11), Warren and Mahoney have ensured that many of their new buildings have a strong sustainability focus, stating that "sustainability is integral to the achievement of architectural excellence" (Warren and Mahoney, 2016).

Jasmax is another large practice in New Zealand, focused on the Auckland market, that includes architecture and interior design. Their output is also strongly ESD-focused. Jasmax's recent completion of the Tuhoe building Te Uru Taumatua in the Bay of Plenty, puts the firm at the forefront of sustainability in terms of New Zealand design, with NZ's first completed entrant to the Living Building Challenge (see Chapter 18) (Jasmax, 2014).

However, it was another architectural practice, Studio Pacific Architecture, which helped set the ESD scene for NZ back in 2005 with the first Green Star-rated new building in the country: the Meridian building (Studio Pacific, 2007). Studio Pacific Architecture is a younger, smaller practice (at the time of the design of Meridian it was about 30 strong), but is one with a strong design sense and increasingly strong ESD credentials. The Meridian building was the company's first NZ project to focus on ESD as the key design requirement for a commercial building and featured an aspirational "shopping list" of desirable ESD features (noted below) from which the design team could pick and choose strategically in order to fit the budget. The work in the ESD field was driven alongside their association with consulting engineers Beca, the key ESD advisor on the Meridian project (Beca, 2016).

The building project featured substantial detailing input by Peddle Thorp Architects and was constructed by Fletcher Construction, NZ's biggest and most well-established construction contractor. The base building (core, frame, services, and building envelope) had a tenant fit-out (internal finishings, etc.) led by Warren and Mahoney that was also ESD-focused, but it is important to note that material selection and design principles of both the base building and the fit-out may at times have had to compete with client cost and budget restraints. You can't always get what you want.

16.2 Project site and design brief

16.2.1 Project site

The Meridian building at 33 Customhouse Quay was, at the time, Studio Pacific's largest project. The site was a bare asphalt-covered slab of reclaimed land at the very edge of the harbor in central Wellington, formerly used as a carpark, and before that, for goods storage. Urban, brownfield sites have fewer restrictions on development than greenfield sites, not having to consider things like environmental and species habitat, for example. Extensive arguments were made to the local Council that rather than car parking, the use of the site as office accommodation was a better use of prime waterfront land, but local lobby groups opposed the development, arguing that any waterfront space should be open space accessible by the public rather than built on. On the other hand, extreme wind conditions at the harbor's edge meant that the site was often simply too cold and windy to use comfortably. Buildings were proposed as a way to break up the wind, provide space for public life, and create places for people to gather. As the building site was highly visible to the public, it was decided by the project team that the building would be focused on sustainability as a leading example of green building for the city and indeed the whole country. To benefit the general public, as much as possible of the ground floor area was to be given over to public amenities such as hospitality (restaurants, cafes, etc.), with commercial uses only situated above the ground floor.

16.2.2 Design brief

Having gained Resource Consent (planning permission) for the building, the design team had to ensure that once completed it could meet the aspirational target of being the "greenest" new building they could build. A "shopping list" of features was drawn up by Alan Barbour (ESD specialist at Beca), and included a heat recovery wheel, exposed concrete for thermal mass, chilled beam ceiling technology, solar hot water heating, photovoltaic panels, low volatile organic compound (VOC) finishes, and natural ventilation wherever possible. The aim of using 60% less energy and 70% less water compared to a standard NZ office building at that time was also set (Marriage and Barbour, 2007).

Material choices were also important aspects to consider in the design of the building and the origin of these materials was considered. The choice of any materials is dictated by aesthetics, suitability, cost, performance, and long-term maintenance requirements, as well as lifecycle assessment and carbon emissions generated in production or transportation. Travel miles via sea are typically not an overly large part of the cost equation, as sea transport is relatively low on carbon use and proportionally more carbon is created in trucking to and from the ports than from the shipping itself (Cenek et al., 2012). European statistics show emissions per tonne-kilometer from maritime shipping and rail are similar (rail is slightly higher) with those from road travel and inland shipping being three times greater or even more (European Environment Agency, 2015).

The subsequent leasing of the building to Meridian Energy (a New Zealand company specializing in producing and selling only 100% renewable sources of energy, mainly hydroelectric and wind power) further emphasized the importance of the building's sustainability credentials and introduced another green requirement: bicycle parking only (no car parking was required by Meridian). These requirements were finalized into a Development Agreement (not publicly released) which tied the design team to the energy and water use targets noted above and also to a set of materials that could be used.

A lengthy evaluation of materials and potential methods to reduce energy use was then taken to assess which of the targets were achievable within the budget. The potential embodied energy of the various building materials was examined and the use of these materials to help reduce energy consumption, both during construction and after the building was in use, was also undertaken. This process ran alongside the birth of the Green Star certification process at NZGBC, as the Meridian project was already underway while Green Star was being established. The building effectively set the 5 Green Star standard for new-build office buildings, and this certification process has been followed by all subsequent NZ commercial buildings seeking a Green Star sustainability rating.

16.3 Materials selection

In designing for a strongly seismic environment, choice of building materials is crucial for achieving safety. The structural requirements of NZ's seismic design have tended to focus on either a steel frame, or a massively strong system of reinforced concrete. Recently, those two systems have been joined by the increasing use of massive timber, in the form of laminated veneered lumber (LVL) columns and beams, along with compressed laminated timber (CLT) floor slabs. For these uses, the perceived environmental advantages of timber over concrete or steel have been promoted, but the current reality is that cost and engineering concerns are always more likely to take precedence in the selection of material for these elements.

As an example the Meridian building is sited right at (or extending over) the water's edge, on reclaimed land, and so any building on that site needed to be built on long piles that extend deep down to the bedrock below. Reinforced concrete is the only realistic material choice for the size and depth of these foundations right next to (and almost directly into) the ocean, despite cement production being associated with creating greenhouse gases (see Chapter 2). The development team noted the ironies of producing a low-energy building that uses a lot of energy to create, but the role of the Meridian building has been to lead New Zealand buildings to a better, lower-energy future.

16.3.1 Steel structure

In the case of the Meridian building, the central core is composed of thick reinforced concrete walls resting on the piled foundations and exposed concrete floors (which brings the advantage of thermal mass into the building), but generally the rest of the structural solution is steel. This was the obvious choice for Studio Pacific Architects and engineers Dunning Thornton Consultants (DTC). While energy-intensive to manufacture (see Chapter 3), steel columns and beams offer the chance to create large column-free spaces and are quick to assemble on site.

Although in NZ steel is made from arc-forging raw iron sand from NZ's black sand beaches on the West Coast or from recycled steel scrap, as world steel markets continue to be highly and increasingly competitive, much of the country's structural steel is currently imported. Resisting this trend, in the case of Meridian the steel was sourced from within NZ and was fabricated by Stevensons, a New Zealand firm. Circular steel columns were also used as conduits for wiring before the columns were filled with concrete for fire-rating purposes. Steel's inherent strength in both tension and compression allowed a relatively simple structural solution to the daring structural feats, such as a 6-m cantilever out over the water's edge.

16.3.2 Glass cladding

In terms of cladding, glass has and probably will remain a consistently popular choice for external cladding of commercial buildings. It permits the facade cladding to be largely installed by just one lead subcontractor, and has some strong environmental credentials (see Chapter 2). The transparency of glass, most often seen as an asset on a waterfront position suchas possessed by this building, can also be a drawback, in that office buildings are frequently overglazed, and admit too much solar radiation along with excessive light.

While many construction projects simply resolve their solar overheating issues with the use of heavily mirrored glass, this has an anonymizing effect on the urban surroundings and is actively discouraged in Wellington city by the District Plan (Wellington City Council, 2001) Instead, tinting and low-emissivity coatings can help the glass to remain largely transparent while also meeting strict energy efficiency requirements. While for many years NZ's glass was "home grown" by melting silica-rich sand into glass, nearly all glass for commercial buildings in NZ is now imported, as local glass-manufacturing facilities tend not to have the ability for making the quality and size of panes that projects this size now require. This importation of glass has a minor effect on the embodied energy of the material as shipping uses little energy (Vale and Vale, 2007). However, the time taken for the shipping needs to be factored into project timelines and now has a major effect on building programs.

While broken glass and offcuts of glazing are generally shipped to Auckland and used as slag, i.e., downcycled for the creation of drink bottles and products like fiberglass batts, all-glass facades have a far greater effect on the environment through the encouragement of wanton use of energy. All-glass facades typically admit massive amounts of solar radiation, often making interiors of buildings far too hot. Extensive HVAC facilities are then typically used to cool the building down again. In winter, the large expanses of glass allow heat to be transferred rapidly out of the building again, so heating and cooling becomes a big issue for servicing and energy use.

The Meridian building introduced a radically different approach to the glazing systems, with New Zealand's first double-skin facade being employed on the east and west facades, consisting of a double-glazed internal facade coupled with a second layer of facade glass situated 600 mm outside the first layer. This enabled a solution that could channel airflow in both summer and winter and allow natural ventilation to occur throughout the building without risking irksome airflows (Wellington is a notoriously windy city). While a double-skinned facade uses three times more glass than a standard single-glazed facade, the size of the mechanical HVAC plant required is reduced substantially and the long-term energy load is also greatly reduced. Indeed, it was only via the application of the double-skin facade principle that the 70% reduction of energy use was achievable.

16.3.3 Louvre selection

Notable on the Meridian project is the facade to the western annex, which has clear glass cladding but is also covered externally with projecting timber louvres (refer to Fig. 16.1). These louvres have the effect of both visually disguising the building as a semisolid object, as well as strongly limiting the amount of natural heat and light that enters through the facade. As the annex faces west, the louvres are adjustable (computer-controlled by the building management system) so as to be able to block a maximum amount of solar ingress when needed. Computer modeling showed that the number of louvres required for shading would be extensive.

There was no obviously ideal solution to the selection of a material for the louvres. Due to the harsh marine environment on the waterfront, the use of steel for louvres was immediately ruled out, and the choice was down to aluminum (strong, nonrusting, and relatively light weight), or timber—either locally grown or imported. The most obvious choices still had substantive problems. Aluminum may not rust, but it still oxidizes and corrodes, is heavily energy-intensive to manufacture, and is mainly made offshore. "Sustainably grown" (meaning that forests are replanted when cut down), locally produced softwood timber has a probable short life with the potential to warp, twist, and rot if exposed, while exotic hardwood with excellent

Figure 16.1 The Meridian building, on Wellington's waterfront, showing the extensive external timber louvreing to the western Annex of the building.
Source: author's photograph.

projected longevity may potentially be harvested from endangered rainforests. What other options could the team pursue?

Aluminum was relatively quickly ruled out as the answer due to the energy-intensive nature of its production. The sharp, crisp, metallic nature of the aluminum also worked against its selection as the architects wanted a more organic, softer form for the Annex louvres, and extensive amounts of aluminum had already been specified for the mullions and transoms of the glazing system and the small area of louvres to the north. The choice had to be timber—but what species, and where from?

In considering the specification of the timber, a major restraint on available timbers was that they had to come from a Forestry Stewardship Council (FSC) source that could warrant that the forests logged were also sustainably managed. Proposed material for these timber louvres went through a lengthy evaluation process, weighing up factors such as use of local material versus imported material, and durability versus sustainability credentials (NZ Wood, 2016).

Although the logging of NZ native forests was once widespread and the use of the excellent native timbers was common for building, native forest logging is now widely considered to be a highly unethical and very nonsustainable practice. Locally grown native hardwoods and softwoods have exceptionally long growth periods and all but a few privately owned native forests have been closed for logging since 2002. As a consequence, it is now very difficult and uncommon to use NZ-grown native timber for construction, unless it is secondhand.

A faster-growing alternative exotic tree species has been extensively planted throughout the country: radiata pine. Locally sourced *Pinus radiata* planks, readily available and certified as grown in a sustainably managed plantation (although see issues with plantation-grown timber in Chapter 1), were a definite option due to their low cost, but were ruled out due to material unsuitability (their fast growth leads to a potential to warp and twist, with increased susceptibility to rot).

Overseas sourcing of timber supplies revealed several excellent timbers with low warping, rot resistance, and purported sustainable harvesting, but these claims could not all be verified. Most of all, in considering material selection, the issue of timber species selection came down to weight. Eucalyptus species (such as spotted gum and blue gum) from Australia were considered, but the density, hardness, and weight of the timber, while having a long life, would produce considerable issues of weight restraint on the louvre assembly structure. The eventual final choice was made to use cedar planks, certified as FSC compliant, which are endemic in Canada and readily imported to NZ. These have good longevity and most importantly weigh up to a third less than the gum species (Fig. 16.2).

16.3.4 Concrete

As a material, concrete can be argued to be both extremely sustainable in terms of its longevity (for instance, the concrete roof of the Pantheon in Rome is nearly 2000 years old) and extremely unsustainable in terms of its environmental impacts (cement production uses vast amounts of energy, water, and produces large amounts of CO_2 (see Chapter 2 and 3). However, when it comes to a structural material for foundations and floor slabs, the choice is extremely restricted: in reality, foundations for large buildings are concrete. For this project, the effort was made to reduce the amount of CO_2 produced by the pouring of the upper floor slabs, via the incorporation of blast furnace fly-ash Duracem, as a partial substitute for more readily available Portland cement Concrete, 2016). While the finished product has similar characteristics to standard concrete, the fly-ash makes the concrete "sticky" and very much more difficult to pour.

The highly public, waterfront nature of the site meant that alkaline spills from concrete pours had to be completely eliminated, as Wellington harbor is relatively unpolluted by world standards and local sea life abounds, with fish, penguins, seals, and orca all being sighted near the site. A "sticky" substance is less likely to run into the sea, but regardless, secure bunding of the site had to be maintained to avoid any chance of concrete spillage.

The ethics of material selection must be continuously reevaluated: the fly-ash (a byproduct of coal burning, which produces CO_2) is imported from Australia as NZ

Figure 16.2 Enlarged area highlighting the Meridian building's cedar louvres arrayed around the southwest corner of the building.
Source: author's photograph.

does not produce this product—but is it better to use "home-grown" cement or imported fly-ash? There is no clear single black-and-white answer to what material is best: there are instead many shades of gray reflecting the many factors in selecting materials in architectural practice. This assessment of the lifecycle of building materials is discussed in more depth in Chapter 4.

16.3.5 Floor coverings: rubber, plastic, and wool

Floor coverings are an area which have high visibility as well as high wear and tear, so the selection of floor coverings is always a hot topic in discussions with the client. Materials such as natural stone are not that commonly used in New Zealand buildings, especially as our local stone is not high quality. Imported materials are more likely, and a wide range of flooring materials exist, with porcelain tiles, carpet tiles, and timber veneers often selected as the final covering.

For the Meridian building, an innovative material was specified for the flooring in the main circulation stair through the building, with the concrete structure of the stair covered and complemented by recycled rubber flooring. This material was selected at least partially for its acoustic qualities, meaning that footfalls are almost inaudible on the open-plan staircase. It was also chosen due to the fact that it was made from old car tyres and other rubber products minced up and reconstituted into new flat stair treads,which was particularly appealing. While the VOC content of the recycled rubber matting was not anticipated to be that high at first, the smell of this new rubber flooring was strong to start with. With the situating of the main air extract ducts directly over the stairs this problem was minimized and over time the rubber flooring has emitted steadily fewer odors.

By contrast, due to leasing requirements and tenant preference, the general office floors were to be clad in carpet. The selection of the carpet was surprisingly complex and an extensive evaluation process was made with the client and the fit-out architect to reach the final selection. NZ's natural locally grown carpet material is sheep's wool, and NZ has a long history of manufacturing and exporting high-quality virgin wool carpets to the world.

At first, the option seemed clear: surely a naturally grown wool carpet would be the best, most sustainable material. The alternative carpet specification was for nylon carpets woven into a plastic backing, involving nylon fibers which are a product of the oil industry and therefore are a contributor to creating unwanted greenhouse gases.

On face value, a wool carpet seems far more environmentally friendly than a nylon carpet, especially in NZ where the sheep roam freely year-round in open green fields. However, by the time you consider that the growth of the wool entails considerable numbers of methane-emitting sheep whose wool requires chemical scouring, creating severe pollution concerns, the material selection question is complicated. Lifecycle analysis indicated that nylon carpets are competitive in ESD terms, as long as they are recycled.

Coupled with the fact that wool carpets eventually go to landfill, and that Interface, the USA manufacturer of nylon carpet tiles, vows to take back used carpet and recycle it into more new carpet, the decision was taken to specify nylon carpet tiles (Interface, 2016). This decision was not taken lightly and still creates doubts in the design team's minds: what if the manufacturer goes bust in future years and cannot recycle the tiles? Is the export and import of new and used carpet tiles really a sustainable practice? Is the specification of nylon carpets, perceived by many as less desirable than wool carpets, responsible for the continued decline and seemingly inevitable collapse of the wool carpet manufacturing sector in New Zealand?

16.3.6 Wall coverings

The most common internal wall surface in NZ commercial building construction is plasterboard, acting either as part of a fire-rated wall to the building core, or acting merely as a partition cladding to enclose office space. It is a versatile material that can be skimmed over lightly with plaster, and sanded and painted to become a smooth surface for a wall. As such, it is virtually ubiquitous throughout a building site, but it is also frequently compromised by breakage.

Typically there is much wastage of this material on the job and, in the past this waste material was consigned straight to landfill. Contamination of materials, necessitating their dumping, typically happens with the stages of fixing or final finish. While the plasterboard is screw-fixed to the studs (which is reversible), it is also glue-fixed for strength (which is not reversible). Other finishes are often fixed to the outside of the plasterboard, such as tiling, acoustic lining, and paint. In all cases, these involve largely plastic-based adhesion of the substrate to the surface, which has the effect of causing difficulties in separating the materials out again at point of disassembly, and inevitably, this causes waste.

On the Meridian project, recycling of uncontaminated plasterboard offcuts was enabled by consigning them to a composting scheme where the paper outer layer, and gypsum inner core can break down organically. This was the first time it had been trialed in New Zealand with a major contractor such as Fletcher Construction.

16.4 Waste minimization

While the use of concrete, steel, and glass was largely unavoidable on a commercial project of this size, aspects of sustainability also had to be considered when dealing with the offcuts of these elements. Key amongst these was the desire to create a minimal amount of waste, as this is traditionally one of the less desirable side effects of building construction, with normally a large amount of mixed wastes in a series of large bins being shipped off site. Waste disposal is expensive, and construction waste makes up a large proportion of overall council waste, so this is an area where real progress can be made.

On this project, the contractor on site, Fletcher Construction, set up a series of much smaller, labeled bins, and piloted a scheme where the offcuts of plasterboard (used in the internal wall finishes) were recycled for compost, untreated timber was collected for firewood, and only treated timber consigned to landfill. Glass, aluminum facade cladding, metal ceiling grid offcuts and copper wiring offcuts were also sent to be recycled, each in their own collection bin (Rodio and Allen, 2012). While some construction workers on site initially found this "presorting" of waste products irritating, the larger workforce soon found this was not an onerous restriction, and followed it with enthusiasm. Fletcher Construction now employ similar waste recycling schemes on all their large building projects (Fletcher, 2016).

16.5 Conclusion

There is no doubt that the adherence to careful consideration of materials choice and use may have the effect of pushing up the building cost in some areas. The end result has considerable financial rewards however, as the subsequent sale price of the building showed a healthy profit for the developer concerned, although the consultants and contractors involved did not get to share in this. Considerable extra work was required of both Beca and Studio Pacific, as well as Fletcher Construction, but this is (unfortunately) not a factor that is financially reimbursed on building completion.

The "shopping list" of desired features was achieved with the exception of photovoltaic cells (the decision was made to let the efficiency rate of the panels continue to improve before installation) and the underfloor ventilation (which would have necessitated a very different and more expensive construction system). The ethics of material selection and perceived continual "greening" of our building stock is just one part of what makes our modern cities more or less sustainable. The ethics of workforce employment also has an influence—the luxury of being able to ponder and calculate the CO_2 equivalent of a delivery by sea from Australia versus a delivery by truck from Auckland is a first-world scenario. In poorer countries, the eventual selection of materials is largely dictated by budget and availability, with less ability to consider the selection of one material over another in terms of sustainability (or other) criteria.

The success of the Meridian building was evident in that it gained 5 Green Stars, which is at the upper end of the Green Star rating scale. There have subsequently been other projects which have gone on to achieve 6 Green Stars, but generally the aim now for most reputable property developers is for all new buildings to achieve a minimum of 4 Green Stars or above (NZGBC, 2016). The general interest in sustainability as a marketing tool has resulted in a boom of new "green" buildings within the corporate office sector in NZ (Prins, 2013). Green Star rating has now been enlarged to include not just new design, but also "in use" ratings and now covers other building categories such as educational, industrial, existing building reuse, and more recently domestic architecture (Home Star) rating (NZGBC, 2016).

The market has now permanently changed in NZ, with many developers keen to see buildings imbued with at least a green tinge to ensure healthy returns, but the NZGBC's Green Star rating is still seen as an aspirational benchmark for buildings. As an example of a building to help set up and test the system, the Meridian building was a fundamental part of that test and this chapter has shown some of the ways that material selection can be influenced by practical issues in a real-life situation.

References

Beca, 2016. Meridian Energy Building Kumutoto Wellington Waterfront. Available from: http://www.beca.co.nz/projects/buildings/commercial_offices/meridian_energy_building.aspx (accessed 15.06.16.).

Cenek, P., Kean, R., Kvatch, I., Jamieson, J., 2012. Freight Transport Efficiency: A Comparative Study of Coastal Shipping, Rail and Road Nodes. Opus International Consultants, Lower Hutt. Available from: http://www.nzta.govt.nz/assets/resources/research/reports/497/docs/497.pdf (accessed 15.06.16.).

Concrete 3, 2016. Meridian Energy Building - Wellington Waterfront Concrete Core to Sustainable Building Design. Available from: http://www.sustainableconcrete.org.nz/page/meridian-energy-building-wellington.aspx (accessed 15.06.16.).

European Environment Agency, 2015. Specific CO_2 Emissions per Tonne-km and per Mode of Transport in Europe, 1995–2011. Available from: http://www.eea.europa.eu/data-and-maps/figures/specific-co2-emissions-per-tonne-2 (accessed 2.08.16.).

Fletcher Construction, 2016. Meridian Building Site 7 Kumutoto. Available from: http://www.fletcherconstruction.co.nz/projects.php?action=search&id=265 (accessed 15.06.16.).

Interface, 2016. Environmental Accreditations. Available from: http://www.interface.com/APAC/en-AU/about/mission/Environmental-accreditations-en_AU (accessed 15.06.16.).

Jasmax, 2014. Te Uru Taumatua. Available from: http://www.jasmax.com/work/te-uru-taumatua/sectors/sustainable-design/2929 (accessed 15.06.16.).

Marriage, G., Barbour, A., 2007. A Green Building for the Roaring Forties. In: Proceedings of 2nd PALENC Conference and 28th AIVC Conference on Building Low Energy Cooling and Advanced Ventilation Technologies in the 21st Century. PALENC. Crete, Greece, pp. 106–110.

Ministry for the Environment, 2013. Deforestation Mapping 2012 – Final Report. Available from: https://data.mfe.govt.nz/document/101–2012-deforestation-mapping-final-report-indufor-asia-pacific/ (accessed 23.09.16.).

Ministry for the Environment, 2015. Environment Aotearoa 2015. Available from: http://www.mfe.govt.nz/sites/default/files/media/Environmental%20reporting/environment-aotearoa.pdf (accessed 23.09.16.).

NZGBC (2015) History of Green Star. Available at https://www.nzgbc.org.nz/Category?Action=View&Category_id=293, accessed 18 Jan 2017.

New Zealand Green Building Council, 2016. Office Certified Projects. Available from: https://www.nzgbc.org.nz/Category?Action=View&Category_id=194 (accessed 15.06.16.).

Prins, H., 2013. Project - The attributes of Wellington's Green Star rated buildings - Phase 1. Available from: http://wellington.govt.nz/~/media/about-wellington/research-and-evaluation/built-environment/2013-wellingtons-green-star-rated-buildings.pdf?la=en (accessed 15.06.16.).

Rodio, J., Allen, E., 2012. The Meridian Building. University of San Diego. Available from: http://www.josre.org/wp-content/uploads/2012/10/Merdian-Building-in-Wellington-New-Zealand-Near-Net-Zero-Compatibility-Mode.pdf (accessed 15.06.16.).

Studio Pacific Architecture, 2007. Meridian Building. Available from: http://www.studiopacific.co.nz/projects/2007/meridian-building/ (accessed 15.06.16.).

Vale, R., Vale, B., 2007. Time to eat the dog? the real guide to sustainable living. Thames and Hudson., London.

Warren and Mahoney, 2016. Who We Are. Available from: http://www.warrenandmahoney.com/en/practice/ (accessed 15.06.16.).

Wellington City Council, 2001. District Plan. Available from: http://wellington.govt.nz/~/media/your-council/plans-policies-and-bylaws/district-plan/volume01/files/v1chap01.pdf?la=en (accessed 15.06.16.).

NZ Wood, 2016. Case Studies - The Meridian Building. Available from: http://www.nzwood.co.nz/case-studies/the-meridian-building/ (accessed 15.06.16.).

Sustainable and healthy building practice in Germany

17

Rainer Hirth
Hochschule Coburg, Coburg, Germany

17.1 Introduction

Unhealthy building materials are not a modern problem, but rather are an old issue with broad negative effects (see also Part III). One late 19th century example of such a problem was the use of an intense green color, *scheeles grün* or *schweinfurter grün*, which was very popular for wallpaper and plaster topcoat. The color contained copper arsenate and people were poisoned by arsenic gases coming out of the walls of their homes (Szladja, 1999). It took decades to arrive at a clear understanding of the relationship between these deaths and recent house redecorating. Some scientists believe that the serious mental problems the famous painter Vincent van Gogh faced, along with other artists, might have been related to this green color. Van Gough used this in combination with other, equally dangerous pigments containing heavy metals, such as lead-white, tin-white, chrome-orange, and chrome-yellow. Unfortunately, the use of *scheeles grün* was only banned in the 1940s, almost a whole century after its popularization (Estaugh et al., 2007).

When it comes to similar issues with unhealthy components in building materials there is still a long way to go. In 1985, Wulf-Dietrich Rose published a substantive and influential book in German on avoidance of indoor toxins (Rose, 1985). It provided a summary of key information on the health performance of building materials. Although Rose's writing was highly polemical, it seems to have stood the test of time over subsequent years. Rose was sued many times by manufacturers of the products he criticized, but remarkably enough, he was never convicted, or ordered by the courts to either correct the text or pay compensation for damages. His last challenge was the trial against T-Mobile Austria in which Rose asserted that mobile waves are damaging genes and causing cancer (Rose, 1987). In 2001, Rose won this case in the highest Austrian court. Unfortunately, these court cases clearly show the extent of resistance to employing healthier practices in all human activities.

A number of other serious scandals helped increase awareness of such issues in Germany and inspired improvements in the regulations. One such example is the timber finishes Xylamon and Xyladecor manufactured by Desowag. These were very popular and widely used in the 1970s and 1980s (Der Spiegel, 1997). Unfortunately, these products were associated with a persistent release of a very toxic substance, PCB/lindane, which was emitted even decades after it was installed. It took many years between the first cases of health problems to the final banning of these products. One of the most disturbing facts of this particular case

Materials for a Healthy, Ecological and Sustainable Built Environment: Principles for Evaluation.
DOI: http://dx.doi.org/10.1016/B978-0-08-100707-5.00017-4

is that, when it was presented in court, it emerged that the Desowag company had a policy of hiding information. The list of similar events is long; from the dust from early Rockwool insulation (associated with causing lung cancer), to several formaldehyde issues. It appears that many manufacturers have tried to hide suspected known adverse impacts for as long as possible.

More recently, a change to such resistance to disclosing known risks seems to have finally arrived. Under longlasting public pressure, the European Union (EU) regulations have started to change. As part of this process, within the EU several toxic components of building materials have been removed. For example, after almost 20 years of clear evidence, PCB has been removed under the Stockholm Convention, which focuses on removal of persistent organic pollutants (POPs) (Stockholm Convention, 2016; signed in 2001). Similarly, there has been a decrease in the expectation that chemical treatment of loadbearing timber is needed, which since 2013 has not been required if the timber is kept dry (DIN EN 335-06 – 2013). Within these changes, not even the battens under tiles are required to be chemically protected. However, lists, such as the Stockholm list of POPs, are still continually being extended (Stockholm Convention, 2016). One of the advantages of these changes is that construction workers are also experiencing an improvement in health, due to decreased exposure to toxic sawdust.

Nowadays, it is easier in the EU to get a clearer picture of what is contained in a particular building material. Materials normally come with an Environmental Product Declaration, which gives information about embodied energy, resources used, and the extent to which the product contributes to the greenhouse effect, air pollution, erosion, etc. In addition, such declarations provide estimates about the material's general quality, lifecycle, insulation features, and influence on internal air quality, etc. Amassing such information about materials helps in the *ökobilanz* or lifecycle assessments of a building. These systematic analyses include a lifecycle assessment, waste production, and other aspects, that make it possible to get a clearer picture of the total ecological influence of the building.

17.2 Sustainable and healthy construction in Germany

Currently, in Germany, many considerations of sustainability are discussed separately from the impact building materials might have on human health. This is clearly shown by the organization of the existing certification systems.

17.2.1 Certified sustainability in Germany

Although there are numerous certification systems around the world, the most commonly used system in Germany is administered by the German Sustainable Building Council (DGNB) (2016). Introduced in 2007, currently DGNB has a market share of more than 80% but it is a small market as only a small number of buildings are certified. A portion of buildings in Germany use other certification systems such as BREEAM (from the United Kingdom) and LEED (from the

United States), while the Living Building Challenge (see Chapter 18) is almost completely unknown.

The DGNB certificate takes into account ecological, economic, technical, and procedural aspects of building, together with sociocultural quality. The system is complex, with 28 different points which are individually evaluated. Their scores form varying proportions in the overall result. However, this variation in attributed proportional significance means that some aspects are not recognized as being as important as others. For example, the "site factor" (access to public transport, bike stations, reuse of wasteland, etc.) gets a very modest number of points, while the lifecycle assessment of building materials and "commercial viability" are recognized through points as being very important. Perhaps an unusual feature of the DGNB certificate is that the building investors, building owners, and proprietors are recognized as being equally important as the architects and engineers (DGNB, 2016). This has both positive and negative implications. On one hand, the DGNB certificate is well recognized in the free real estate market, where the value of the buildings definitely increases with the certificate. Similarly, landlords appear to appreciate that the DGNB allows them to get a clear picture of the money they will need to spend to maintain the building. However, the complexity of the system is so great and costly that obtaining the DGNB evaluation can cost up to 4% of the building construction costs. Jointly these features indicate that the system is still in need of further improvement.

The German government has tried to take a leading role in the implementation of sustainable construction for its own buildings. In 2015, the Federal Ministry for the Environment, Nature Conservation, Building and Nuclear Safety (BMUB) published the updated "Leitfaden Nachhaltiges Bauen" (Guideline for Sustainable Building, also available in English, BMUB, 2016). These are compulsory for buildings of the state, with smaller local municipalities being free to build without restraints. The BNB is close to the DGNB, but it goes further, as it aims for the comprehensive consideration of the entire lifecycle of a building, taking into account the ecological, economic, technical, and procedural aspects and the sociocultural quality. The focus is mainly on office buildings, but guides for schools and other public buildings are also appearing. However, these systems only tangentially mention the impact building materials might have on human health.

17.2.2 Sustainability: avoiding health problems

The essence of sustainable construction would benefit from the inclusion of careful consideration of the healthiness of building materials. It is encouraging to see that really toxic and unhealthy chemicals are increasingly disappearing from products on the market. On the other hand, new problems come up in terms of intolerances and allergies (issues discussed in Part III). Allergenic building materials are an increasingly relevant problem. Current estimates suggest that about 25%–30% of the German population have some form of allergy, which is a very large proportion. Evaluations of the allergenic contribution of building materials are complex and difficult to quantify. An additional problem is that building materials are part of a combination of many different stimuli influencing allergic reactions.

In Germany, the Technischer Überwachungs Verein (TÜV) (TÜV Rheinland Group, 2016) is a large multidisciplinary institution evaluating a wide variety of technical issues associated with products, including car warrants of fitness and atomic power plant safety. Allergenic impact from building materials is also within TÜV's responsibilities. TÜV also provides independent surveys of the indoor climate for buildings and, if up to standard, awards a ToxProof certificate for the building. Although voluntary, some prefabricated house companies acquire ToxProof certificates. This certificate does not include any consideration of sustainability, but rather exclusively focuses on the absence of toxicants in indoor air.

When it comes to evaluations of the toxicity of indoor air, the ToxProof certification goes a step beyond other existing EU legal requirements. As part of its certification, evaluations of 11 groups of building materials and components are undertaken (floor coverings and suspended ceilings from timber, other timber materials, doors, sockets, paint, parquet overcoat, sealing materials, body filler, tile adhesive, gypsum boards, plaster, concrete parts, and movable dividing walls). These are checked against the safe threshold value for dozens of chemicals and, based on this, a summative value of volatile organic compounds is calculated. It should be noted that in the context of buildings and especially indoor air, normally it is the sum of all irritants that causes health problems and allergic reactions, rather than a single material on its own. The ToxProof certificate provides a reasonably good level of security for people living with allergies. However, in the majority of contemporary cases, the focus on safer and less allergenic materials means an increase in final building cost for the clients, but not necessarily, and architectural outcome that is more sustainable.

What this introductory discussion shows, is that a designer has to navigate through a number of separate organizations to develop a good response both in terms of ecological sustainability and the impact of building materials on human health. Greater integration would clearly be helpful. In practice, such integration is rarely requested by clients. The differences between the following two examples highlight some of these issues.

17.3 Example 1: A semidetached healthy residence

The clients were a family of five who in 2001 commissioned Hirth Architects to design a semidetached house on a pleasant, south-facing site. They were interested in an energy-saving, passive solar house which used natural materials. In the first meeting with the architects, they mentioned that one of the family members had allergies. This influenced a number of design decisions because even some completely natural materials, such as wool insulation, can cause allergic problems (Fig. 17.1).

The impact of buildings and building materials on human health should concern everybody, not only those with allergies, especially given that there are other

Figure 17.1 Residence M. View from southeast with the timber façade.

known health impacts of concern from natural materials. One such example is radon gas which can enter homes from the ground, but also from building materials such as bricks, stone, and earth extracted from areas naturally high in radon. Radon is a tasteless and odorless radioactive gas and is a known problem in South Germany, Austria, and Northern Italy (Winter, 1993), creating potential adverse impacts for users of buildings constructed with materials extracted from these areas. Another reasonably well-known issue with natural materials is high levels of essential oils in new timber coming from pine varieties, such as Swiss Stone Pine (*Pinus cembra*). Some timber oils have been associated with generating allergic reactions when timber is still fresh.

In this case, the clients were interested not only in building their house using healthy materials and avoiding allergenic ones, but also the family wanted to integrate this with sustainable building practices. In particular, they wanted to achieve a healthy internal climate, in terms of thermal comfort, with a lot of natural light and good use of passive solar gains.

After a long discussion with the architects on the advantages and disadvantages of different construction approaches, the clients chose a highly insulated timber-framed house with a high thermal mass brick wall inside. This wall was curved and in many ways operated as the core of the house (Fig. 17.2). The form of the wall can be seen as providing an embracing gesture while playing an important role in organization of the circulation within the house. The wall also creates a light shaft leading daylight through the whole house from the roof to the ground floor. Through such gestures this very important element for structure and heating of the house is also architecturally celebrated in the space.

Figure 17.2 The central loadbearing wall works like a vertical heater.

The wall was a load-bearing structure supporting all-timber floor beams, and provided thermal mass for heat storage in spring and autumn, and during the heating period was also the main source of heating in the house. Thin copper pipes were placed on the constructed brick wall and covered with a special ductile lime plaster (Fig. 17.3). The wall was painted with an old-fashioned lime paint, which does not interfere with the heating system. This combination of thermal mass with additional active heating provided low-temperature radiant heating. In contrast to the more common radiant floor heating, with this approach, the whole body receives heat, rather than just the feet. The seminal work of Danish scientist Ole Fanger from the 1960s is still used in Germany as an important source for defining a good internal thermal climate. Among other conditions, Fanger explained the importance of radiant heating and the surface temperature of the walls for this (Sieber and Welter, 1993; Fanger, 1970). Therefore, while the human body loses heat through convection, radiation, and evaporation, radiation often plays an important role in the sensation of warmth, which is why radiant heat in a space is experienced as a very pleasant form of heating.

One of the advantages of radiant heat is that the air temperature in a room can be set lower without any loss of comfort, because the body receives the warmth in the form it needs. A fresh spring day with a decent amount of sun is a good example of how people can feel warm even when the air temperature is low. This is significant, because the greatest loss of heat is through the external walls and increases with the temperature difference between the outside and the inside.

Figure 17.3 The main part of the heating system: Copper pipes before getting plastered.

Thus, even a modest decrease in internal temperature greatly reduces the heat loss during the heating period. This approach saves operational energy used for heating, and can provide a superior sense of thermal comfort.

Achieving the elements of passive solar architecture did not come without challenges. The local district plan (Bebauungsplan) instructed the clients to build a compulsory garage on the sun-facing south side. This requirement was in clear conflict with the needs of passive solar architecture. Fortunately, the rules did not define the floor level for the garage, which is why in this house the garage was lowered half a storey into the ground, leaving enough of the facade for passive solar use.

This move also challenged the design treatment of the basement. Basements have been traditionally common in Germany and are still found in the majority of newly built houses, although they often tend to be underutilized. Constructing basements is also costly. This residence has no basement for two reasons: the clients wanted to spend their money on other things, and once presented with the facts on the high environmental impacts of concrete, they did not want to build with more concrete than absolute necessary. The end result is a modest and strategic use of a simple concrete slab on a pebble bed. The slab area was minimized to reduce the amount of steel reinforcement required.

While both the curved wall feature inside and the concrete slab facilitated collection of solar gain, timber was the main material for the rest of the house (Fig. 17.1). All timber used was completely untreated. Spruce was selected for all construction and structural purposes, and larch for the façade cladding. No finish was used on the outside, with the European larch slowly becoming gray like historic houses in the Alps. Because this process will greatly change the appearance of the building over time, the clients were fully informed in advance about the process and they accepted the "imperfect" appearance of larch.

The cavity in the external walls was filled with 20 cm cellulose insulation. This choice was made in an effort to avoid plastic and formaldehyde-emitting insulation materials. The cellulose insulation product used here was a fully recycled material from old newspapers and cardboard, treated with boron salt, which acts as a fire retardant and increases resistance to insects. It comes in soft plates like Rockwool or in flakes which are blown into the wall cavity using a type of inverse vacuum cleaner. Its insulation value is comparable with Rockwool but it gives a somewhat better performance in summer due to a higher specific weight.

In the interior, beech parquet was used as flooring throughout, with a linseed oiled finish. Two bedrooms were originally carpeted with an expensive eco carpet. This is a high-quality product made from the untreated fine hair of Cashmere goats and the assumption was that this careful selection would not interfere with the health sensitivities of the inhabitants. Unfortunately, the carpet caused an immediate allergic reaction with the client, which was a noticeable disappointment for both clients and architects. Unfortunately, some things can only be discovered through trial and error, especially when allergies, which are unpredictable, are involved. Fortunately, after replacing the carpet with the beech parquet the problems disappeared immediately. No other similar problems arose.

Although currently sustainability and indoor toxicity are considered as separate by the labeling and regulative context, this example clearly shows that the two easily complement each other and should be considered together. One great reward of this project was that once the clients settled into this house, the health of the family improved over time in comparison to their former dwelling. This begs the question: Wouldn't it be a very good outcome if all architecture could help to improve the health of its inhabitants while also minimizing the negative impact on the planet?

17.4 Example 2: Energy-efficient water residence

The clients were a couple who in 2008 commissioned Hirth Architects to design a house on a good site close to the River Rhine in Germany. The site had great advantages including being close to nature but also had some very significant disadvantages because it was on a floodplain which floods regularly every 5–10 years.

It was also exposed to busy traffic at weekends caused by the public enjoying the same natural features. In response, the architects decided to work with water as the main feature of the site and the core aspect of the *genius loci*, the spirit of the place. The building clearly had to be able to handle regular flooding, but rather than embracing water only for those periods, the design proposed using water as a permanent feature in the form of a pond. On a conceptual level the flood and pond were expressions of each other in different forms (Fig. 17.4).

As a result, the building cantilevers over the water surface, and within the building a negative volume was created to form an inverse winter garden which embraces the water. It is possible to close this winter garden with a large sun blind, effectively making the pond part of the living room. In this set up, reflections of the sun on the water could be seen on the ceiling of the living room, while the space was more private. The form of the building consists of two cubes, placed to form an opening to the south. This opening enabled cross-ventilation and the evaporative cooling effect of the pond to be used on hot summer days. The basement and bottom parts of the building were made using waterproof concrete walls. The ground level was prepared for flooding to exceed predicted levels by as much as 60 cm.

The client, a scientist, was fascinated by energy-saving technologies and renewable energy and was willing to use new technology in the house. The hot water central heating system used two components for heating the water: solar roof panels and a fully automated wood pellet burner. The pellet burner used standardized pellets from waste wood material, generated when tree trunks are barked and processed in sawmills. Pellets for a whole year could be stored in the basement.

The main reasons for the very low energy demand of the building were the very highly insulated external walls and roof, and the triple-glazed windows.

Figure 17.4 Water Residence L. View from the west with the pond.

The external walls consisted of a load-bearing sand-lime brick of 17.5 cm and a composite system of 28 cm of polystyrene and plaster. The polystyrene composite system (*Wärmedämmverbund Systeme*) is a construction system currently common in as much as 80% of all new buildings in Germany. The high demand for this construction system reflects its proportionally low costs for the achievable energy savings. However, this tends to be an uncomfortable compromise because while it increases the thermal performance of the building, it also transfers some environmental sustainability problems to the future. This insulation system presents real problems when it comes to the end of its use and the demolition of buildings, because it is difficult to separate the different materials. Furthermore, the polystyrene currently used in wall insulation is flammable. Because of this it becomes designated as special waste in the disposal stage. Unfortunately, currently there are few alternative choices when it comes to insulating massive walls.

With EU regulations forcing the building market to increase the recycling of all waste coming from building sites to 70% by 2020, and bearing in mind that 52% of all waste by weight stems from the building sector (Statistische Ämter des Bundes und der Länder, 2016) issues with the disposal of polystyrene insulation are only going to increase. Finally, the core problem is that many architects see polystyrene facades as a paradoxical dead end: saving oil (used for heating) by using oil (for polystyrene manufacture). In the end, this residence easily covers all of its operational energy demand by using renewable energy sources, which might to a certain extent balance the oil used for producing the polystyrene insulation. Fortunately, better alternatives are currently appearing on the market, using materials such as wood fibers or hemp, which could offer an insulation material with better overall ecological performance over its lifecycle. A few companies have recently been granted public licences (the certificate from the Deutsches Institut für Bautechnik (DIfB)), which it is the entry point for enabling the professional use of such materials. Therefore, there is hope that polystyrene facades and wall systems might not be a permanent feature of sustainable buildings in the EU.

In this example, consideration of indoor air chemical composition and allergies did not figure at all. The building aspired to achieve sustainability by engaging with energy-saving practices. The insulation, solar water-heating system, and the pellet burner all have high embodied energy but these features were critical to achieving a highly energy-efficient building.

17.5 Conclusion

This chapter shows that at that this point in time the concern for healthier indoor environments is only partly integrated into other existing protocols for sustainable construction in Germany. The regulations and recommendations concerned with human health in terms of indoor air quality and the impact that the construction industry has on the natural environment still operate to a large extent as separate from each other. However, the first example discussed here shows that it is very possible to address both of these concerns and give them reasonably equal priority. With the increases in incidents of allergies, being able to provide healthy indoor environments is likely to remain an important issue. This is why it is important to keep including the consideration of human health in all architectural design efforts.

The second example highlights that when sustainable building in terms of energy efficiency is the aim, and when both architects and clients give this priority, finding materials that satisfy performance and cost considerations without causing negative impacts on the natural environment can be problematic. Considered together, these examples signal the importance of a broader definition of a "sustainable" building, and illustrate that current common building practices and labeling systems struggle to support the inclusion of a wider range of sustainability considerations into building design.

References

Bundesministerium für Umwelt, Bau und Reaktorsicherheit (BMUB), 2016. Leitfaden Nachhaltiges Bauen. Available from: http://www.nachhaltigesbauen.de/fileadmin/pdf/Leitfaden_2015/LFNB_D_final-barrierefrei.pdf (accessed September 2016.).

"Der Spiegel," 1997. (Magazin) Issue 50 et al.

DGNB, 2016. Deutsche Gesellschaft für nachhaltiges Bauen/German Sustainable Building Council. Available from: www.dgnb.de (accessed September 2016.).

DIN EN 335-06 – 2013. Dauerhaftigkeit von Holz und Holzprodukten - Gebrauchsklassen: Definitionen, Anwendung bei Vollholz und Holzprodukten; Beuth Verlag (EU Regulation).

Estaugh, N., Walsh V., Chaplin, T., Sidall R., 2007. Pigment compendium: a dictionary of historical pigments, p. 122.

Fanger, O.P., 1970. Thermal Comfort: Analysis and Applications in Environmental Engineering. Danish Technical Press, Copenhagen.

Rose, W.-D., 1985. Wohngifte erkennen und vermeiden. Edition Wandlungen, Oldenburg.

Rose, W.-D., 1987. Elektrosmog – Elektrostreß. Kösel-Varlag, München.

Sieber, H., Welter A., 1993. Architektur unter der Sonne, Deutsche Bauzeitschrift, Issue 6, pp. 907–912.

Statistische Ämter des Bundes und der Länder, 2016. Umwelt–Entsorgung von Abfällen nach Abfallkategorien 2014. 23 August. Stuttgart, Germany. Statistisches Landesamt Baden-Württemberg.

Stockholm Convention, 2016. Stockholm Convention – Protecting human health and the environment from persisting organic pollution, Available from: chm.pops.int (accessed September 2016.).

Szladja, F.M., 1999. Artists' Pigments 1780–1880: History and Uses. Available from: http://www.lilinks.com/mara/history.html (accessed September 2016.).

TÜV Rheinland Group, 2016. Toxproof certificate. Available from: http://www.tuv.com/de/deutschland/gk/produktpruefung/pruefzeichen_zertifizierungen/toxproof/toxproof.html (accessed September 2016.).

Winter, M., 1993. Webelements - University of Sheffield. Available from: www.webelements.com (Accessed September 2016.).

The Bullitt Center: A "Living Building"

18

Adele Leah
Victoria University of Wellington, Wellington, New Zealand

18.1 Introduction

The Bullitt Center is a six-storey commercial office building, with more than 4150 m^2 (44,700 ft^2) of net rentable office space, located in the Central District of Seattle (Fig. 18.1) (International Living Future Institute, 2015b, Scully, 2015). The Center is owned and operated by the Bullitt Foundation, a 60-year-old philanthropic institution, whose mission is to promote environmentally responsible human activities and sustainable communities in the Pacific Northwest (The Bullitt Foundation, 2014). The vision of the President and CEO of the Bullitt Foundation, Denis Hayes, was to build "the greenest urban office building in the world." The Bullitt Foundation would occupy and operate the building throughout its 250-year lifespan, and tenants would rent the additional office space (Scully, 2015). It would be a physical demonstration of the foundation's commitment to urban ecology (Bena, 2014). Construction of the Bullitt Center began in July 2011, and the building was completed in March 2013 (Scully, 2015). The materials selection process for the construction of the Bullitt Center was lengthy and rigorous, and stringent requirements were met in terms of the chemical composition of the specified products and the location of the manufacturer. This case study outlines the requirements of the Living Building Challenge (LBC) certification program, with particular focus on the materials used in construction, the materials vetting process, and a selection of the specified materials.

18.2 Living Building Certification

The idea for the LBC was developed by Bob Berkebile and Jason F. McLennan as a result of their work together at BNIM on the EpiCenter (Montana State University's Green Building Project) in Bozeman, Montana. McLennan, who currently serves as the CEO of the International Living Future Institute, presented the LBC Standard version 1.0 to The Cascadia Green Building Council in August 2006, and it was formally launched to the public 3 months later (International Living Future Institute, 2015a). Denis Hayes selected the LBC building certification program as an alternative to the US Green Building Council's Leadership in Energy and Environmental Design (LEED) certification. At the time, the LBC standard

Materials for a Healthy, Ecological and Sustainable Built Environment: Principles for Evaluation.
DOI: http://dx.doi.org/10.1016/B978-0-08-100707-5.00018-6

Figure 18.1 Exterior of the Bullitt Center. Image courtesy of Nic Lehoux for the Bullitt Center.

(version 2.0) most closely resembled the scientific, ecology-based philosophy of the Bullitt Foundation (Scully, 2015) and Hayes considered it to be the more rigorous scheme in terms of sustainability, requiring that the building actually perform as designed, and meet all the criteria during a full year of operation before it can obtain "Living Building" certification. In contrast, LEED certification is based on preconstruction information and predicted energy performance, and can be obtained prior to the building's first day of operation (Bena, 2014). Since the construction of the Bullitt Center, the LBC standard has evolved (the current version in 2016 is 3.0). The LBC is made up of seven performance categories called "Petals": Place, Water, Energy, Health and Happiness, Materials, Equity, and Beauty (International Living Future Institute, 2015d) and each "Petal" comprises of a number of "Imperatives" (Table 18.1).

The criteria which the Center were required to meet to attain LBC 2.0 certification included the following: the building must not use more energy than it produces; all water required must be collected from rainwater, and all wastewater and graywater must be managed and treated on site; and every building user should have access to operable windows providing fresh air and daylight (Scully, 2015). The Bullitt Center has a 1300 m^2 (14,000 ft^2) photovoltaic array which produces

Table 18.1 LBC 3.0 "Petals" and Imperatives

	Place		Materials
01. 02. 03. 04.	Limits to Growth Urban Agriculture Habitat Exchange Car Free Living	10. 11. 12. 13. 14.	Red List Embodied Carbon Footprint Responsible Industry Living Economy Sourcing Net Positive Waste
	Water		**Equity**
05.	Net Positive Water	15. 16. 17. 18.	Human Scale + Humane Places Universal Access to Nature & Place Equitable Investment JUST Organizations
	Energy		**Beauty**
06.	Net Positive Energy	19. 20.	Beauty + Spirit Inspiration + Education
	Health & happiness		
07. 08. 09.	Civilized Environment Healthy Interior Environment Biophilic Environment		

Source: http://living-future.org/

all of its electricity on site, and geothermal heating and cooling are also utilized. A variety of methods are used in the building to conserve and manage water, which include rainwater harvesting, a green roof, composting toilets, and a bioswale located on strips of land near the footpaths (which typically would normally just be grass) to treat graywater (Scully, 2015).

The project team had to be receptive to new ideas and be capable of working together in a creative, innovative, and productive manner. In contrast to the traditional, sequential process from architect to engineer to contractor, the design process was deeply integrated, with all problems identified and costs estimated before breaking ground. Team members were chosen, in most cases, due to their skills and knowledge, and also their good working relationship with other team members. The project architect, Miller Hull Partnership, sought to integrate sustainability and design, and processes were guided by the performance goals (Scully, 2015).

It is beyond the scope of this chapter to discuss each of the LBC "Petals" and the response to these in the design and construction of the Bullitt Center. Consequently, the remainder of this chapter focuses on the imperatives in the Materials "Petal," in particular Imperative 10, the Red List.

18.3 Materials "Petal"

> *The Living Building Challenge envisions a future where all materials in the built environment are regenerative and have no negative impact on human and ecosystem health.*
>
> *International Living Future Institute (2014)*

The Imperatives in the Materials category of the LBC Standard (3.0) aim to remove the worst known materials and practices (in terms of their impact on the health of ecosystems, climate, and people) and direct industry towards a responsible materials economy (International Living Future Institute, 2014).

18.3.1 Red List (Imperative 10)

The Red List Imperative is a list of chemicals and compounds that are rigorously screened out of any materials used in an LBC-certified building. Table 18.2 contains a list of the chemicals found in the Red List and includes formaldehyde and lead. The requirements stipulate that wet applied products must have low levels of volatile organic compounds (VOCs). The materials vetting process for the Bullitt Center is described later in this chapter.

Declare is a building product label associated with the International Living Future Institute (International Living Future Institute, 2015c). The label takes complex chemical analysis and raw material source location information and presents this as an "ingredients" label for building materials.

Table 18.2 LBC 3.0 Red List

Alkylphenols	Halogenated flame retardants (HFRs)
Asbestos	Lead (added)
Bisphenol A (BPA)	Mercury
Cadmium	Polychlorinated biphenyls (PCBs); Perfluorinated compounds (PFCs)
Chlorinated polyethylene and chlorosulfonated polyethlene	Phthalates
Chlorobenzenes	Polyvinyl chloride (PVC)
Chlorofluorocarbons (CFCs) and hydrochlorofluorocarbons (HCFCs)	Polyvinylidene chloride (PVDC)
Chloroprene (neoprene)	Short-chain chlorinated paraffin
Chromium VI	Wood treatments containing creosote, arsenic, or pentachlorophenol
Chlorinated polyvinyl chloride (CPVC)	Volatile organic compounds (VOCs) in wet applied products
Formaldehyde (added)	

Source: http://living-future.org/

> *By offering a platform for public disclosure that surpasses any other materials label, Declare rises above the greenwash and enables a deep connection between suppliers and consumers.*
>
> *International Living Future Institute (2015c)*

The online database is freely available to consumers and specifiers and helps materials specifiers to meet the LBC's strict materials requirements by streamlining the process. The materials in the database typically meet the requirements of the LEED V4 Building product disclosure and optimization—material ingredients credit (LEED, 2016). The current LBC Standard (3.0) requires projects to use one *Declare* product for every 500 m^2 of gross building area. Project teams are also required to send information about *Declare* to at least 10 manufacturers not currently using the building product label (International Living Future Institute, 2014).

18.3.2 Embodied Carbon Footprint (Imperative 11)

The Embodied Carbon Footprint Imperative requires that the total embodied carbon (tCO_2e) impact from construction be accounted for through a one-time carbon offset in the Institute's new Living Future Carbon Exchange or an approved carbon offset provider (International Living Future Institute, 2014).

18.3.3 Responsible Industry (Imperative 12)

The Responsible Industry Imperative requires that raw materials (stone and rock, metal, minerals, and timber) be extracted as sustainably as possible, and using fair labor practices. All timber used in construction must be certified to Forest Stewardship Council (FSC) 100% labeling standards or be sourced from salvaged sources. The intentional harvest of timber on-site in order to clear the site, or to restore or maintain the continued ecological function of the on-site bionetwork is also acceptable (International Living Future Institute, 2014). The FSC Project Certification for the Bullitt Center is discussed later in this chapter.

18.3.4 Living Economy Sourcing (Imperative 13)

The Living Economy Sourcing Imperative stipulates that the project must contribute to the expansion of a regional economy rooted in sustainable practices, products, and services. Place-based solutions are to be incorporated, and there are a number of restrictions not only in terms of the location of materials and product manufacturers, but also the location of consultants and subcontractors.

18.3.5 Net Positive Waste (Imperative 14)

The Net Positive Waste Imperative requires the project team to seek to conserve natural resources and identify ways to integrate waste back into an industrial (recyclable) or natural (biodegradable) nutrient loop. A pre building audit must be completed on each site with existing infrastructure to record available materials for reuse or recycling. Projects must either reuse an existing structure, or use at least one salvaged material per 500 m^2 of gross building area. The project team has to produce a Material Conservation Management Plan to explain how the project optimizes materials in the Design Phase, Construction Phase, Operation Phase, and End of Life Phase, which includes a plan for deconstruction or adaptive reuse. During construction, a percentage (by weight) of the resulting construction waste material must be integrated back into either an industrial or natural nutrient loop. The percentages are as follows: metal 99%; paper and cardboard 99%; soil and biomass 100%; rigid foam, carpet and insulation 95%; and all others (combined weight average) 90%. Every LBC project requires a dedicated infrastructure for the collection of compostable food scraps and recyclables while in use (International Living Future Institute, 2014).

18.4 Forest Stewardship Council Project Certification

The Bullitt Center was the first commercial building in the United States (US) to earn FSC Project Certification (Kahn, 2013). FSC certification (which is separate from Living Building Certification) ensures that timber products come from responsibly managed forests that provide environmental, social, and economic benefits to local communities (Forest Stewardship Council, 2016). Designed to last 250 years, the heavy timber beams used in construction serve as a carbon sink for the life of the building, as well as structural supports, whilst accentuating and defining the interior spaces (Scully, 2015). The FSC Project Certification required an independent assessment, and Soil Association Woodmark conducted the onsite audits. Schuchart Corporation, the general contractor for the construction of the Bullitt Center, procured timber products from a variety of sources including: Matheus Lumber in Woodinville, Glulam timbers from Calvert Glulam, dimensional lumber from Potlatch, small timbers from Elk Creek Forest products, structural plywood from Roseburg Forest Products, and recycled lumber from Rhine Demolition (Kahn, 2013). The Glulam timbers were industrial grade, rather than architectural or appearance grade to reduce the cost. Each glulam member was precision cut and waste was minimal (Smith-Gardiner, 2012).

The lagging boards, used to temporarily shore the basement walls while the concrete foundation walls were being installed, were preused. The project team worked with a local salvage yard to procure sections of timber beams that were taken from a local automotive warehouse that had recently been deconstructed. Hollow metal doors were specified in the service areas of the building (basement and stairwells) due to their durability. Joseph David from Point32 recalls:

> *After those doors had been specified, there weren't really that many additional doors in the building. One issue that we faced at the time was that it was difficult to find a locally made, 100% FSC certified door that was 100% formaldehyde free.*

I believe that has changed since we completed the project a few years ago, but it was a challenge that ultimately led us to using hollow-metal throughout.

David (2016)

18.5 Material selection

The materials vetting process was undertaken collaboratively by the developer, contractor, and their subcontractors, and was managed by Joseph David from Point32. Products recommended by the subcontractors and installers were prescreened, manufacturers were contacted directly, and Material Safety Data Sheets and product questionnaires were completed (International Living Future Institute, 2015b). The LBC Red List and the VOC content of products were the main drivers of the product selection process, as the LBC requires indoor air testing for VOCs (Smith-Gardiner, 2013). It was also important to identify products that were produced domestically.

The most important thing that I learned through the material vetting process was the value of creating a collaborative conversation with manufacturers. Clear documentation and disclosure was often the barrier for a product to be specified in the project. Instead of reaching out to manufacturers with an ultimatum, we had much greater success explaining to them what we were trying to accomplish, what information we required, and inviting them to participate in the process. Some declined, but those who chose to participate were very forthcoming with the information that we needed.

David (2016).

Products were selected based on the information that was made available to the project team by manufacturers. The catalogue of products used throughout the construction of the Bullitt Center was made available to the public in January 2014 in keeping with the philosophy of the Bullitt Foundation that the process be transparent and educative (The Bullitt Center, 2014). At the beginning of the catalogue, the author states that not all products listed comply with the LBC 2.0 Standards, as in a number of instances, market restrictions, local building code requirements, and buildability resulted in temporary exceptions being granted by the LBC to use noncompliant products. One example of this includes the Glulam timbers which used a formaldehyde-based binder, as there was no acceptable alternative (International Living Future Institute, 2015b). The author also writes: "Due to 'proprietary' and 'trade secret' claims that are prevalent in the industry, it is possible that the project team inadvertently used non-compliant products" (The Bullitt Center, 2014). These disclaimers highlight the difficulty of the materials selection process, and the challenge of identifying and specifying 100% compliant materials. Joseph David from Point32 reflects on the material selection process:

The most difficult and challenging products to select were those that were either extremely complex (hundreds of distinct parts) or those cases where the industry was not willing to fully disclose the chemicals used in their products. In these cases, we had to work extremely hard to fully document our research efforts, and ultimately make an informed decision based on the most reliable information available. In instances where the manufacturer was forthcoming with the details of their product's make-up, we were able to quickly vet the product, and move toward purchase and installation.

David (2016).

The catalogue of products is extensive, and consequently, the remainder of this chapter focuses on a selection of the specified materials, focusing on those which were the most difficult to specify, and includes: the air and water barrier system, coatings, kitchen and bathroom fittings, carpets, floor finishes, and furniture.

18.5.1 Air and water barrier system

Prosoco R-Guard FastFlash was selected as the vapor-permeable air and water barrier system for the building envelope (The Bullitt Center, 2014) and has since received an LBC Declare label for this and a number of other products (Prosoco Inc., 2016a). Prosoco reformulated their product specifically for the Bullitt Center project, by replacing the phthalate plasticizers contained in their air- and water-resistive coatings and sealants with a low-toxicity alternative. Prosoco's Sustainability and Environment Director states:

Prosoco is conducting an ongoing, rigorous review of building envelope, concrete finishing and concrete/masonry protective products. . .Our process involves discussions through the supply chain coupled with direct control over raw materials sourcing and quality control at our Lawrence, Kansas manufacturing facility.

Fuhlhage (2016)

The FastFlash system has recently been utilized in the R.W. Kern Center, Massachusetts, which officially opened in April 2016, and is currently the largest LBC-certified building in New England (Prosoco Inc., 2016b). Prosoco's Consolideck LS and Consolideck LSGuard were also utilized to seal and protect the R.W. Kern Center's concrete floors, and Prosoco's Natural Stone Treatment WB Plus was utilized as a water repellent to protect the exterior masonry. This example demonstrates how through the LBC certification process, materials specifiers are able to influence and ultimately affect the properties of products.

18.5.2 Coatings

One of the aims of the building was to have the structure itself serve as the surface finish internally, thus limiting the number of finishes (Smith-Gardiner, 2013). Fig. 18.2 shows the interior of Suite 300. Rubio Monocoat Oil, a zero VOC hardwax oil produced in Belgium, was selected to seal the internal timber, glulam

Figure 18.2 Interior of Suite 300. Image courtesy of Nic Lehoux for the Bullitt Center.

beams, and staircase treads (The Bullitt Center, 2014). Rubio Monocoat Oil Plus does not contain water, and its components are mainly natural products (Rubio Monocoat, 2011). The oil adheres with the first microns of the wood by molecular bonding and only one layer is required. Safecoat Acrylac, a clear, high-gloss, water-based replacement for high-solvent content lacquer was used on cabinetry where more durability was required (The Bullitt Center, 2014). Acrylac has low VOC content and is LEED-approved (American Formulating & Manufacturing, 2014).

Acro Pure Interior primer and paint by Miller Paint was selected due to it being low VOC and LBC Red List compliant (Miller Paint, 2015). It was more difficult to identify a paint for the exposed steel structure, as this needed to be applied over an intumescent fire-proofing paint. The product selected was Sher-Cryl High Performance Acrylic by Sherwin-Williams, an exterior grade paint with a higher VOC content (Smith-Gardiner, 2013, The Sherwin-Williams Company, 2016). This shows that the complete removal of VOCs from large commercial buildings is currently extremely difficult, even with the very best of intentions. As demand increases for low-VOC and LBC Red List-compliant products, manufacturers will need to consider and research the reformulation of their products, which will effect change in the building materials industry.

18.5.3 Kitchen and bathroom fittings

The kitchen cabinetry is made with PureBond Hardwood Plywood using FSC-certified materials (Smith-Gardiner, 2013) and finished with Safecoat Acrylac. PureBond is manufactured exclusively in North America using a formaldehyde-free, soy-based assembly technology. The technology was developed by Dr. Kaichang Li, who discovered that soy proteins can be modified to perform similarly to byssal threads, the proteins secreted by mussels which enable them to keep their grip on rocks that are constantly pounded by waves. Soy proteins deliver phenomenal adhesion, and also offer exceptional water resistance. Columbia Forest Products have been manufacturing formaldehyde-free hardwood plywood since 2005 (Columbia Forest Products, 2016). The bench tops, Trinity Glass, are made locally from 75% recycled glass and low-carbon cement (Smith-Gardiner, 2013, The Green Building Center, 2016). In 2014, a new federal law regarding lead content in plumbing products that deliver water for human consumption took effect throughout the United States, and as such, faucets and fixtures are manufactured lead-free (Westerling, 2013). ECAST, from Chicago Faucets, was selected for the kitchen and bathroom fixtures. ECAST are durable, high-quality brass faucets and fittings that are designed and manufactured with less than one quarter of one percent (0.25%) total lead content by weighted average (Chicago Faucets, 2016). The sink is stainless steel. For appliances, reasonably priced energy-efficient options were sought.

Daltile ceramic tiles were utilized in the bathrooms (The Bullitt Center, 2014). Daltile have a number of sustainability initiatives which include recycling, reusing, and reducing solid materials and water used in their manufacturing process, and conserving energy (Daltile, 2016). The selection of the partitions for the bathrooms was challenging, as a number of options contained formaldehyde or a foam core with flame retardant. High-density polyethylene was compliant, but was not considered to be durable enough, or have the right aesthetic. The solution was to use FSC-certified wood panels painted with an exterior grade hard enamel paint (Smith-Gardiner, 2013). This example demonstrates how compromises had to be made in order to specify a suitable product in terms of durability and aesthetics. It also highlights an area in the building materials industry where a gap exists for a new, less chemically sensitive product.

18.5.4 Carpet

The requirement to avoid polyvinyl chloride (PVC) and various other chemicals, together with the need for domestic manufacturing, left the project team with only one option. Patcraft carpet was selected due to Patcraft's ability to provide details about their carpet ingredients together with the fact that they are a Seattle-based company (Patcraft, 2016a). A selection of Patcraft carpet products contain no chemicals on the LBC Red List and are registered with the *Declare* program as LBC-compliant (Patcraft, 2016b). Interface carpet, which is manufactured in Europe, is also PVC-free and under their Cool Carpet program Interface offset all the greenhouse gas (GHG) emissions associated with the manufacture, maintenance, and end-of-life management of their carpet, by investing in renewable energy and other projects that prevent GHG emissions from entering the atmosphere (Interface, 2016). If Interface carpet tiles are preused, then they meet the requirements of the LBC, despite being manufactured in Europe, and the office space of the Integrated Design Lab on Level 2 of the Bullitt Center is carpeted with preused Interface carpet tiles (Smith-Gardiner, 2013). This highlights the challenge of identifying LBC Red List-compliant products that are sourced and manufactured locally.

18.5.5 Floor finishes and furniture

Floor finishing strategies were determined by the tenant renting the floor. Joseph David from Point32 recalls that in one case the tenant chose to use the vetted Consolideck products, another tenant chose to grind the floors and apply a wax polish, and others used approved carpet products. Fig. 18.3 shows the interior of the co-working space (Suite 400).

Version 2.0 of the LBC Standard did not review furniture, leaving it to the discretion of the building owner to set parameters. This was changed in version 3.0 of the standard, with furniture systems being included in the Materials “Petal” (Stamats Communications Inc., 2016). The Bullitt Center project team set guidelines for appropriate furniture which was to contain no Red List materials and also meet California Air Resources Board (CARB) Phase 2 standards. CARB is part of the California Environmental Protection Agency (CEPA), and once a substance is identified as a toxic air contaminant (as formaldehyde was in California in 1992), the CARB is required by law to determine if there is a need for controls (California Air Resources Board, 2012). In 2007 CARB approved an airborne toxic control measure (ATCM) to reduce formaldehyde emissions from composite wood products (including hardwood plywood, particleboard, and medium-density fiberboard) and also furniture and other finished products made with composite wood products. The ATCM which was approved by the Office of Administrative Law in 2008, set a

Figure 18.3 Interior of Suite 400 coworking space. Image courtesy of Nic Lehoux for the Bullitt Center.

threshold for formaldehyde emissions from furniture (California Air Resources Board, 2016). Greenguard Certification from UL Environment, a business unit of UL (Underwriters Laboratories), an independent, not-for-profit testing laboratory, was also considered to be an acceptable standard, along with preused furniture (Smith-Gardiner, 2013, UL Environment, 2016). Three manufacturers worked with the project team to provide proposals for furniture layouts, these were Herman Miller (Herman Miller Inc., 2016), Watson (Watson, 2016), and Teknion (Teknion, 2016). Their work was considered valuable in illustrating to prospective tenants how the spaces could be utilized in terms of layouts of desks, access to daylight, and views. Tenants selected furniture according to their individual needs.

18.6 Product transparency

> *Transparency is vital; as a global community, the only way we can transform into a truly sustainable society is through open communication and honest information sharing, yet many manufacturers are wary of sharing trade secrets that afford them a competitive advantage, and make proprietary claims about specific product contents.*
>
> *International Living Future Institute (2014)*

Human illness, habitat and species loss, pollution, resource depletion, and waste are some of the issues to which building materials contribute during their lifecycle. Currently, it is extremely difficult to measure the environmental impact and toxicity of the built environment, or even individual buildings, due to a lack of specific product information. Joseph David from Point32 reflects:

> *As the healthy material industry advances, standardizing the way in which we pass information from manufacturer to project team must rapidly change. At the time of the project, we created our own project specific questionnaire for manufacturers to complete. This was a very time consuming process for both the team and the manufacturer (some companies were receiving hundreds of project specific forms – each slightly different).*
>
> *David (2016)*

The Bullitt Center case study illustrates how an individual project can encourage and drive innovation in industry to create more healthy materials and products. As discussed previously, the air and water barrier system (Prosoco R-Guard FastFlash) was chemically altered to enable it to be specified in the Bullitt Center, and there were a few cases where manufacturers substituted components in their products to meet the criteria, e.g., using EPDM rubber gaskets in lieu of the standard neoprene in a common plumbing coupling (David, 2016). By demanding a high standard, the project team have effected change in the manufacturing industry, and have positively contributed to the phenomena of rapidly disappearing toxic compounds in materials which has been referred to as the "sustainable chemical revolution" (Stoneham Knott, 2014).

18.7 Conclusion

The Bullitt Center provides a source of inspiration for others who might wish to walk a similar path in the creation of a certified "Living Building." The catalogue of products which has been made available to the public provides an excellent starting point for other project teams working in the same region, and materials lists for three other Living Buildings are currently publicly available (International Living Future Institute, 2015e). If there is to be real change in the materials that are used to construct the built environment, it will be essential to build upon the work of others, acknowledging those who have gone before, and making the way clearer for those who follow. It will also be essential to educate people about building materials and products, as consumer-driven industry innovation may be the fastest route to a more responsible chemical management system in buildings and therefore lead to a healthier built environment.

References

American Formulating and Manufacturing, 2014. Safecoat Clear Finishes, American Formulating & Manufacturing. Available from: http://www.afmsafecoat.com/products.php?page=3 (accessed 15.08.16.).

Bena, R.B., 2014. Living Proof: The Bullitt Center. University of Washington Center for Integrated Design, Washington.

The Bullitt Center, 2014. Core & Shell As-Built Product List, Seattle, Washington, The Bullitt Foundation. Available from: http://www.bullittcenter.org/building/red-list-compliant products/ (accessed 4.08.16.).

The Bullitt Foundation, 2014. Mission, Seattle, Washington, The Bullitt Foundation. Available from: http://www.bullitt.org/about/mission/ (accessed 4.08.16.).

California Air Resources Board, 2012. Rulemaking Process. California Environmental Protection Agency, California. Available from: https://www.arb.ca.gov/toxics/compwood/process.htm (accessed 20.09.16.).

California Air Resources Board, 2016. Composite Wood Products ATCM. California Environmental Protection Agency, California. Available from: https://www.arb.ca.gov/toxics/compwood/compwood.htm (accessed 20.09.16.).

Chicago Faucets, 2016. ECAST® Meeting The New Lead-Free Standard, Chicago Faucets. Available from: http://www.chicagofaucets.com/catalog/catalog.php?feature = ECAST (accessed 19.08.16.).

Columbia Forest Products, 2016. PureBond Hardwood Plywood, Columbia Forest Products. Available from: http://www.columbiaforestproducts.com/product/purebond-classic-core/ (accessed 19.08.16.).

Daltile, 2016. Greenworks/LEED Sustainability, Daltile. Available from: http://www.daltile.com/greenworks/sustainability (accessed 19.10.16.).

David, J., 2016. Email dated 8th September, RE: Bullitt Center Response. Typed to Leah, A.

Forest Stewardship Council, 2016. Certification, United States, Forest Stewardship Council. Available from: https://us.fsc.org/en-us/certification (accessed 23.08.16.).

Fuhlhage, D., 2016. Living Building Challenge Red List Reviewed Products, Prosoco Inc. Available from: http://www.prosoco.com/media/2502/living-building-challenge-red-list-conformant-products-summary-01-26-2016.pdf (accessed 15.08.16.).

The Green Building Center, 2016. Trinity Glass, The Green Building Center. Available from: http://www.greenbuildingcenter.net/products/countertops/trinity-glass/ (accessed 19.08.16.).

Herman Miller Inc., 2016. Reuse, Reduce, Recycle: Designing Products and Processes for Sustainability, Herman Miller Inc. Available from: http://www.hermanmiller.com/research/topics/all-topics/reuse_reduce_recycle_designing_products_and_processes_for_sustainability.html (accessed 19.08.16.).

Interface, 2016. Zeroing Out Greenhouse Gas Emissions, Interface. Available from: http://www.interface.com/APAC/en-AU/about/modular-carpet-tile/Cool-Carpet-en_AU (accessed 16.08.16.).

International Living Future Institute, 2014. Living Building ChallengeSM 3.0. Seattle, Washington.

International Living Future Institute, 2015a. A Brief History of The Living Building Challenge, The International Living Future Institute. Available from: http://living-future.org/ilfi/about/history-0 (accessed 20.09.16.).

International Living Future Institute, 2015b. Bullitt Center, Seattle, Washington, International Living Future Institute. Available from: http://living-future.org/bullitt-center-0 (accessed 4.08.16.).

International Living Future Institute, 2015c. Declare, International Living Future Institute. Available from: http://living-future.org/declare (accessed 9.08.16.).

International Living Future Institute, 2015d. Living Building Challenge 3.0, International Living Future Institute. Available from: http://living-future.org/lbc/about (accessed 4.8.16.).

International Living Future Institute, 2015e. Project Team Resources, International Living Future Institute. Available from: http://living-future.org/lbc/support (accessed 22.08.16.).

Kahn, B., 2013. Bullitt Center Earns FSC Project Certification, Living Proof Blog. Available from: http://www.bullittcenter.org/2013/02/21/bullitt-center-earns-fsc-project-certification/ (accessed 22.07.16.).

LEED, 2016. Building Product Disclosure and Optimization - Material Ingredients, Washington, USGBC. Available from: http://www.usgbc.org/credits/new-construction-core-and-shell-schools-new-construction-retail-new-construction-healthca-24 (Accessed 9.08.16.).

Miller Paint, 2015. Acro Pure Interior, Miller Paint. Available from: http://www.millerpaint.com/miller-interior-paints/232-acro-pure.html (accessed 16.08.16.).

Patcraft, 2016a. Bullitt Center Project Profile, Patcraft. Available from: http://www.patcraft.com/content/LiteratureCenterPDFs/BullittCenter_ProjectProfile_web.pdf (accessed 16.08.16.).

Patcraft, 2016b. Living Building Challenge, Patcraft. Available from: http://www.patcraft.com/html/SustainabilityLivingChallenge (accessed 16.08.16.).

Prosoco Inc., 2016a. Sustainability. Some Manufacturers Talk Sustainability & Transparency. We live it, Prosoco Inc. Available from: http://www.prosoco.com/sustainability/ (accessed 15.08.16.).

Prosoco Inc., 2016b. What a Living Building is made of, Prosoco Inc. Available from: http://www.prosoco.com/what-a-living-building-is-made-of/ (accessed 15.08.16.).

Rubio Monocoat, 2011. Rubio Monocoat. Available from: http://www.rubiomonocoatusa.com/ (accessed 15.08.16.).

Scully, J., 2015. Bullitt Center Case Study, ULI Case Studies, Urban Land Institute.

The Sherwin-Williams Company, 2016. Sher-Cryl High Performance Acrylic, The Sherwin-Williams Company. Available from: http://protective.sherwin-williams.com/detail.jsp?A=sku-26028%3Aproduct-6795 (accessed 16.08.16.).

Smith-Gardiner, N., 2012. FSC Wood as a Responsible Building Material, Living Future Blog. Available from: http://www.bullittcenter.org/2012/05/30/fsc-wood-as-a-responsible-building-material/ (accessed 22.07.16.).

Smith-Gardiner, N., 2013. Finishes and Indoor Air Quality at the Bullitt Center. Living Proof Blog. Available from: http://www.bullittcenter.org/2013/12/05/finishes-and-indoor-air-quality-at-the-bullitt-center/ (accessed 22.07.16.).

Stamats Communications Inc., 2016. Update to Living Building Challenge: Learn about key changes in Version 3.0, Stamats Communications Inc. Available from: http://www.buildings.com/article-details/articleid/17851/title/update-to-living-building-challenge.aspx (accessed 22.08.16.).

Stoneham Knott, A., 2014. The Chemical Revolution, United Kingdom, Miramar. Available from: http://www.miramar.global/news/chemical-revolution-andrew-stoneham-knott (accessed 22.08.16.).

Teknion, 2016. HPD Collaborative, Teknion. Available from: http://www.teknion.com/ap/inspiration/sustainability/hpd-collaborative (accessed 19.08.16.).

UL Environment, 2016. About Greenguard Certification, Underwriters Laboratories. Available from: http://greenguard.org/en/about.aspx (accessed 21.09.16.).

Watson, 2016. Practical Environmentalism, Watson. Available from: http://www.watsonfurniture.com/company/practical-environmentalism (accessed 19.08.16.).

Westerling, K., 2013. Are You Ready For 'Lead Free' 2014? Water Online. Available from: http://www.wateronline.com/doc/are-you-ready-for-lead-free-0001 (accessed 19.08.16.).

Conclusion

Emina Kristina Petrović, Brenda Vale and Maibritt Pedersen Zari
Victoria University of Wellington, Wellington, New Zealand

It is doubtful that any built environment professional inadvertently wants to cause the degradation of ecosystems, climate change to continue, or for people to become ill because of their decisions about building materials. Unfortunately, this is the current norm. This book set out to discuss the principles and issues behind evaluating building materials for sustainable construction, and has explored a broad variety of issues, discussed new and established methodologies, and used case studies to illustrate how decision-making processes might change to reflect shifting requirements and ethics. The size and complexity of the book clearly indicate that this is neither a simple nor straightforward process. Rather, there is much to be considered. The most important overall message of this book is that each material should be considered according to the numerous criteria outlined here and a recommendation of whether or not it can contribute to making sustainable buildings will vary depending on many parameters. These include where in the world a particular material is to be used, where the material comes from, how it is to be used in the building, and whether or not the building users have particular health requirements that might be affected by the indoor air quality. No material is good in all aspects and the best (and for designers most unpalatable) advice for sustainability is not to build.

Where building is unavoidable the most important objective in relation to mitigating climate change, is to reduce the nonrenewable operating energy to zero, whether this is energy for heating or cooling or running the other services. When it comes to sustainable design, the materials used to achieve this goal have to be selected based on their ability not to reduce local ecosystem services, but if possible to improve and support these. The material also needs to be considered for its effects on the health of its users. However, materials are only part of a building—the design, form, and systems of the latter also have to be appropriate for its purpose. Thus, choice of materials and overall building design need to be synchronous. What this book argues is that this complex consideration of materials needs to be part of making sustainable buildings.

In the past this was all part of local vernacular traditions, where the choice of materials had been honed over many years and through many experiences. This approach has been lost with industrialization, but new ways of thinking about sustainable materials also require building professionals to think about the local context, climate, ecosystems, and health impacts. A building that lasts a long time

Materials for a Healthy, Ecological and Sustainable Built Environment: Principles for Evaluation.
DOI: http://dx.doi.org/10.1016/B978-0-08-100707-5.00031-9

with minimum maintenance may be a good strategy for sustainability and this again means thinking about climate. Building in earth, a low embodied energy material, may not be appropriate for all locations or climates. This is where vernacular traditions have much to teach, because they evolved in response to climate, locally available materials, and cultural context to make buildings that would have a minimum environmental impact. The big difference between such traditions and now is that expectations about the building process have also changed. It is no longer acceptable to wait a year for an earth wall to dry out before occupying a building. Rather, numerous pressures demand accelerated construction, which can conflict with sustainability considerations.

To be sustainable, the built environment has to be part of a sustainable society that operates in a sustainable way. Of great importance is employing strategies to decrease (or eliminate) nonrenewable operating energy, when possible to construct with low embodied energy materials that will be easy to maintain, and to design spaces that are flexible enough to be adapted to new uses over the life of the building. Careful consideration of the impacts that extraction and manufacture of materials have on ecosystems, and the effects of materials on the health of building users is also of paramount concern. This book asserts the importance of considering all of these parameters when choosing building materials.

This book has brought together knowledge from many fields and applied it to built environment design, asserting that such an integrated or holistic approach is essential for complete evaluation of materials. Some sustainability considerations, such as those discussed in Parts I and III, are only in the early stages of being included in architectural decision making. Others such as lifecycle energy assessment (discussed in Part II) are more common but still not universal. This book further proposes that transdisciplinary considerations of building materials are essential for a more complete understanding of their total implications and impacts, on people, ecosystems, and the climate. What is needed is powerful individual drive from building professionals to look for a greater level of detail and to find information on all relevant facets of materials. This in turn might require rethinking many aspects of traditional architectural construction, to include these broader and more complex considerations if sustainable buildings are to be the result.

Index

Note: Page numbers followed by "*f*" and "*t*" refer to figures and tables, respectively.

C

F

M

R

Kolibri 360 GmbH
Ferdinand-Jühlke-Straße 7
99095 Erfurt, Deutschland
produktsicherheit@kolibri360.de